Riveted Lap Joints in Aircraft Fuselage

SOLID MECHANICS AND ITS APPLICATIONS
Volume 189

Aims and Scope of the Series

The fundamental questions arising in mechanics are: *Why?*, *How?*, and *How much?* The aim of this series is to provide lucid accounts written by authoritative researchers giving vision and insight in answering these questions on the subject of mechanics as it relates to solids.

The scope of the series covers the entire spectrum of solid mechanics. Thus it includes the foundation of mechanics; variational formulations; computational mechanics; statics, kinematics and dynamics of rigid and elastic bodies: vibrations of solids and structures; dynamical systems and chaos; the theories of elasticity, plasticity and viscoelasticity; composite materials; rods, beams, shells and membranes; structural control and stability; soils, rocks and geomechanics; fracture; tribology; experimental mechanics; biomechanics and machine design.

The median level of presentation is the first year graduate student. Some texts are monographs defining the current state of the field; others are accessible to final year undergraduates; but essentially the emphasis is on readability and clarity.

For further volumes:
http://www.springer.com/series/6557

Andrzej Skorupa • Małgorzata Skorupa

Riveted Lap Joints in Aircraft Fuselage

Design, Analysis and Properties

Springer

Andrzej Skorupa
Faculty of Mechanical Engineering
and Robotics
AGH University of Science and Technology
Al. Mickiewicza 30
30-059 Kraków
Poland

Małgorzata Skorupa
Faculty of Mechanical Engineering
and Robotics
AGH University of Science and Technology
Al. Mickiewicza 30
30-059 Kraków
Poland

ISSN 0925-0042
ISBN 978-94-007-4281-9 ISBN 978-94-007-4282-6 (eBook)
DOI 10.1007/978-94-007-4282-6
Springer Dordrecht Heidelberg New York London

Library of Congress Control Number: 2012941830

Printed on acid-free paper

Springer is part of Springer Science+Business Media (www.springer.com)

Preface

Fatigue of the pressurized fuselages of transport aircraft is a significant problem for all builders and users of hundreds of thousands of aircraft, servicing millions of air travellers and cargoes. Primary facets of this issue are: How to assure a sufficient lifetime for each and every component of each and every aircraft. How to determine adequate safety features compatible with differing structures. How to formulate and enforce inspection procedures commensurate with the demands of individual aircraft. These aspects are all addressed in various formal protocols for creating and maintaining airworthiness, including damage tolerance considerations. In most transport aircraft, fatigue occurs in joints, and more especially in lap joints between sheets of aluminium alloys, sometimes leading to circumstances that threaten safety in critical ways. This fact was recognized as early as the 1950s. Since then, fatigue of lap joints has become a fundamental design question and an increasingly detailed inspection problem for the aircraft operator. The aircraft industry and the airlines are both interested in the use of thin sheet material in order to achieve weight reduction. Moreover, the economics of aircraft production, the increasing sophistication of inspection procedures and the need for lengthened lifetimes of aircraft already in service must all be considered. The problem of fatigue of lap joints has been considerably enlarged by the goal of extending aircraft lifetimes which may already exceed, e.g., 20 years.

Fatigue of riveted lap joints between aluminium alloy sheets, typical of the pressurized aircraft fuselage, is the major topic of this book. Bonded lap joints are not included. These joints appear to have attractive fatigue properties. However, no major breakthrough has occurred due to questions associated with production techniques, quality assurance and inspections. One newcomer to the industry is using weldable aluminium alloys in integral structures with stiffeners, which offer a means of replacing lap joints by butt joints. Long time service experience with these structures is not yet available and this type of connection is also outside the present survey.

Fatigue response of a riveted connection depends on the integrated effect of a large number of variables related to joint design and production and applied loading conditions. In view of that dependency, numerous research programs on fatigue of riveted lap joints have been carried out and reported in the literature, as discussed in this book. The book consists of ten chapters. Chapter 1 gives basic information on structural design solutions for fuselage skin joints and loading conditions. Although the stress distribution in a fuselage lap joint is of a complex character, a great majority of experimental studies reported in the literature were carried out in laboratory conditions on simple small lap joint specimens under uniaxial tension. The relevance of such results to riveted joints in a real structure is considered in Chap. 2. The fatigue behaviour of riveted lap joints shows a considerable dependency on factors associated with the production process. In Chap. 3, the following production variables are taken into account: sheet material, rivet type and material as well as the manufacturing process, including riveting techniques, rivet hole imperfections, surface treatment of the sheets and the squeeze force. The latter is a major factor that influences the fatigue behaviour of riveted joints. In Chap. 4, the dependence of joint fatigue performance on various design parameters is addressed. Specifically, the effects of the number of rivet rows, rivet row spacing, rivet pitch in a row, rivet pattern and sheet thickness are accounted for. An analytical solution and experimental results on load transmission in lap joints with mechanical fasteners are considered in Chap. 5. Special attention is paid to the experimental and theoretical determination of fastener flexibility and to friction between the faying sheets in view of their importance for load transfer. Eccentricities occurring in the overlap region of a joint induce a so-called secondary bending. Estimates of secondary bending by means of simple analytical models, FE computations and measurements are presented in Chap. 6. Also, implications of secondary bending for joint fatigue performance are considered. The nucleation and shape development of fatigue cracks in longitudinal lap joints is covered in Chap. 7. Issues given special attention are the influence of the squeeze force on the mode of failure and the significance of fretting for fatigue crack initiation. A characteristic and very dangerous form of fatigue damage in longitudinal riveted lap joints is the so-called multi-site damage (MSD). In Chap. 8, passenger aircraft catastrophic accidents due to MSD are described first. Next, an overview of experimental investigations into MSD performed on full scale fuselage panels and riveted lap joint specimens is offered. Chapter 9 is devoted to fatigue crack growth and fatigue life prediction methodology for riveted lap joints, including the MSD problem. Models and codes most commonly used for that purpose are outlined and stress intensity factor solutions appropriate for cracks at rivet holes are presented. A particular consideration is given to the equivalent initial flaw size concept due to its significance for the prediction quality. Residual strength predictions for riveted lap joints in a fuselage structure are addressed in Chap. 10. Failure criteria and crack growth directional criteria are thoroughly considered. Various computational approaches to estimate residual strength of panels with riveted connections are presented and reported comparisons between predictions and experimental results are reviewed. Structural risk analysis methodology applicable to riveted joints with MSD is overviewed. The major issues of each chapter are recapitulated in the last section. The material presented in the book is richly illustrated.

Acknowledgements

Though the authors of the book appear to be the only contributors to its contents, in reality it is an achievement made possible due to the help and efforts of other persons whom we would like to thank.

First of all we want to express our deepest thanks to Prof. Jaap Schijve from the Delft University of Technology, our friend and collaborator for many years, for his encouragement, enduring help and interest in this work. Without his suggestions and support, in particular providing us with a number of literature sources to which it was difficult to gain access, we would have never been able to prepare this book.

We want to express our appreciation to Johannes Homan, M.Sc., from Fatec Engineering for his useful comments and information on some selected problems.

Our co-workers Dr. Tomasz Machniewicz and Adam Korbel, M.Sc., from the AGH University of Science and Technology in Kraków were always available for discussion. They also deserve our particular thanks for their great help in preparing the figures in a printable form.

We gratefully acknowledge the thorough work done by Mr. Edwin Beschler, an English language copy editor provided by Springer to improve the text of the book.

With respect to publishing the book by Springer, we appreciate the nice and effective cooperation with Ms. Nathalie Jacobs and Ms. Anneke Pot as well as Ms. Arulmurugan Pavitra responsible for the typesetting of the book.

An incentive to write this book was the authors' participation in projects concerning fatigue of riveted joints in aircraft structures. Financial support from the associated governmental research funds within the years 2009–2012 is acknowledged.

Acknowledgments

Contents

Nomenclature

Symbols

a	depth of part through crack
c	crack length
D	rivet driven head diameter
D_o	rivet shank diameter
d	rivet hole diameter
$\mathrm{d}c/\mathrm{d}N$	fatigue crack growth rate
d_e	expanded hole diameter
E	sheet material elastic modulus
E_f	fastener material elastic modulus
e	load path eccentricity
F_{cl}	clamping force
F_{sq}	squeeze force
F_u	shear strength
f	rivet flexibility, or load frequency
H	rivet driven head height
H_o	initial rivet height (protruding)
h	countersink depth
he	hole expansion
K	stress intensity factor
K_b	bending factor
M	bending moment
N_f	fatigue life
P	applied tensile load
p	rivet row spacing, or pressure
R	stress ratio, or fuselage radius
R_{TR}	load transfer ratio
r	rivet hole radius
S	applied or nominal stress

S_b	nominal bending stress
S_{LO}	lift-off stress
S_{sq}	squeeze pressure
S_u	ultimate tensile strength
S_y	yield stress
s	rivet pitch in row
T_{BP}	bypass load
T_{BR}	bearing load
T_{FR}	friction force
T_{TR}	transfer load
t	sheet thickness
w	bending deflection, or specimen width
δ	rivet deflection
μ	friction coefficient
ν	Poisson's ratio
ψ_c	critical CTOA
σ_b	local bending stress
$\sigma_{LINK\text{-}UP}$	applied stress at which crack link-up occurs
σ_r	residual radial stress, or residual strength
σ_θ	residual tangential stress

Subscripts

a	amplitude
m	mean
max (min)	maximum (minimum) level of S or K in fatigue cycle
op	crack opening level of S or K in fatigue cycle

Abbreviations

CA	constant amplitude
CGR	fatigue crack growth rate
CTOA	crack tip opening angle
CW	cold working
EIFS	equivalent initial flaw size
EPFM	elastic–plastic fracture mechanics
FE	finite element
LEFM	linear elastic fracture mechanics
MSD	multiple-site damage
POF	probability of failure
PSC	plane strain core

PZL	plastic zone link-up
SB	secondary bending
SIF	stress intensity factor
SSD	single-site damage
SY	strip yield
TTCI	time to crack initiation
VA	variable amplitude

Units and Conversion Factors

1 ft = 0.3048 m
1 in. = 25.4 mm
1 ksi = 6.8948 MPa
1 psi = 6.8948 kPa
1 lb = 0.4536 kg
1 ksi$\sqrt{in}$ = 1.099 MPa $\sqrt{m}$

Chapter 1
Riveted Lap Joints in a Pressurized Aircraft Fuselage

1.1 Constructional Solutions of the Fuselage Skin Structure

The contemporary transport aircraft fuselage is a skin structure supported by frames and stringers, Fig. 1.1. The skin and the stiffening elements carry the flight loads including those due to cabin pressurization. The stringers are joined by riveting, bonding or spot welding directly to the skin, as exemplified in Fig. 1.2. In order to connect the skin with the frames, two different solutions can be applied. One of these, addressed as the shear-tied frame, involves attaching the frame directly to the skin and tear straps, Fig. 1.2. Another possibility, illustrated in Fig. 1.3, is to connect the frames only to the stringers by means of stringer clips. This solution is referred to as the floating frames. The internal tear straps, Fig. 1.4, located at and sometimes also between the frame stations, force longitudinal skin cracks (parallel to the fuselage axis) to turn circumferentially. If this situation takes place between the tear straps, a segment of the skin bends back creating an opening. This phenomenon, described as "flapping", is a safe failure mode that limits the failure to the affected bay only (Kosai et al. 1992). A thin skin will flap more easily than a thick one (Maclin 1991). More information on flapping one can find in Swift (Swift 1987). Some aspects are also considered further on in this book. The straps can carry the load of the cracked skin. These crack stopper bands, although undesirable from the production point of view, are applied in several types of aircraft. Sometimes, instead of the fail-safe straps, waffle pattern doublers bonded to the skin are used, Fig. 1.5. Different types of crack stopper bands (integral, riveted and bonded made of the 2024-T3 and 7075-T6 Al alloys, a Ti alloy and ARALL) and their capability to cause crack growth retardation are reported by Schijve (1990).

A. Skorupa and M. Skorupa, *Riveted Lap Joints in Aircraft Fuselage*, Solid Mechanics and Its Applications 189, DOI 10.1007/978-94-007-4282-6_1,

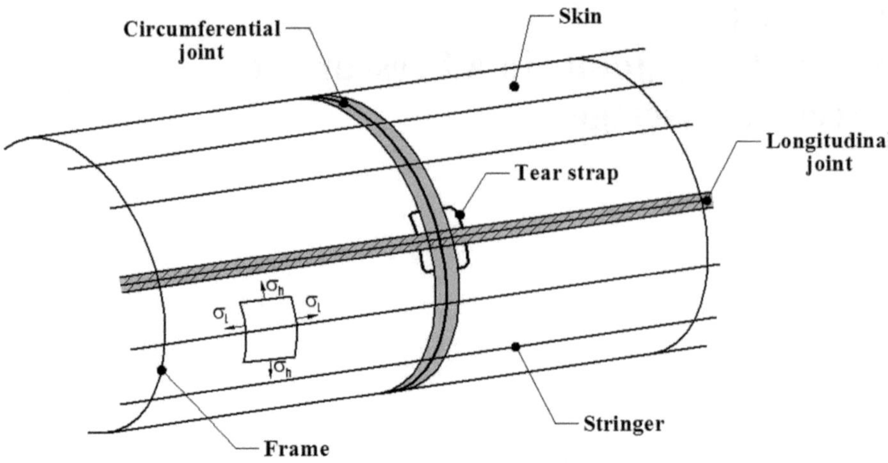

Fig. 1.1 Components of the fuselage skin structure

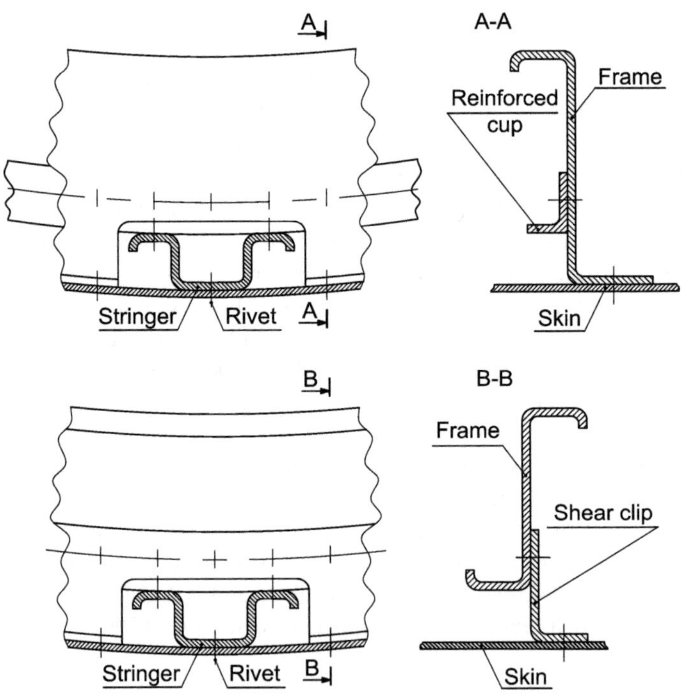

Fig. 1.2 Shear tie connection between the frame and skin (shear-tied frame)

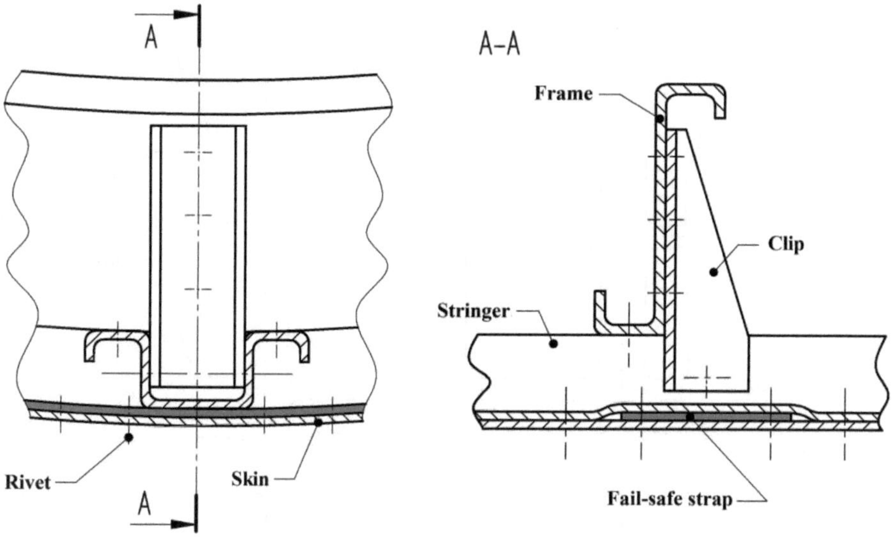

Fig. 1.3 Stringer clip connections between the frame and stringer (floating frame) with a fail-safe strap

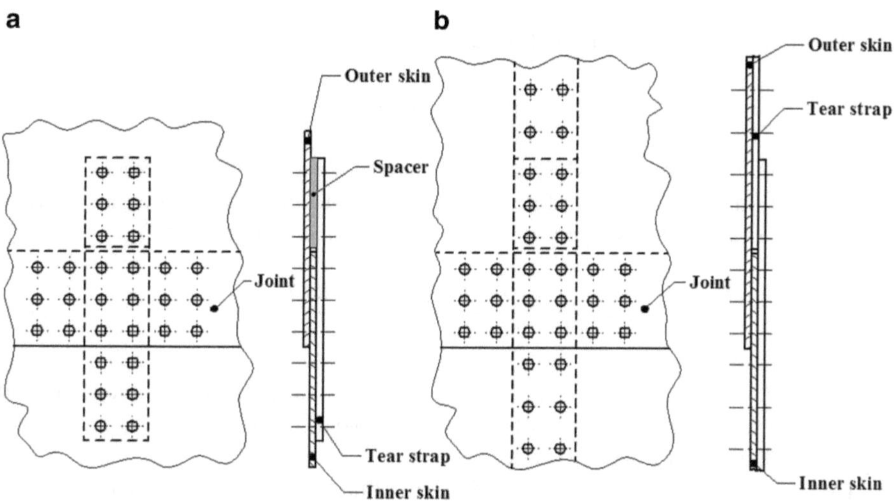

Fig. 1.4 Tear strap: (**a**) short; (**b**) continuous

1.2 Loading Conditions for a Longitudinal Lap Splice Joint

Due to cabin pressurization, membrane stresses are generated in the fuselage skin, see Fig. 1.1. According to the theory of thin-wall pressure vessels, a primary loading for the skin is hoop tension. For an unstiffened vessel the hoop (circumferential)

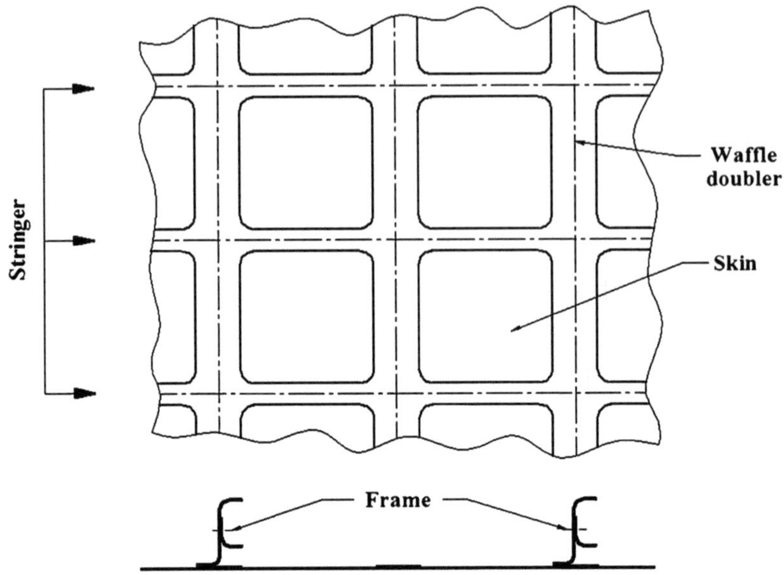

Fig. 1.5 Bonded waffle doubler

stress can be calculated from the formula

$$\sigma_\mathrm{h} = p \cdot R/t \tag{1.1}$$

where p is internal pressure, R is radius of curvature and t is wall thickness.

Taking into account circumferential and axial stiffening elements, an average uniform hoop stress across the bay can be calculated as (Flugge 1952)

$$\sigma_h = \frac{t_x pR + \nu(t_\varphi - t)pR/2}{(1-\nu^2)t_\varphi t_x + \nu^2 t(t_\varphi + t_x - 1)} \tag{1.2}$$

with $t_x = t + A_\mathrm{L}/S$ and $t_\varphi = t + A_\mathrm{F}/L$, where A_L and A_F are the longeron and frame cross section area respectively, S and L denote the longeron and frame spacing respectively, and ν is Poisson's ratio.

The longitudinal stress σ_l equals 0.5 σ_h. For that reason, a longitudinal skin joint is more critical than a transverse (circumferential) joint. A typical solution for the longitudinal connection is the lap-splice joint. The overlaps of the mating skins are generally joined together using permanent fasteners, but also a bonded or riveted-bonded design can be applied. The riveted longitudinal joint can consist of two, three and even four rivet rows, Fig. 1.6. However, for a better smoothness of the fuselage outer surface, some longitudinal lap splices are replaced by single-strap joints, as presented in Fig. 1.7. A more typical application of the latter type joints are, however, the circumferential skin connections.

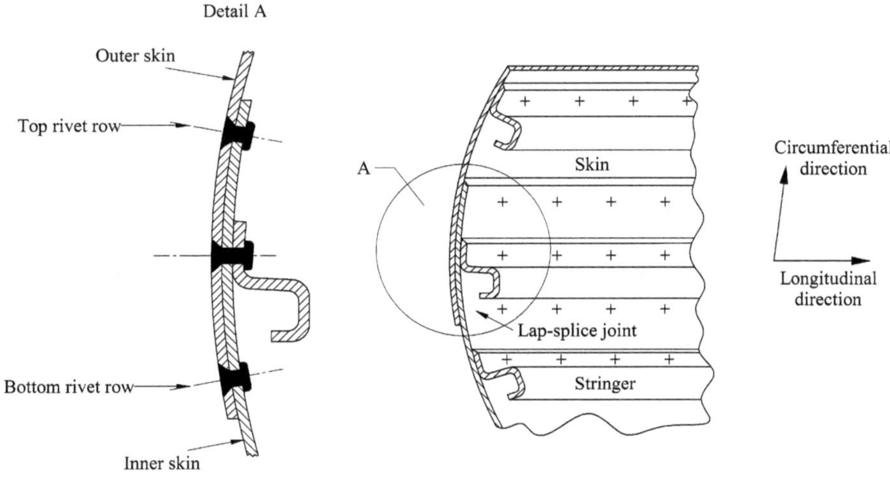

Fig. 1.6 Typical fuselage longitudinal riveted lap splice joint

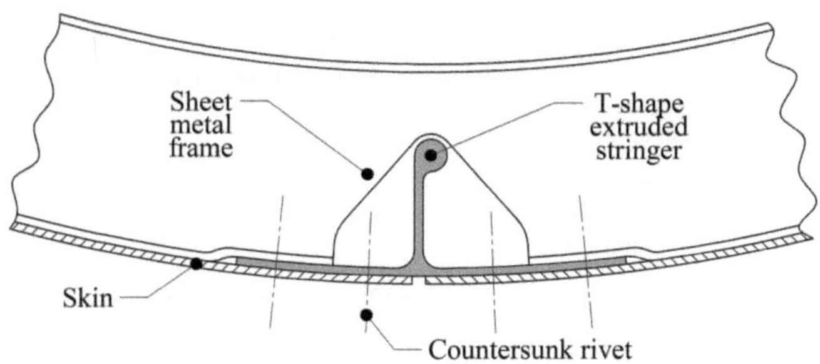

Fig. 1.7 Longitudinal single-strap joint from Falcon 7X, Dassault Aviation (Le Telier and Repiton 2007)

In general, most commercial aircraft transport cabins are designed for an 8,000 ft (2,438 m) altitude pressure. At this altitude the atmospheric pressure is 10.92 psi (0.075 MPa). The values of a differential pressure Δp between the atmospheric pressure p_a at different altitudes and the cabin pressure $p_c = 10.92$ psi are given in Fig. 1.8. For example, a typical Boeing's wide body airplane has a fuselage radius R of 127 in. (3.23 m) and a skin thickness t of 0.053 in. (1.6 mm). In that case, for the altitude of 45,000 ft (13,715 m) Eq. 1.1 yields the nominal hoop stress of 17.4 ksi (120 MPa). According to Niu (1999), the σ_h stress in the skin varies between around 80 and 110 MPa, depending on the aircraft type.

The longitudinal splice can additionally experience bending or shear stresses resulting from air turbulences (predominantly during the climb and descent period at low altitudes) and maneuvers (take-off, landing), depending on the circumferential location of the splice and the position along the fuselage, Fig. 1.9.

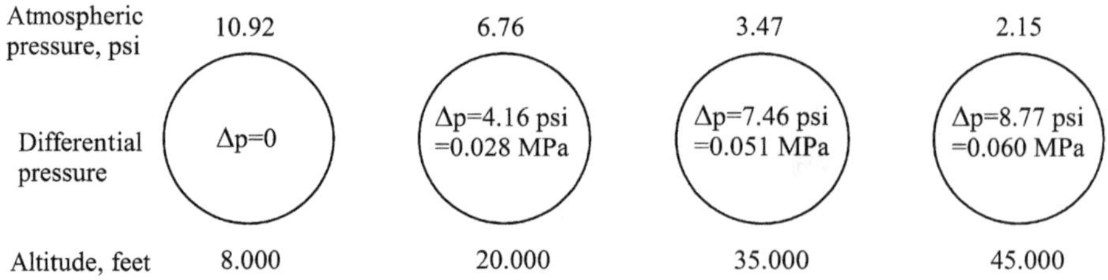

Fig. 1.8 Differential pressure at different altitudes (Niu 1999)

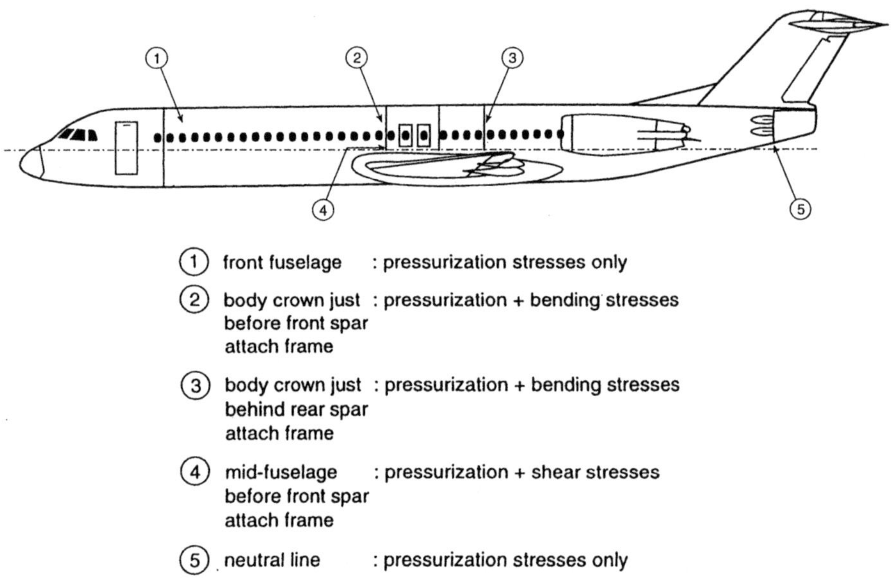

Fig. 1.9 Transport aircraft fuselage loading conditions (Wanhill 1996)

Lap joints and single-strap joints are the connections with eccentricities occurring in the overlap which introduce the so-called secondary bending. The magnitude of the maximum bending stress at locations shown by arrows in Fig. 1.10 can be as large as the applied tensile stress (Schijve 2009b). Consequently, the fatigue performance of the joints with eccentricities is inferior to that of symmetric joints, like the double-strap joints for which the secondary bending does not occur. This is demonstrated in Fig. 1.11 presenting results of constant amplitude (CA) fatigue tests by Hartman and Schijve (1969) which indicates that, for the same rivet row spacing, the fatigue strength at 2×10^7 cycles for the double-strap joint is about twice that for the lap joint.

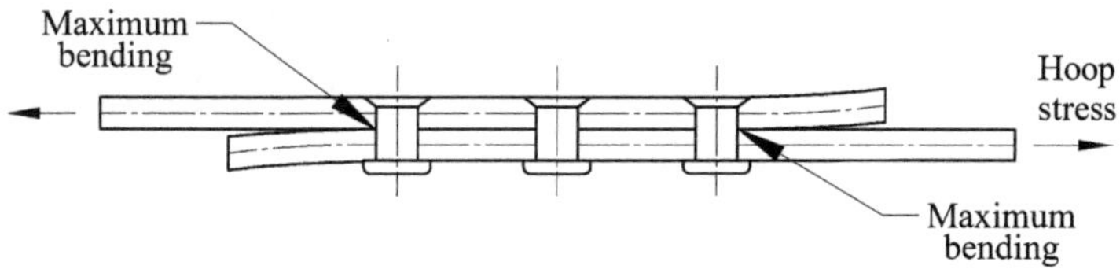

Fig. 1.10 Secondary bending in a lap splice joint

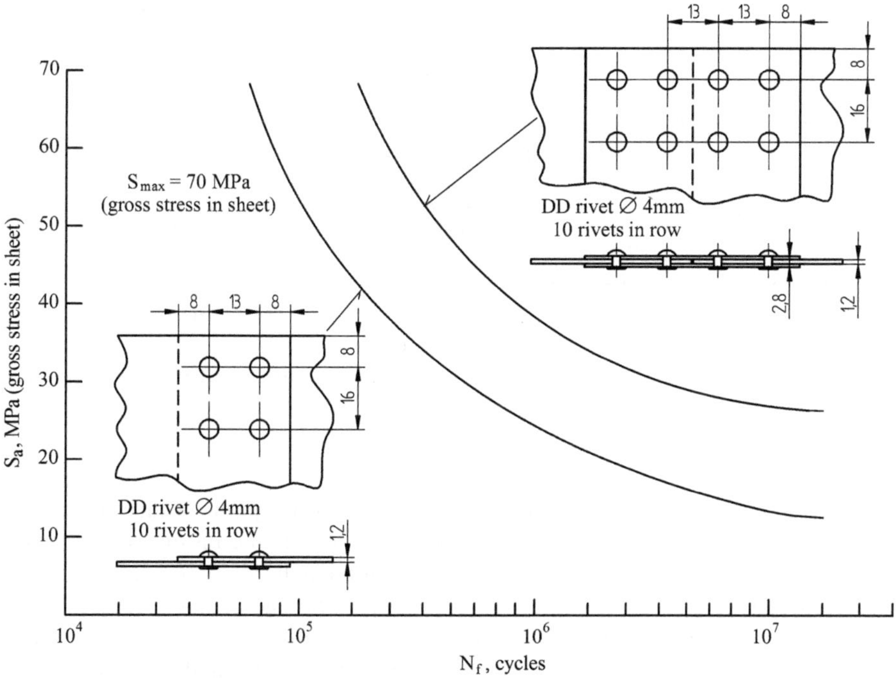

Fig. 1.11 Fatigue curves for different types of riveted joints. Material 2024-T3 Alclad (Hartman and Schijve 1969)

1.3 Bonded and Riveted-Bonded Lap Joints

Because riveted-bonded and adhesive-bonded joints are outside the scope of the present work, only limited information on their fatigue properties in comparison to riveted joints is presented below. A combination of riveting and adhesive bonding is applied to improve the strength characteristics of joints. Replacing the riveting with the adhesive bonding of the Al-alloy sheet material in a lap joint, or combining both processes, is attractive for several reasons. In a bonded joint, due to the continuous attachment of the overlapping sheets, stress concentration does not occur. On the

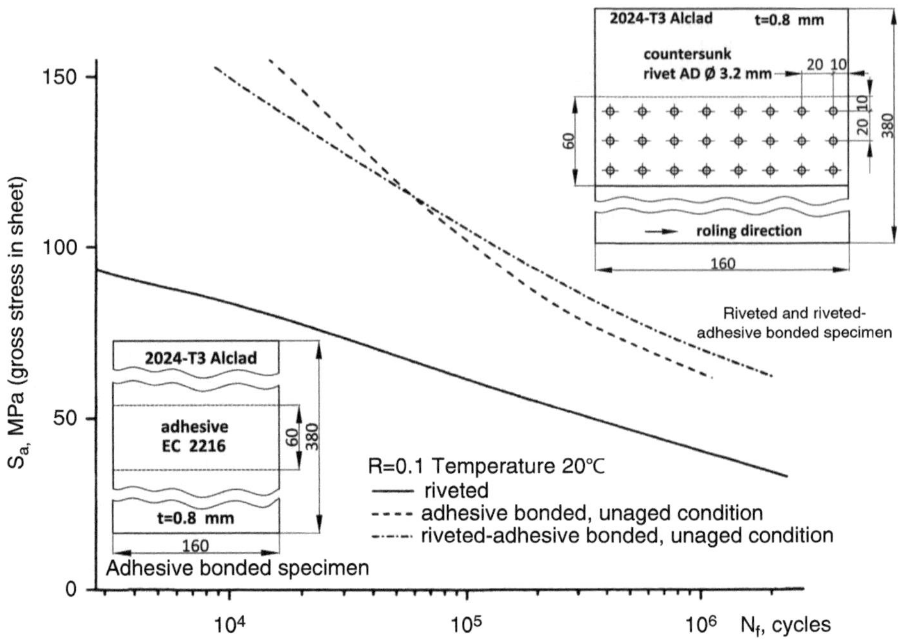

Fig. 1.12 Fatigue curves for riveted, adhesive bonded and riveted-adhesive bonded joints (Hartman1966)

contrary, a severe stress concentration does occur in riveted joints because the sheets are attached to one another at discrete points only, i.e. by the fasteners. Unlike for the riveted joints, metallic contact is absent from the adhesively bonded joints and thus fretting is eliminated (Schijve 2009a). From the fatigue point of view, adhesive bonding is therefore a preferred joining method. A confirmation has been already offered due to early experiments by Hartman and Duyn (1952) and Hartman (1966). In the latter investigation the fatigue tests were carried out on single-lap joints of 2024-T3 Alclad sheets 0.8 mm thick with three rows of countersunk AD type (2117-T4 alloy) rivets 3.2 mm in diameter. Similar specimens were additionally adhesive bonded with AW 106 (manufactured by CIBA) and EC 2216 (3M) adhesives. Neither adhesive requires curing at elevated temperatures, which is very convenient from the practical point of view. A curing cycle for some days at room temperature is applied instead. The test data shown in Fig. 1.12 indicate significantly improved fatigue properties of the joints when adhesive bonding or riveting-adhesive bonding is applied.

1.4 Fatigue Damage of Longitudinal Lap Splice Joints

Load transmission in a multiple row lap joint occurs for a larger part by the outer rivet rows, where also maximum moments due to the secondary bending are induced. For this reason these rows are considered fatigue critical, where the cracks are initiated.

For longitudinal lap joints in aircraft fuselage structures a typical failure scenario is the so-called Multi-Site-Damage (MSD). It is a situation when many cracks are initiated at about the same time in a critical row of rivets, grow in a similar manner, and suddenly coalesce to form a single long crack, which can lead to catastrophic failure. Crack initiation and early crack growth prior to the occurrence of a visible crack consume the major part of the joint fatigue life. The period from a first crack link-up to failure takes only a few percent of total fatigue life. On the other hand, because damage development during the first stages of MSD is rather slow, it is possible to detect cracks when scheduled inspections are performed.

1.5 Summary of This Chapter

The contemporary transport aircraft fuselage is a stiffened skin structure that contains stringers (longerons) parallel to the aircraft axis and frames perpendicular to that axis. The stringers are connected to the skin, and the frames can be directly joined with the skin (so-called shear tied frames) or connected only to the stringers by means of stringer clips (floating frames).

In order to force longitudinal cracks in the fuselage skin to turn circumferentially (the so-called flapping), tear straps are located at frames and sometimes also between the frame stations. Due to the tear straps a damage area can be limited to only one bay of the fuselage. Sometimes, instead of the fail-safe straps, waffle doublers bonded to the skin are applied.

A primary loading for the skin of a pressurized fuselage is the hoop (circumferential) stress resulting from a differential pressure between the atmospheric pressure and the cabin pressure. For this reason, longitudinal skin joints are usually more critical than circumferential joints. A typical solution for the skin longitudinal connection is the lap splice joint. Lap joints are the connections with eccentricities occurring in the overlap which introduce the so-called secondary bending with a high magnitude of resulting bending stress.

Overlaps of the mating sheets are generally assembled together using rivets, but riveted-bonded or adhesive bonded joints are also applied. A typical failure scenario for longitudinal riveted lap joints in the fuselage skin is the so-called Multi-Site Damage (MSD).

Chapter 2
Differences Between the Fatigue Behaviour of Longitudinal Lap Joints in a Pressurized Fuselage and Laboratory Lap Joint Specimens

2.1 Stress Distribution and Specimen Geometry

As already said earlier, a primary loading for the fuselage skin is hoop tension. However, the total effect of pressurization is a combination of hoop and longitudinal tension and local out-of-plane bending of the skin, the so-called pillowing. Besides the circumferential pillowing between the stiffeners shown in Fig. 2.1a, pressurization also causes a longitudinal pillowing of the skin and stringers between the frames, Fig. 2.1b. Because the stringers hardly restrain the skin pillowing, this effect will be strongly dependent on the type of the stringer-frame connection, for example the shear-tied frame or "floating" frame, cf. Figs. 1.2 and 1.3. Due to the pillowing, the hoop stress is not uniformly distributed between the frames. The loading complexity is additionally increased by the tear straps. A comparison between the measured and predicted (from FE analysis) membrane hoop stresses in a panel typical for the Boeing narrow-body fuselage airplane is shown in Fig. 2.2. The structure consists of "floating" frames and riveted tear straps. The data are for the location along an axial line adjacent to the lap joint. The stresses are normalized by the nominal hoop stress in an equivalent unstiffened cylinder of the same radius R and lap joint skin thickness t. The maximum stress equal to 80% of the nominal stress occurs midway between the tear straps, while the stresses in the vicinity of the tear straps are much lower.

A non-uniform stress distribution in the skin of fuselage panels was also observed by Terada (2001) and Molent and Jones (1993). Stress measurements by Terada were carried out on a one-third scale model of the B-737 fuselage structure. Figure 2.3, showing his measurement results, indicates that the hoop stress at location A close to the frame is only around 50% of the stress value recorded in the middle of the central bay of the fuselage panel (location D). At the same time, the latter stress is slightly lower than that at location F close to the stringer. Molent and Jones reported a 25% difference in the hoop stress measured at the frames and midway between the frames in a fuselage panel of Boeing 727. For the frames, the measured hoop stress was 63% of that in an unstiffened vessel, whereas the stress

A. Skorupa and M. Skorupa, *Riveted Lap Joints in Aircraft Fuselage*, Solid Mechanics and Its Applications 189, DOI 10.1007/978-94-007-4282-6_2,

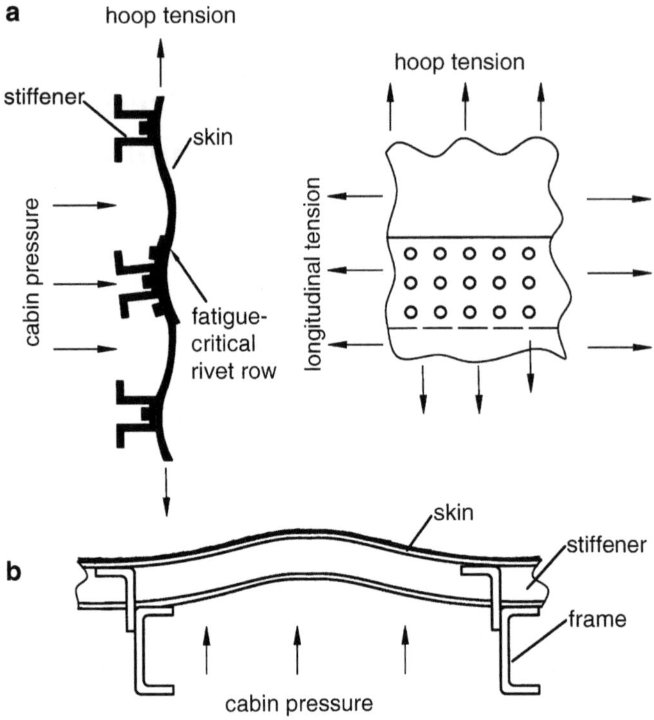

Fig. 2.1 Effect of cabin pressurization on the longitudinal lap splice joint deformation: (**a**) pillowing between stringers; (**b**) pillowing between frames

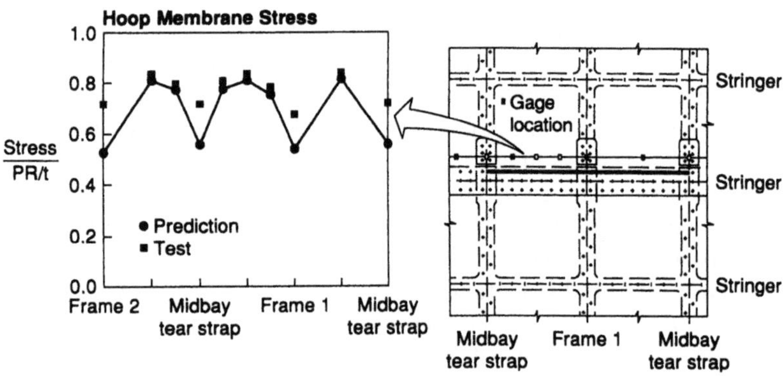

Fig. 2.2 Example of hoop stress distribution between frames (Miller et al. 1994)

recorded midway between the frames was only 17% lower. For Boeing 737 a 27% difference in the hoop stress at the frames and in the middle of the waffles was observed. In general, the hoop stress reduction at the frames can be on the order of

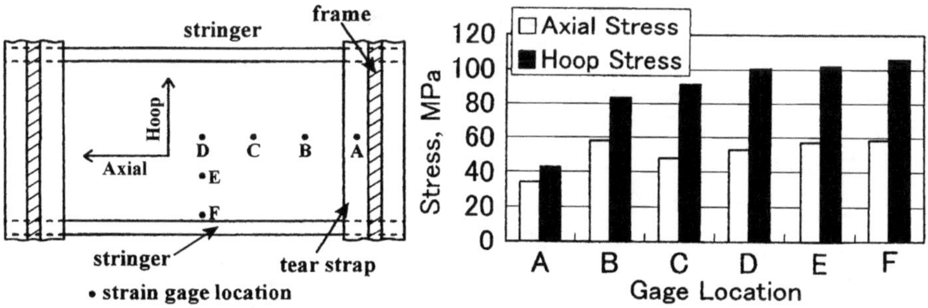

Fig. 2.3 Stress distribution measured at the internal surface of the central bay of a fuselage panel (Terada 2001)

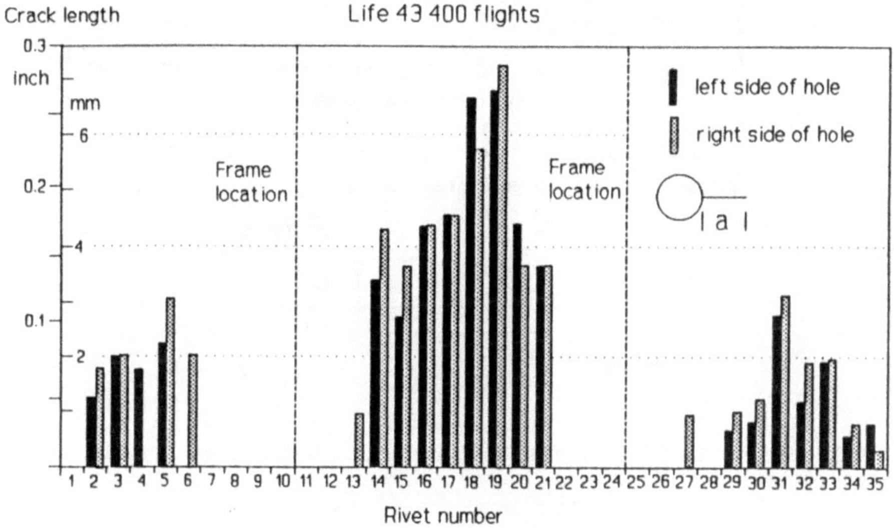

Fig. 2.4 Fatigue cracks in a riveted lap joint of an aging B727 fuselage (Mayville and Warren 1991)

20–30% for a stiff connection between the frame and skin (shear tied frame), and on the order of 10% for a low stiffness frame-skin joining (floating frame) without crack stopper bands at the frames (Müller 1995).

The stress behaviour revealed in the aforementioned investigations is consistent with measurement results on the distribution of crack lengths at rivet holes for an aging Boeing 727 aircraft panel reported by Mayville and Warren (1991) and produced in Fig. 2.4. It is seen that the longest cracks were detected midway between the frames while no cracks were found at the frame locations.

It is obvious that the load distribution along a rivet row must be affected by the frame stiffness, as illustrated by Fig. 2.5 which presents the rivet load distributions in the end row obtained from 3-D FE calculations by Müller (1995) for two ratios

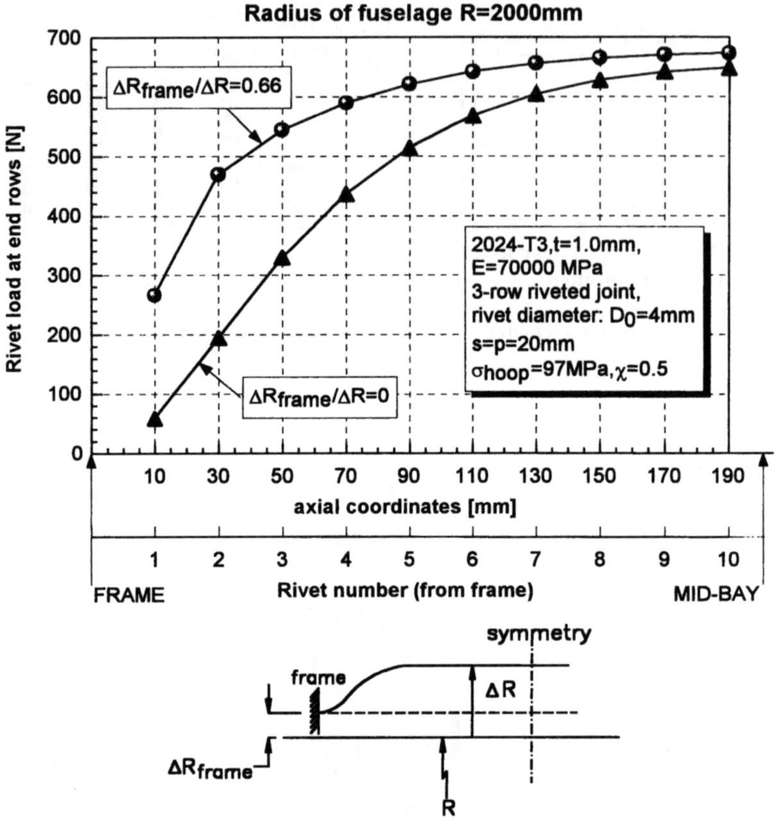

Fig. 2.5 Effect of frame stiffness on the load distribution along a rivet row. Biaxiality ratio 0.5 (Müller 1995)

of the frame radial displacement ΔR_{frame} to the radial expansion of an unstiffened cylinder ΔR,

$$\Delta R = \frac{pR^2}{Et}\left(1 - \frac{\nu}{2}\right), \tag{2.1}$$

where R is radius of the cylinder, p is cabin pressure, t is sheet thickness, E is the Young modulus, and ν is Poisson's ratio.

It is seen that for infinitely stiff frames ($\Delta R_{\text{frame}} = 0$) the load transfer in between the frames is highly non-uniform. If the frame expands $0.66(\Delta R)$, the load transfer becomes already more homogeneous.

Not only the frame stiffness but also the tear strap configuration plays a considerable role in the load distribution in the top rivet row. Furuta et al. (1997) considered two types of flat panels with three rivet rows in a longitudinal joint, as shown in Fig. 2.6a. The only difference between the panels was the tear strap width.

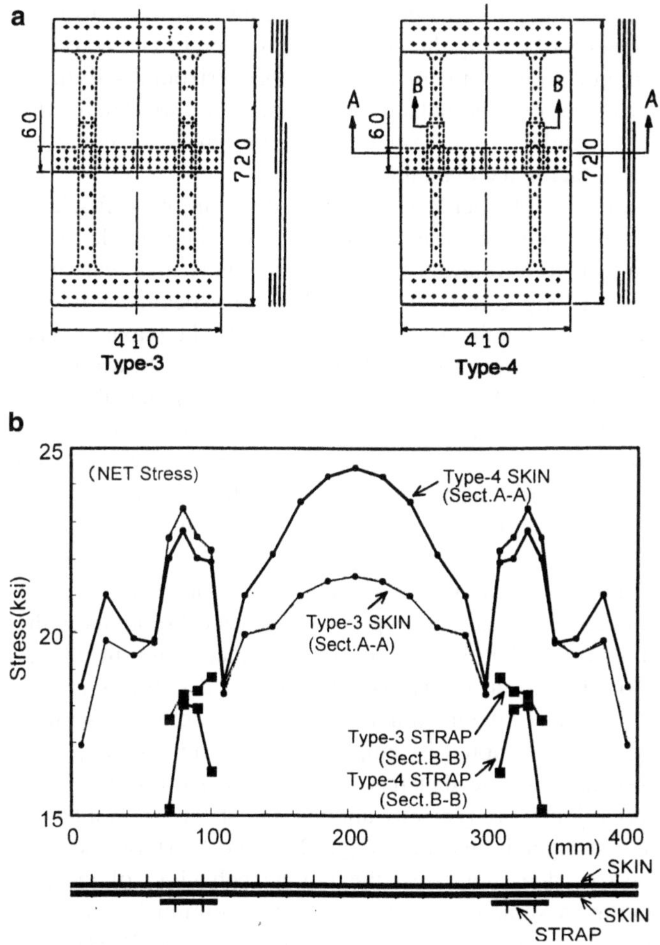

Fig. 2.6 Stress distribution for lap joints in wide panels with tear straps: (**a**) panel geometries; (**b**) FE results (Furuta et al. 1997)

The Type-3 panel had a constant width tear strap. The tear strap width in the Type-4 panel was narrowed compared to the Type-3 specimen except at the spliced area. FE analysis results shown in Fig. 2.6b indicate that the narrowing of the tear strap width causes a substantial increase in the stress in the skin splice.

For an unstiffened pressure vessel the biaxiality ratio (longitudinal to hoop stress) is 0.5. For a stiffened pressure vessel, as for example the fuselage, the biaxiality ratio should be less than 0.5 due to the longitudinal stiffening elements, i.e. the stringers. However, the longitudinal pillowing of the skin between the stringers creates an additional longitudinal stress resulting in an increase of the biaxiality ratio, as seen in Fig. 2.3.

Additional longitudinal stresses in the fuselage skin are also generated as a result of the fuselage bending due to the aircraft weight and aerodynamic loading (specifically the bending moment created by the horizontal stabilizer). Depending on the location of the longitudinal lap splice joint with respect to the fuselage neutral axis, the stress parallel to the rivet line may be tensile or compressive, thereby augmenting or reducing the biaxiality effect.

Despite a complex character of the stress distribution in fuselage longitudinal riveted lap joints, laboratory tests on simple, uniaxially loaded small specimens can provide information relevant to the riveted joints in a real structure. Vlieger (1994) carried out fatigue tests on riveted lap joint specimens under CA uniaxial and "in phase" biaxial loading (biaxiality ratio 0.5). The biaxial loads were applied to the test section by means of load introducing arms, as shown in Fig. 2.7. Fatigue lives of the biaxially loaded specimens were always higher than for the uniaxial loading, Fig. 2.8. A similar fatigue behaviour was observed in tests carried out by Vlieger and Ottens (1997) on three-row countersunk riveted specimens typical for the Fokker fuselage longitudinal lap joint. These results suggest that fatigue tests under uniaxial loading can provide safe results on the fatigue performance of biaxially loaded riveted specimens.

Stiffener doublers, if sufficiently thick and wide, were found long ago to favourably influence the fatigue properties of riveted lap joints with a single row of rivets (Hertel 1969). For this type of joint geometry, the stiffener prevents inner sheet deflection directly at the rivets. The increased stiffness created around the rivet leads to the shift of the maximum secondary bending stress in the inner sheet away from the hole. It should be emphasized that single-row joints represent an extremely poor design case that has not been used for years. Nowadays, when three-row riveted joints are typically applied, the sheets are not bent in the middle row where the stringer is installed, Figs. 2.9a and 2.10. So, adding the stiffener in the middle row does not affect the fatigue behaviour of the joint unless the stiffener rotation is somehow restrained, which diminishes secondary bending, Fig. 2.9b. This conclusion is confirmed by Vlieger's et al. (1994) experiments, who performed fatigue tests on uniaxially loaded panels (cf. Fig. 2.8) with and without a longitudinal stiffener, and did not find any significant effect of the stringer on the panel fatigue life. Furthermore, in uniaxial tests the loading of the stiffener will not be realistic. Due to Poisson's effect, the stiffener will be loaded in compression, which in reality is hardly the case (Vlieger et al. 1994). The above cited works substantiate neglecting the stringers in test specimens.

It should be noticed that the so-called edge effect always occurs when uniaxially loaded test specimens are used. It is caused by differences in the lateral contraction (Poisson's effect) in two overlapping sheets, Fig. 2.11a, due to differences in the stress levels associated with load transmission through the joint. Consequently, edge rivets of each row are additionally loaded by the lateral contraction displacement, which leads to a premature crack initiation at these rivets in the critical rows, Fig. 2.11b. The cracks grow to the specimen edges and cause a general collapse before significant crack nuclei at the inner rivet holes develop. The edge effect is more pronounced for wider specimens and, if not suppressed, it is responsible for the

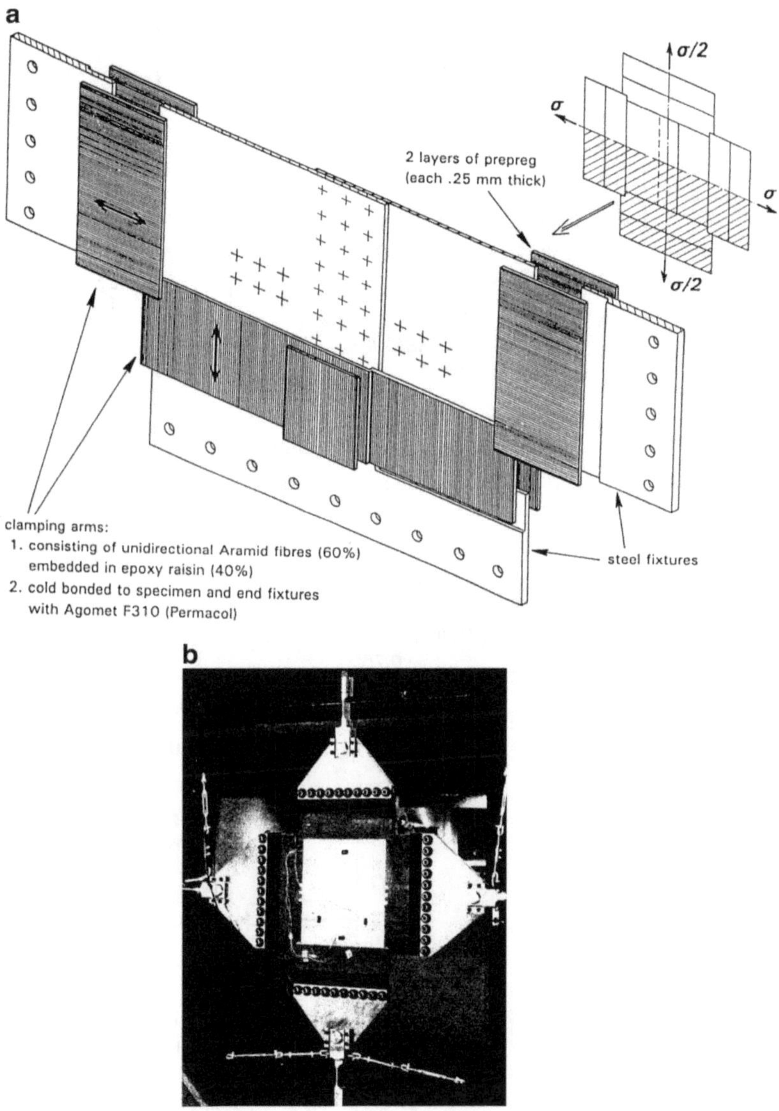

Fig. 2.7 Biaxially tested lap joint specimens: (**a**) specimen design; (**b**) specimen mounted in the test frame (Vlieger et al. 1994)

undesirable dependence of the fatigue life on the specimen width. Measures taken to avoid edge cracking are overviewed by Müller (1995) and Fawaz (1997). According to the present authors' experience, applying a somewhat higher squeezing force when installing the edge rivets in the critical rows can effectively prevent the edge cracking.

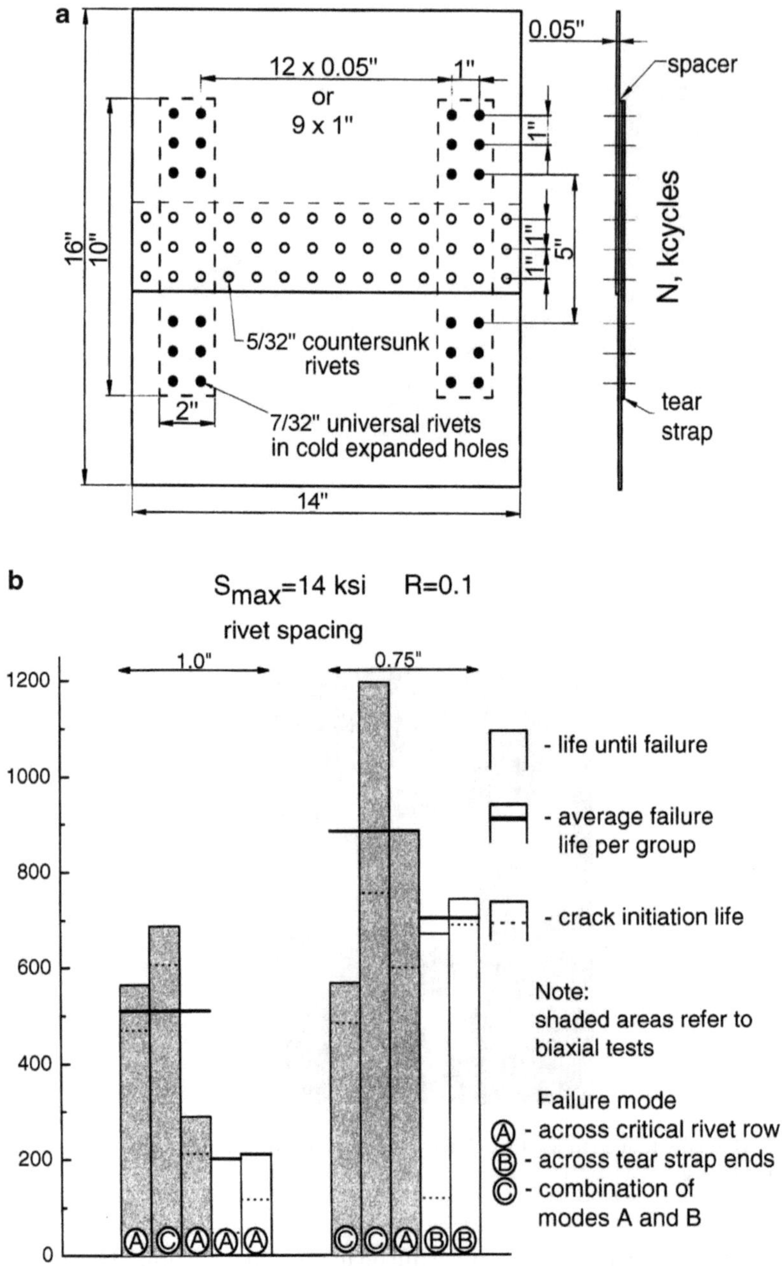

Fig. 2.8 Comparison between uniaxial and biaxial fatigue test results for riveted lap joint specimens: (**a**) specimen geometry; (**b**) results on fatigue lives and failure mode (Vlieger et al. 1994)

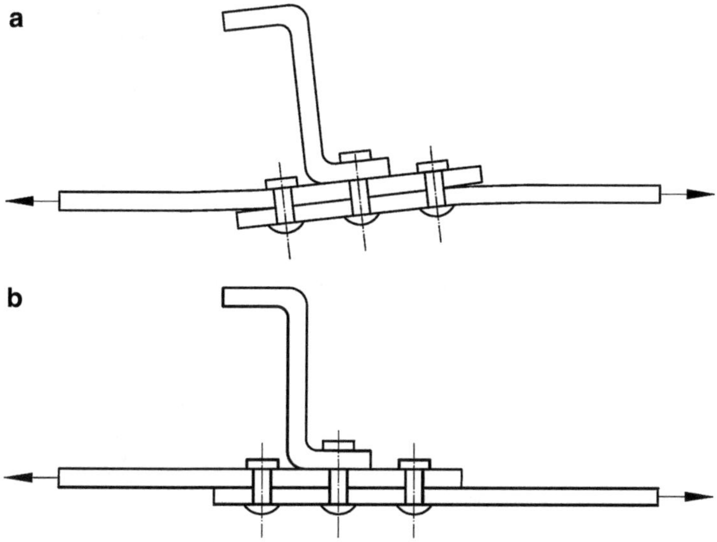

Fig. 2.9 Deflection of a three-row riveted lap joint with a stiffener: (**a**) stiffener rotation possible; (**b**) stiffener rotation restrained

Figure 2.12 presents the fatigue lives for identical lap joint configurations with either 25 or 3 rivets per row corresponding to a specimen width of 500 and 60 mm respectively (Müller 1995). For the wide specimens the edge effect was avoided by applying clamping pieces at the edges. The total fatigue lives were always longer for wide specimens than for narrow specimens. The same conclusion follows from the work by Vlieger and Ottens (1997) who conducted biaxial fatigue tests ($S_{max1} = 96$ MPa, $S_{max2} = 0.5S_{max1}$, $R = 0.1$) on three-row riveted lap joint specimens made of 2024-T3 Alclad sheet 1.2 mm thick. For all specimens the rivet spacing (countersunk rivet 3.2 mm dia) was equal to 20 mm. The only difference was the specimen width, i.e. 120 and 180 mm. The average fatigue life for the narrower specimens was 263 versus 533 kcycles for the wider specimens. The explanation of the specimen width effect on fatigue life is that, unlike for the narrow specimens, link-up of different cracks can occur for the wide test pieces without immediately leading to the net section yielding and final fracture. The above observations support the use in laboratory tests of possibly wide specimens which can provide a more realistic representation of the fatigue damage development for a real fuselage structure. However, the narrow specimens can still be useful for parametric studies.

Lap joints in the fuselage structure are curved. For example, Boeing's typical narrow body and wide body aeroplanes have a radius of curvature of 74 in. (1.88 m) and 127 in. (3.23 m) respectively. However, small specimens typically used in laboratory tests are flat. Müller (1995) employed a special test set-up to apply a uniaxial fatigue loading on bent riveted lap joint specimens. He observed fatigue

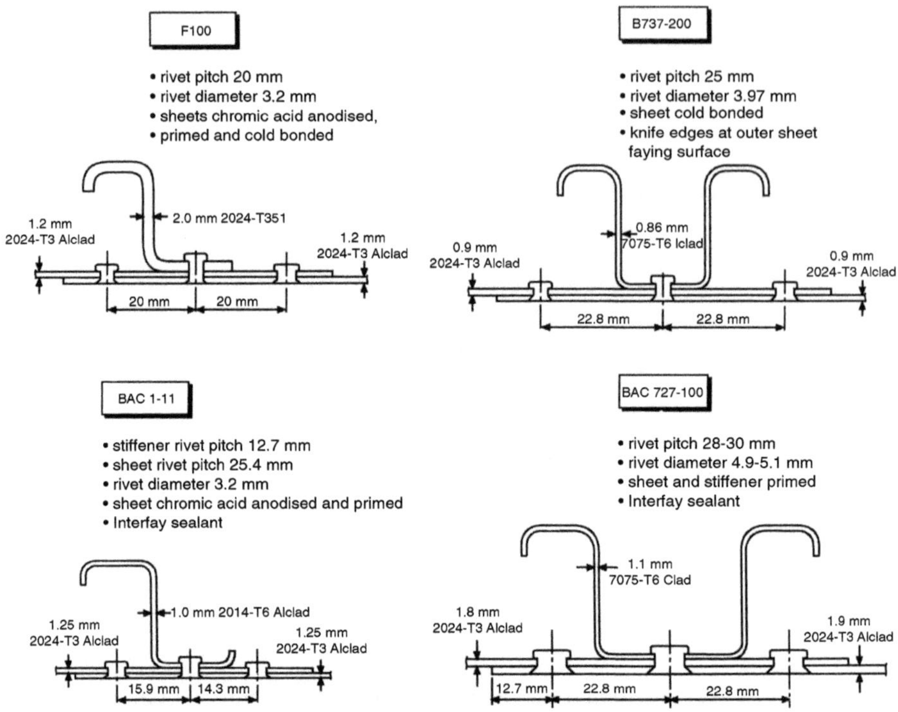

Fig. 2.10 Examples of fuselage lap joints with stringers (Wanhill and Koolloos 2001)

lives for the curved specimens to be significantly shorter than for the flat specimens of the same configuration (three-row riveted lap joint made of 2024-T3 Alclad sheet 1 mm thick, countersunk rivet NAS1097AD5-5). Also crack growth in a barrel test (biaxial loading) was considerably faster than in the flat specimens due to the bulging of the skin caused by the crack. This effect is recognized by the aircraft industry when considering test results for relatively small, simple, lap joint specimens and properties of the real fuselage structures. The discrepancies can be partly associated with the influence of the stiffening elements' geometry on the pillowing and bending of the fuselage skin and with the effect of the internal pressure (biaxial loading). The curvature forming method (elastic bending or rolling), the squeeze force, and possible imperfections of the riveted joint in a structure resulting from the manufacturing process should also be taken into account.

According to Müller (1995), in order to obtain the same fatigue life for a simple laboratory specimen as for a joint in the fuselage, the stress applied on the specimen should be 1.2 times larger than the nominal stress in the fuselage skin. Such an increase in the stress on the specimen typically yields an order of 2 decrease in its fatigue life.

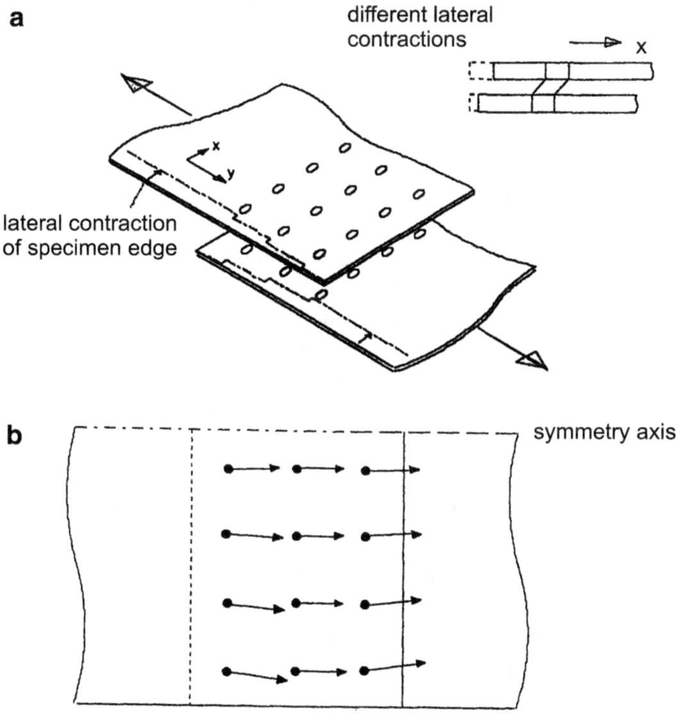

Fig. 2.11 Edge effect in a uniaxially loaded lap joint: (**a**) differences in lateral contraction within the overlap; (**b**) rivet force direction deviations due to the edge effect (Müller 1995)

Fig. 2.12 Influence of specimen width on fatigue life. Edge effect prevented (Müller 1995)

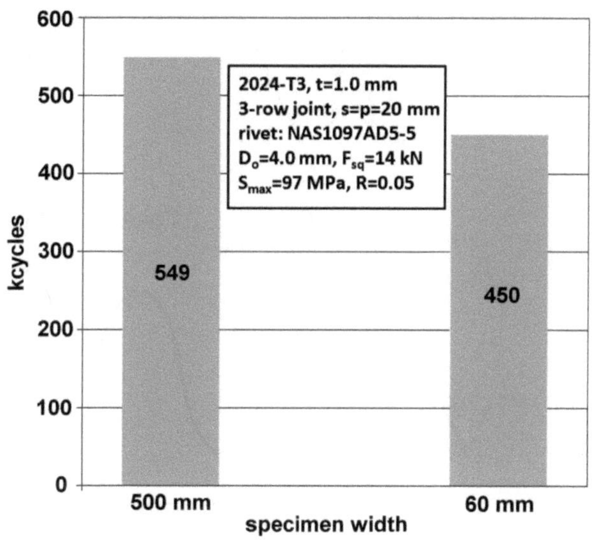

2.2 Effect of the Load Frequency and Environmental Conditions

For a fuselage lap joint one load cycle is equal to the pressure cycle of the cabin. The period of this pressure cycle can range from about half an hour to ten or even more hours. A part of the pressure cycle relevant to crack growth is the climb period (thicker line in Fig. 2.13a). Tests performed by Schijve (2009a) on 7074-T6 alloy specimens submerged in salt water have shown that the crack growth rate becomes higher for a longer rise time, i.e. for a slower load increase. The duration of a cruise period does not eminently influence the crack extension. For this reason, the simplified pressure cycle seen in Fig. 2.13a can be replaced by a saw tooth wave shape shown in Fig. 2.13b, provided that the rise time *t* for both wave shapes is the same. But in laboratory tests a sinusoidal wave shape is usually applied, Fig. 2.13c. Because of the accelerated crack extension in the upper part of the sinusoidal waveform compared to the saw tooth wave shape (resulting from higher stresses, Fig. 2.13d), the sinusoidal waveform leads to more conservative test results.

Especially in the case of parametric studies, when the fatigue tests are performed to compare the endurances for various options, load frequencies must be much higher than those resulting from the duration of the climb period (typically on the order of 10 min). A beneficial influence of increasing the load frequency on the fatigue behaviour of riveted lap joints seems to be confirmed by some experiments.

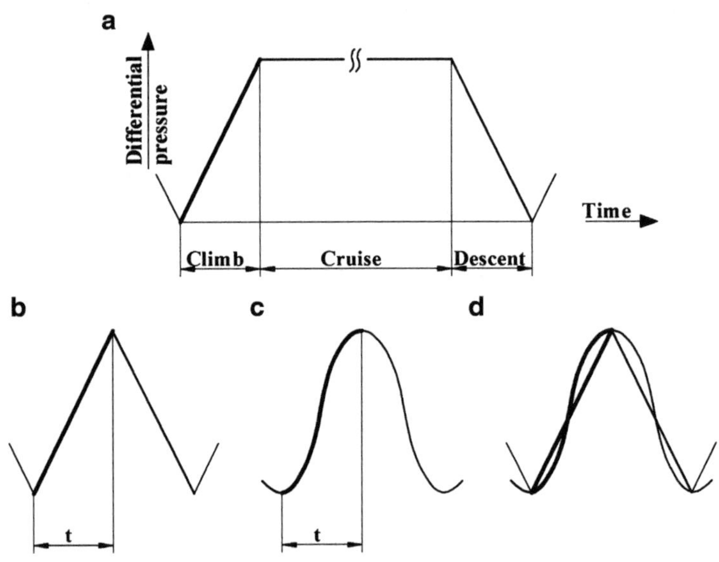

Fig. 2.13 Simulation of cabin pressurization cycle in fatigue tests: (**a**) simplified cabin pressurization cycle; (**b**) saw tooth waveform; (**c**) sinusoidal waveform; (**d**) comparison between both waveforms

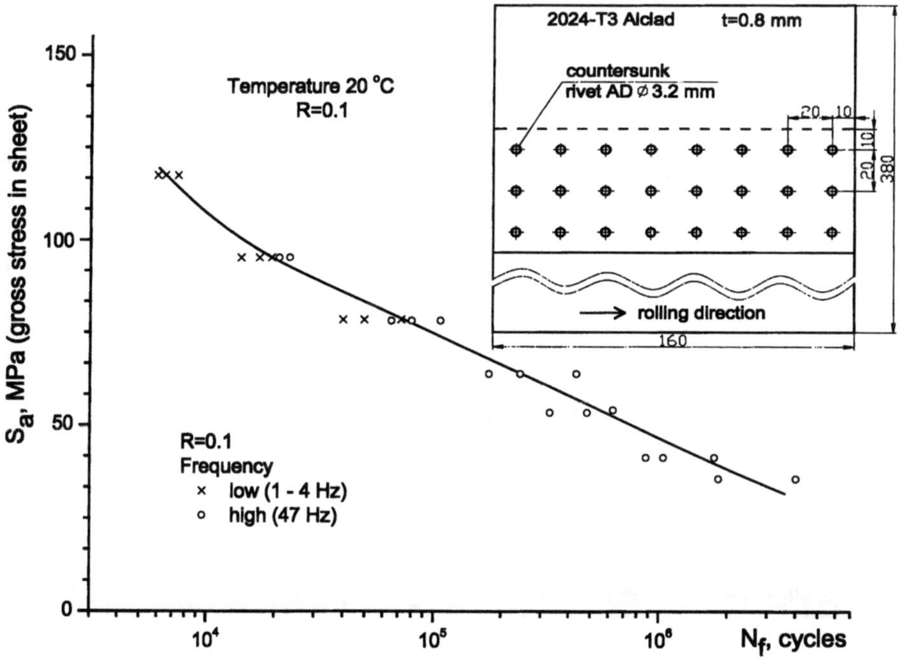

Fig. 2.14 Influence of test frequency on lap joint fatigue lives (Hartman 1966)

The frequencies used by Klaassen (1955) were 0.15 and 40 Hz, and the data revealed no frequency effect for the 2024-T3 alloy, but for the 7075-T6 material a higher endurance at the higher frequency was observed. On the other hand, experiments carried out by Hartman (1966) implied a fairly small beneficial effect of increasing the loading frequency for the 2024-T3 aluminium alloy, Fig. 2.14. The same tendency was noted by Schütz (1963) for three-row dimpled lap joints tested at frequencies of 0.5 and 30 Hz. Furthermore, he observed a decreasing scatter of the specimen fatigue life with an increasing frequency. In CA fatigue tests carried out by Müller (1995) on 2024-T3 Alclad material three-row lap joint specimens two different load frequencies of 1 and 10 Hz yielded practically the same fatigue lives.

The present aircraft generation is designed for an economical operational life up to 90,000 and more flight cycles. Therefore, corrosion of metallic materials is one of the most important processes affecting the life of an aeroplane structure. It is obvious that corrosion damage depends first of all on the in-operation environmental conditions. Furuta et al. (1997) investigated the effect of a corrosive environment (3.5% Na Cl solution simulating sea water) using spliced panel specimens representative for the fuselage structure of B-737 and shown earlier in Fig. 2.6a. Sheets made of the 2024-T3 Clad aluminium alloy 1 and 2 mm thick were joined with three rows of countersunk rivets. No sealant or adhesive was applied during the manufacturing of the panels. The specimens were fatigued under CA loading ($S_{max} = 96$ MPa,

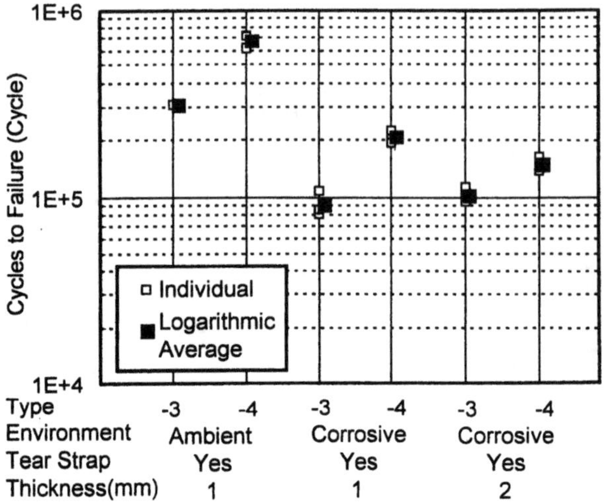

Fig. 2.15 Influence of the corrosive environment of fatigue lives of wide panel specimens (Furuta et al. 1997)

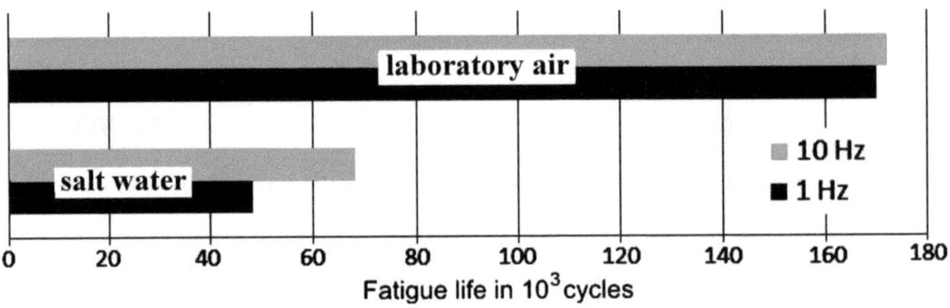

Fig. 2.16 Influence of environmental conditions on fatigue life of lap joint specimens (Müller 1995)

$R = 0.125, f = 5$ Hz). As can be seen in Fig. 2.15, fatigue lives of the specimens tested in the corrosive environment were about 1/3 of those observed in the ambient condition. The skin thickness effect was insignificant.

A similar decrease in the lap joint fatigue life caused by the corrosive environment was observed by Müller (1995). The specimens with three rows of countersunk rivets joining 1.8 mm thick 2024-T3 sheets were immersed in salt water (5% Na Cl) and tested under CA loading ($S_{max} = 120$ MPa, $R = 0.1$). No interfaying sealant was applied prior to riveting. As shown in Fig. 2.16, the fatigue life in the salt water was about three times shorter than in the laboratory air. A frequency effect can also be observed in the considered figure. The low frequency of 1 Hz caused a 25% reduction in life in salt water compared to the results for 10 Hz. No influence of frequency was, however, revealed in the laboratory air. A similar effect of the frequency (1 and 30 Hz) in the salt water was observed for small specimens by Furuta et al. (1997), but the difference in their fatigue lives was less evident than in Müller's tests referred to above.

On the other hand, experiments performed by Daimler Benz Aerospace Airbus (DA) on specimens representing a fuselage longitudinal lap joint (skin material 2024-T3) revealed no influence of different environments, i.e. sodium chloride (3.5% solution) and laboratory air (Schmidt and Brandecker 1997). However, in contrast to the tests by Müller and Furuta et al. addressed above, the DA specimens were manufactured according to production standards, i.e., with the surface treatment (CA anodizing, primer, top coat, wet assembling and wet riveting). Evidently, laboratory tests on small specimens can be useful in research on corrosion damage only if during the manufacturing and assembling of the test pieces the production standards are maintained.

The pressurized aircraft fuselage is exposed in flight to a low temperature on the order of −50°C, whereas laboratory tests on riveted specimens are performed at room temperature. Information on the fatigue properties of riveted joints in low temperatures is lacking in the literature. Schijve (2009a) refers to tests carried out by Broek on the 2024-T3 and 7075-T6 Al alloys which indicate that crack growth becomes slower at lower temperatures. A lower water vapour content in air at low temperatures can be responsible for this trend. It is expected that the same favourable effect of a low temperature will also apply to the fatigue strength of riveted joints. It should be noticed that fracture toughness of the 2024-T3 alloy remains stable down to a temperature on the order of −50°C. However, laboratory tests are not capable of simulating environmental effects in a realistic way. The test duration is too short to cause corrosion damage in the lap splices and the exterior of the test article is not cooled to cruise altitude temperatures during each pressurisation cycle. This implies that any transpiration is unlikely to result in moisture condensation and entrapment within the lap splices.

The influence of environmental conditions on the fatigue properties of aluminium skin materials is considered in Sect. 3.1.

2.3 Summary of This Chapter

Due to complex loading conditions of the skin (a combination of the hoop and longitudinal stress, local out-of-plane bending) a non-uniform stress distribution in the skin is observed. In general, the hoop stress at the frames is lower than in the middle of the bay. A hoop stress reduction at the frames depends on stiffness between frame and skin and fail-safe strap design.

Despite a complex character of the stress distribution in fuselage longitudinal lap joints, laboratory tests on simple, uniaxially loaded small specimens can bring useful information relevant to the riveted joint in a real structure. Especially, it concerns the influence of production variables and design parameters on the fatigue behaviour of joints.

Fatigue properties obtained from tests on flat specimens are superior to those observed for curved specimens (barrel tests). This effect should be accounted for when transferring test results for flat specimens to a real fuselage structure.

Laboratory tests on small specimens can be useful in research on corrosion damage only if during manufacturing and assembling of the test pieces the production standards are maintained. However, laboratory tests cannot fully account for environmental effects.

Fatigue tests under uniaxial loading provide conservative results on the fatigue performance of flat, biaxially loaded riveted specimens. When uniaxially loaded specimens are tested, the so-called edge effect (difference in the lateral contraction in two overlapping sheets) occurs. Because this effect leads to a premature crack initiation in edge columns of rivets, these rivets should be installed using a higher squeeze force. If MSD in lap joints is studied, possibly wide specimens should be used.

Loading frequency in constant amplitude tests seems to be of little importance for the test results.

Chapter 3
Production Variables Influencing the Fatigue Behaviour of Riveted Lap Joints

A large number of production variables can affect the fatigue behaviour of a riveted joint. Among these, the most important are the sheet material, the fastener type and material, and the manufacturing process. The influence of the above factors on lap joint fatigue properties is reviewed in this chapter.

3.1 Sheet Material

Two basic types of sheet material are used for the fuselage skin, namely 2024-T3 and 7075-T6 aluminium alloys (this present day nomenclature has replaced the older one, namely 24 ST and 75 ST respectively). The minimum mechanical properties required of all sheets by the US specification are compiled in Table 3.1. Either sheet material can be provided with a soft aluminium layer (pure Al for 2024 alloy and Al + 1% Zn for 7075 alloy) at both sides of the sheet, thickness about 5% of the total thickness each. The cladding layers are applied for corrosion protection (Schijve 2009a).

Static properties, such as yield stress and tensile strength of both materials, are somewhat better when tested parallel to the grain direction, but the elongation does not show a noticeable dependence on the rolling direction. The loading direction effect is larger for yield stress than for ultimate strength. For example, the directionality of the 2024-T3 sheet materials from seven different manufacturers, illustrated by the ratio between the yield stress in the transverse and the longitudinal direction, equalled 0.9 on average (Schijve 1968). Tensile tests of lap joints with AD or DD rivets (see next section) prove that the maximum load for specimens oriented perpendicular to the rolling direction can be even 10% lower than for specimens cut out parallel to the rolling direction (Hartman et al. 1965).

A distinct directionality effect is also observed under fatigue loading conditions. For 2024-T3 Alclad sheet material the crack propagation life is about 40% longer if the material is loaded in the rolling direction compared to the specimens loaded in

A. Skorupa and M. Skorupa, *Riveted Lap Joints in Aircraft Fuselage*, Solid Mechanics and Its Applications 189, DOI 10.1007/978-94-007-4282-6_3,

Table 3.1 Mechanical properties of sheet materials

Sheet material	Thickness (mm)	S_y (MPa)	S_u (MPa)	δ (%)	Specification
2024-T3	0.25–1.57	276	414	15	AMS-QQ-A-250/5
Alclad	1.60–3.25	289	427	15 $l = 2''$	MIL-HDBK-5H October 2001
7075-T6	0.30–0.99	482	538	7	AMS-QQ-A-250/13
Clad	1.06–3.18	482	551	8 $l = 2''$	MIL-HDBK-5H October 2001
D16Cz	0.50–1.90	270	405	13	OST1 90070-92
	1.90–10.5	275	425	11	

Note: Loading direction across rolling direction

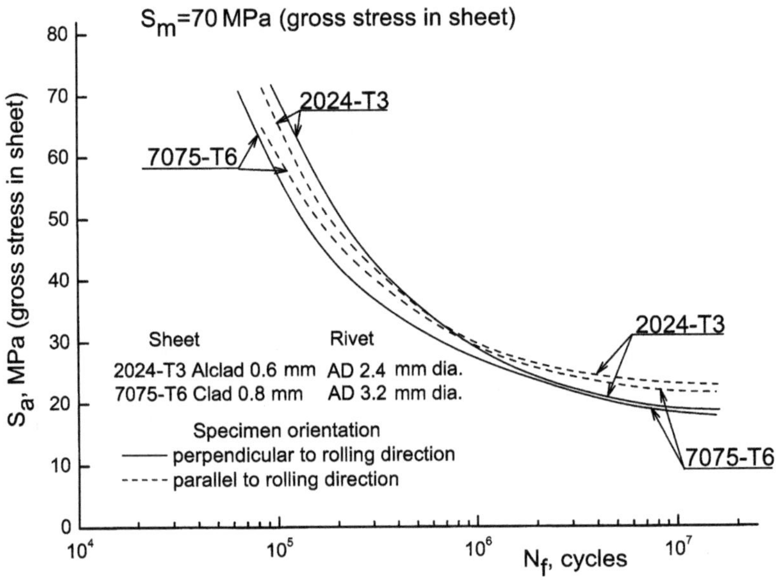

Fig. 3.1 Effect of specimen orientation relative to rolling direction on the fatigue life of double row lap joints (Hartman et al. 1965)

the transverse direction (Schijve and de Rijck 1966). However, CA tests revealed no consistent differences in fatigue life for both cases of the riveted specimen orientation, Fig. 3.1 (Hartman et al. 1965).

In general, 2024-T3 Alclad is slightly superior to the 7075-T6 clad alloy with respect to their fatigue properties, but its static strength is lower. If a high static tensile strength is required for a component, the relatively high S_y and S_u of 7075-T6 can be advantageous compared to the lower values for 2024-T3.

Experimental results that back up the opinion that the 2024-T3 fatigue properties are better than those of 7075-T6 have been reported in several works, e.g. (Hartman and Klaassen 1956; Schijve 1956; Hartman et al. 1965). The *S-N* curves from CA

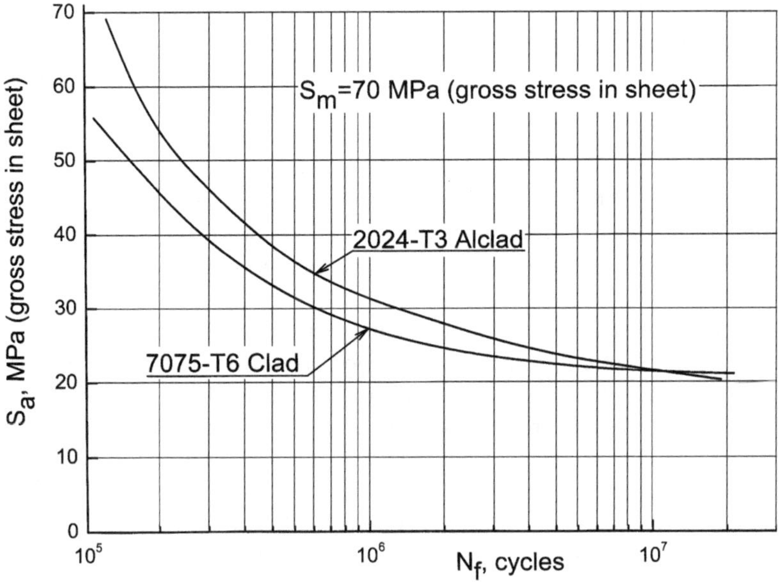

Fig. 3.2 Comparison of mean *S-N* curves for lap joints in 2024-T3 Alclad and 7075-T6 Clad (Hartman et al. 1965)

tests of Hartman et al. (1965) on two-row simple lap joints (sheet thickness 0.8 mm, AD snap rivets 3.2 mm dia) are shown in Fig. 3.2. Programme loading tests (based on a gust spectrum of Bullen) by the latter authors indicated a greater effect of the sheet material compared to the CA tests. The mean endurance of the 2024 specimens was 32 periods against only 21 periods observed for the 7,075 specimens.

It should be noted that in many older investigations on riveted joints, like those reported above, the *S-N* curves were obtained from CA tests carried out at a constant positive mean stress (S_m). Although information obtained in this way is still of interest, it cannot be ignored that the stress ratio close to zero is the most relevant condition. On the other hand, fatigue diagrams generally suggest that the effect of S_m is not large, especially if the number of cycles is high (Schijve 2009a).

Similarly, results of programme tests (low-high-low sequences of blocks of the CA cycles derived from a gust spectrum) should be treated with some caution. Laboratory investigations in later years have shown that significant load sequence effects were not taken into account when the programme loading was adopted to simulate in-service conditions. Ignoring load interaction phenomena can lead to misleading information on real fatigue properties of a tested structure (Schijve 2009a).

In East-European countries a Russian aluminium alloy D16Cz is also applied for aircraft skin structure. The chemical composition of this alloy is similar to that of the 2024 alloy, the major difference being a lower content of Fe and Si in D16Cz (Schijve et al. 2004). The letters Cz imply that it is a high purity alloy. The difference between D16 and D16Cz can be compared with the difference between 2024 and 2524, which is the high purity version of 2024. D16 sheets are covered by an Al cladding layer. Mechanical properties of the considered alloy according to Russian specifications are given in Table 3.1.

Fatigue tests under CA loading with different stress ratios indicated similar crack growth results obtained for sheet specimens made of 2024-T3 and D16 alloys (Schijve et al. 2004; Nesterenko et al. 2009). Also crack growth data from variable amplitude tests under the mini FALSTAFF load history were in a good agreement with similar results on 2024-T3 sheet specimens reported in the literature (Schijve et al. 2004).

Differences in fatigue crack growth properties may occur between materials according to the same specification obtained from different producers, and also between different batches of material from the same producer. Schijve and de Rijk (1966) compared fatigue crack growth data for 2 mm thick 2024-T3 Alclad sheet delivered by various producers from the USA, the Netherlands and other European countries. The crack growth rates (CGRs) over the range of 10^{-8} to 10^{-6} m/cycle differed by up to a factor of 4, which is quite significant.

Also, long-term natural ageing can detrimentally influence fatigue crack growth (Oldersma and Wanhill 1996). Wanhill (1996) cites an investigation by The National Advisory Committee for Aeronautics (NACA) which has revealed that over the CGR range of 10^{-8} to 10^{-6} m/cycle the continued long term ageing, for about 40 years compared to 20 years, has increased crack growth rates by a factor of 1.5. Wanhill points out, however, that this effect is within the batch-to-batch variations observed by Schijve and de Rijck (1966), as referred to earlier.

Because the factors addressed above can be of technical significance for aircraft designers, Schijve (1968) carried out comparative fatigue tests on riveted lap joints, as shown in the inset in Fig. 3.3. The joints were made of 2024-T3 Alclad sheet material delivered by five different manufacturers indicated by capitals A, C, E, F and G. Mechanical properties and chemical composition of all materials were similar. However, material F excelled other materials under fatigue conditions, while material A exhibited the poorest fatigue performance. The specimens were tested under CA loading (frequency 4,000 cycles per minute) and under programme loading based on a gust spectrum. As seen in Fig. 3.3, the width of the scatter band including all individual test results for CA loading is not very excessive, although differences between the *S-N* curves for different manufacturers are visible. In the programme tests the scatter of the mean fatigue life values was on the order of 2. It could be said again that the differences were small. However, a batch-to-batch variation was not studied in these experiments as the specimens were cut from a single sheet. This variation would probably increase the scatter.

Compared to other aerospace alloys, fatigue crack growth in 2024-T3 is relatively insensitive to environmental effects. Feeney et al. (1970) conducted fatigue crack growth tests in normal "dry" (<10% RH) and "wet" (>90% RH) air and 3.5% NaCl

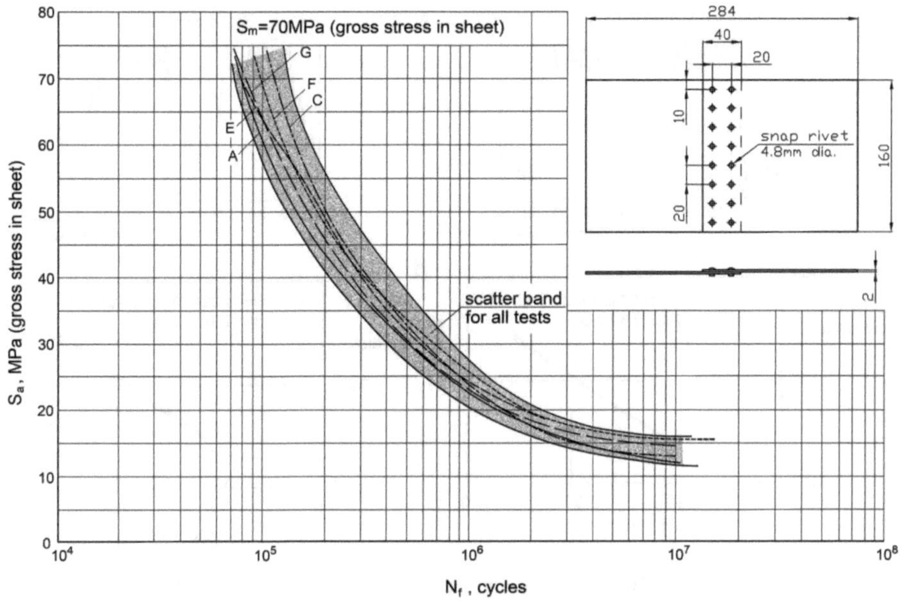

Fig. 3.3 Influence of different sheet material manufacturers on the *S-N* curves for riveted lap joint specimens (Schijve 1968)

solution. At CGRs (d*a*/d*N*) up to 2.5×10^{-7} m/cycle no evident environmental effect was observed for the 2024-T3 alloy. On the contrary, 7075-T6 showed a pronounced acceleration in CGRs in "wet" air and sodium-chloride solution.

The effect of flight simulating loading and of environmental conditions on CGRs for the 2024-T3 Alcad alloy (sheet thickness 1.6 mm) was tested by Schmidt and Brandecker (1997). The flight-by-flight stress-time histories were defined with major cycles of the stress ratio $R = 0.1$ and incremental cycles of $R = 0.8$, Fig. 3.4a, b. The goal of this investigation was to understand the effect of incremental cycles during taxiing and cruise (Spectrum 1, Fig. 3.4a) and during the climb, cruise and descent (Spectrum 2, Fig. 3.4b). Figure 3.4c, d show results of the CGR tests with Spectrum 1 and 2 respectively. As it can be seen, no effect of the different test conditions, namely laboratory air and 3.5% NaCl, on the CGRs has been observed. A similar behaviour of the considered alloy was observed in the same work when the stress-time history applied on the rear fuselage during the Airbus A-330 full scale fatigue test was used in crack growth tests.

The pressurized aircraft fuselage is exposed in flight to temperatures on the order of −50°C. Schmidt and Brandecker (1997) noted that the influence of a low temperature on CGRs for 2024-T3 could be neglected, Fig. 3.5. Also tests carried out earlier by Broek (Schijve 2009a, pp. 491–492) on the 2024-T3 and 7075-T6 Al alloys indicated slower crack growth at low temperatures. A lower vapour content in the air at a low temperature can be responsible for this trend.

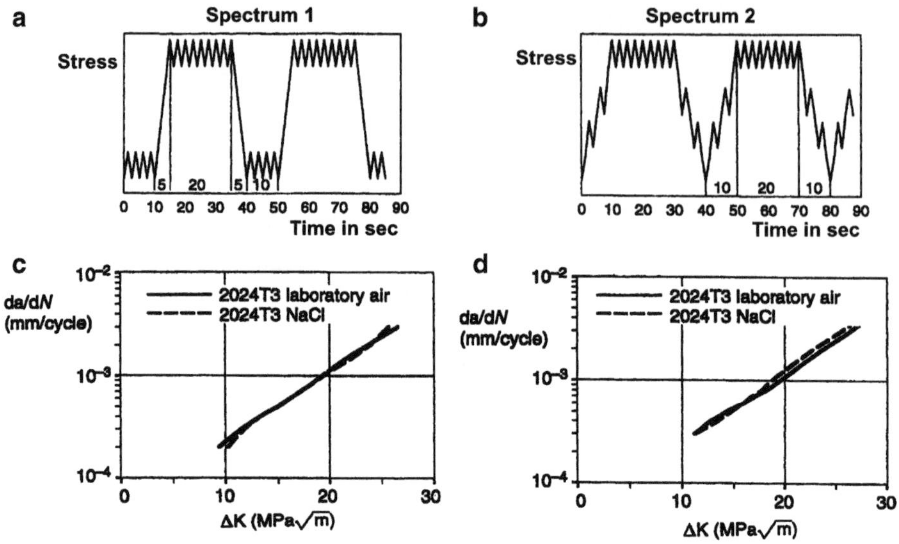

Fig. 3.4 Crack growth for 2024-T3 aluminium alloy under simplified service loading: (**a, b**) load spectra; (**c, d**) test results (Schmidt and Brandecker 1997)

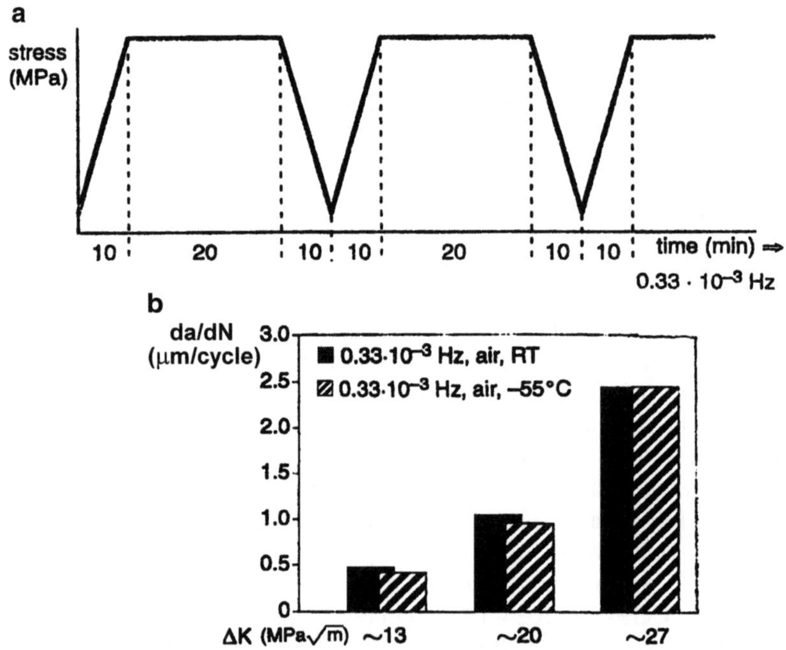

Fig. 3.5 Influence of low temperature on crack growth in 2024-T3 aluminium alloy: (**a**) stress-time history; (**b**) test results (Schmidt and Brandecker 1997)

3.2 Fastener Type and Material

Solid rivets are used when access to both sides of a sheet assembly or structure is possible. Shown in Fig. 3.6 are the most frequently used solid rivets made from aluminium alloys:

- protruding head (snap) rivet, both sheets flat, Fig. 3.6a,
- machined countersunk rivet, outer sheet sunk, inner sheet flat, Fig. 3.6b,
- dimpled countersunk rivet, both sheets dimpled, Fig. 3.6c,
- NACA rivet, outer sheet sunk, inner sheet flat, Fig. 3.6d,
- Briles rivet, outer sheet sunk, inner sheet flat, Fig. 3.6e,
- Allfast crowned rivet, outer sheet sunk, inner sheet flat, Fig. 3.6f.

The protruding head rivet is the one most used in early times. There is a large variety in shapes of the manufactured head for protruding head rivets, e.g. the round head or universal head. The latter is applied for aerodynamic reasons, see Fig. 3.6a. Also the mushroom rivet head with a small protrusion (the so-called compensator) is applied. Due to the compensator, a better rivet hole expansion is obtained, as detailed later. The installation of a protruding head rivet requires only drilling a hole of a diameter typically 0.05–0.1 mm larger than the rivet shank diameter.

When installing light metal solid rivets, the so-called driven head is formed by use of a hand hammer, a compressed air hammer or a riveting press. During this process, which is known as direct riveting, the manufactured head is supported by an anvil (a dolly). With indirect riveting, the hammer is applied to the manufactured head and the driven head is shaped by a holding-up tool.

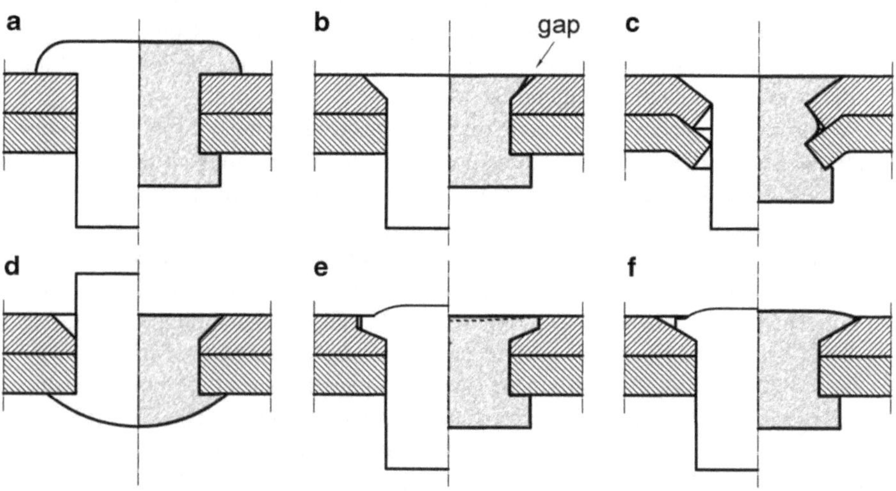

Fig. 3.6 Most frequently used types of solid rivets, explanation in text. *Left side* – before riveting, *right side* – after riveting

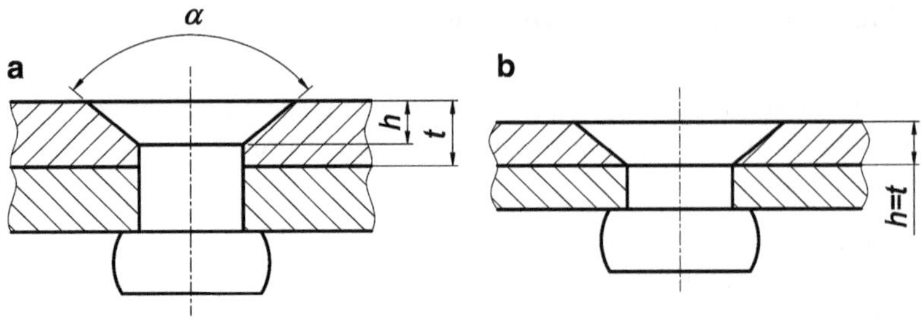

Fig. 3.7 Countersunk sheet geometry: (**a**) recommended geometry; (**b**) knife edge effect

The countersunk rivet has been introduced to obtain a flat outer side of the joint for aerodynamic reasons. When the sheet is thicker than 1 mm, the countersink can be formed by machining away the excess material. According to Niu (1988), the thickness (t) of the countersunk sheet should be equal to at least 1.5 times the depth (h) of the countersunk head of the fastener, i.e., $h/t \leq 0.67$, Fig. 3.7a. On the other hand, Fokker requires $(t-h) \geq 0.2$ mm. If $h = t$, the so-called knife edge effect occurs, see Fig. 3.7b, which is undesired as it contributes to crack initiation in the rivet hole. Bürnheim (ref. XVII/4 in Hertel 1969) postulates that the manufactured head angle α of 120° is favourable. Nowadays, however, $\alpha = 100°$ is a most usual geometry (Müller 1995). For sheet thicknesses $t \leq 1$ mm the countersink is plastically formed in a process referred to as dimpling. Dimpling is preferable to drilling because the original drilled hole wall is too small, which makes the bearing pressure locally abnormal and the joint strength decreases. In the case of a good dimple joint (the dimples of two sheets must make contact), load transfer is mainly over the dimples instead of through the rivets. Rivets applied in the dimple joints are the same as those used in the machined countersunk connections. Countersunk rivets are more costly than protruding head rivets because of the installation requirement to machine or dimple the countersink.

According to generally accepted rules, the protruding head rivets and countersunk rivets are identified by coding the following information: head style, rivet material, shank diameter, rivet length. The code for the rivet material will be described later on. The shank diameter is given in 1/32 in. while the rivet length in 1/16 in. For example, the MS20470AD4-4 rivet would be described as follows: MS20 – catalogue system (Military System), 470 – head style code (in this case it implies the universal head), AD – rivet material code, 4 – shank diameter (4/32 in.), 4 – rivet length (4/16 in.). The countersunk rivets are coded by the Military System as the MS20426 rivet. Other frequently used countersunk rivet types are the reduced head rivet NAS1097 and the intermediate head rivet EN6001. The NAS1097 rivet is suitable for thinner sheets because of the reduced head height, allowing for a less deep countersink. The EN6101 rivet is a compromise between the MS20426 and the NAS1097 rivets (Homan 2008).

After riveting, in the countersunk joint a small gap can be present between the countersunk rivet head and the countersunk hole, Fig. 3.6b. This gap, if not sealed, promotes crevice corrosion and implies that the countersunk rivet head is not perfectly supported by the countersunk hole. Then the load transmission is more concentrated on the cylindrical part of the countersunk hole, which is not beneficial for fatigue resistance.

In order to avoid this gap, the NACA rivet has the driven head in the countersunk sheet which insures a good filling of the countersunk end of the hole. The angle of the countersunk hole is equal to 88°. Because a sufficiently flush outer surface is not always obtained, an extra milling operation on the driven head may be required (so called "shaving").

The Briles rivet is another solution for the gap problem existing in the countersunk joint. The manufactured head end face of the rivet has a special protrusion which, during the riveting process, fills tightly the rivet hole. The Briles rivet requires a special installation process. It should be noted that the Briles rivet has a smaller rivet head compared to the conventional countersunk rivet.

The Allfast crowned rivet is to some extent similar to the Briles rivet. A crown extending from a top surface of the manufactured head deformed during riveting becomes flat and the rivet fills the hole of the worked piece. No post-installation process is required.

In order to reduce the cost of riveting, a slug rivet, which is a certain type of the countersunk rivet, was developed. Instead of individual rivets, rivet wire is used. Consequently, the riveting implies forming the driven heads at both ends. As with the NACA rivet, rivet shaving may be required when the slug rivets are used.

The choice of rivet material depends on the sheet material used. The hardness of the rivet material has to be lower than or equal to the hardness of the sheet material (Oldersma 1992). The potential difference between the rivet and sheet material should be small because of the possibility of galvanic corrosion. Solid aluminium rivets were introduced in 1919 to the German Junkers F13, the first civil, all metal, aircraft. Earlier, steel and brass rivets were used. Nowadays, aluminium alloys commonly used as rivet materials are: 2117-T3, 2017-T3, 2017A-T3, 2024-T3, 7050-T73 (MIL-HDBK-5H).

The AD rivet is manufactured from the 2117-T3 (old notation A17 S) Al alloy (ultimate shear stress $F_u = 30$ ksi ≈ 207 MPa). This medium-hard alloy of the Al-Cu-Mg type contains a rather low percentage of Cu and Mg (Cu 2.5%, Mg 0.3%) and can be riveted in its "as fabricated" condition. The AD rivet is most commonly used.

The D and DD rivet are manufactured from the 2017-T3 (17S) and 2024-T3 (24 ST) hard Al alloy respectively (2017-T3: Cu 4%, Mn 0.5%, Mg 0.5%; 2024-T3: Cu 4.5%, Mn 0.6%, Mg 1.4%). Both type rivets are installed in quenched condition. The heat treatment takes place at a temperature of 500°C for 20–30 min. Afterwards, the rivets are instantly cooled down to room temperature by submersion in water. The rivets are then very soft and can easily be deformed during riveting. To keep the heat treated rivets in the "as quenched" condition, they must be stored at a temperature of 0°C. Storing the rivets under refrigeration is applied to delay the natural age hardening process after the heat treatment. In order to comply

Table 3.2 Rivet materials

Designation	Material	Heat treatment before riveting
AD	2117-T3	No
D	2017-T3	Yes
DD	2024-T3	Yes
E	7050-T73	No
2017A	2017A-T3	Yes

with the specifications, before riveting the quenched rivets should be kept at room temperature no longer than 1 h in order to maintain their soft condition. However, Hartman's (1968) tests revealed that a longer ageing period (up to 1¾ h) prior to the riveting did not have a significant detrimental effect. The advantage of the DD rivet over the D rivet is a superior static shear strength ($F_u = 41$ ksi ≈ 283 MPa and $F_u = 34$ ksi ≈ 234 MPa respectively). Typically, the DD rivets are found installed in high-strength (e.g. 7075 aluminium alloy) primary structures and can be an alternative to the E rivets.

Besides the 2017-T3 material, the 2017A-T3 aluminium alloy is also applied. The 2017A-T3 rivet has the same basic chemical composition as the 2017-T3 rivet, but a bit higher Ti + Zr content and therefore a higher shear strength than the 2017 rivet ($F_u = 38$ ksi ≈ 260 MPa).

Rivets with designation E are made of the 7050-T73 alloy of an Al-Zn type (Zn 6.2% and smaller percentage of other elements) and used when a high static strength is required ($F_u = 46$ ksi ≈ 318 MPa). The E rivet can be squeezed without any heat treatment using ADRM (Automatic Drilling and Riveting Machine) equipment only. For this rivet type, however, cracks in the driven head can occur with already limited squeeze forces (Müller 1995). The rivet materials and their codes are listed in Table 3.2.

In the East-European aircraft industry, rivets made of PA24 and PA25 aluminium alloys are also used. According to the EN 1301-2:1997 standard, the PA24 material ($AlCu_2Mg$) is an equivalent of the 2117 Al alloy. There is no exact Western equivalent of the PA25 alloy ($AlCu_4Mg$) the mechanical properties of which are similar to those of Al 2024. However, the Mg contents for the latter material is higher than for PA25. Both PA24 and PA25 rivets are installed in the "as fabricated" condition.

A comparison of test data for the AD and DD rivets in the 2024 Alclad material, collected by Hartman et al. (1965) and presented in Fig. 3.8, indicates that their fatigue properties are almost the same.

Müller (1995) compared fatigue lives for lap joints riveted with three rows of AD, DD and E rivets. The squeeze force was chosen to obtain the driven head diameters (D) of 1.3, 1.5 and 1.7 times the rivet shank diameter (D_o). According to the Fokker specification, the acceptable driven head dimensions range from $1.25D_o$ to $1.65D_o$. For the typical D/D_o-value of 1.5, the AD and DD rivets show similar fatigue properties at all stress levels and the E rivet surpasses the AD and DD rivets,

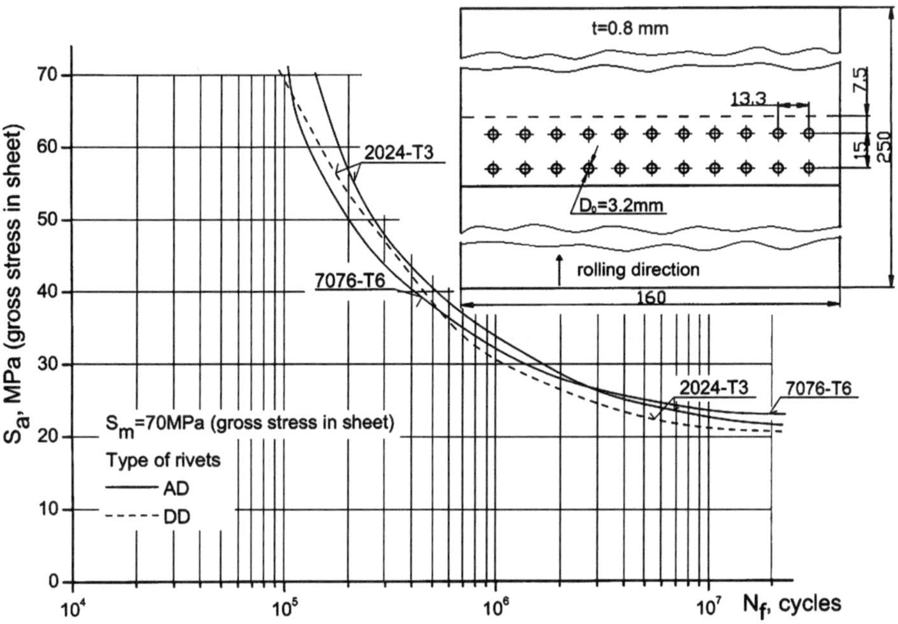

Fig. 3.8 Effect of rivet material on *S-N* curves for lap joints (Hartman et al. 1965)

Fig. 3.9a. As seen in Fig. 3.9b, the same trend was observed for a large squeeze force resulting in $D/D_o = 1.7$. For a low squeeze force ($D/D_o = 1.3$), Fig. 3.9c, the E rivet exhibited poor fatigue properties at higher stress levels due to insufficient hole expansion.

Already Bürnheim (ref. XVII/4 in Hertel 1969) observed that at long fatigue lives ($N_f \geq 10^7$ cycles) the fatigue properties of protruding head and countersunk rivets are the same. Results of similar tests conducted by Hartman (1954) on single lap joints with the protruding head, countersunk and NACA rivets (rivet material 17 S) confirm that the effect of the rivet type is only revealed at relatively high loads, the NACA rivet proving to be superior. These results further show that differences between the various types of rivets become more significant for thicker sheets. Hartman's *S-N* data for 1.6 mm thick specimens are plotted in Fig. 3.10 in comparison with the scatter band for the lap joints of 0.8 mm thick sheets and snap rivets. The author concluded that the NACA rivet should be preferred in cases of a high ratio of stress amplitude to static strength. Also Müller and Hart-Smith (1997) noted that the fatigue lives of three-row lap joint specimens with NACA rivets were longer compared to joints with typical countersunk rivets. The better filling of the countersunk hole in the outer sheet by the NACA method of riveting, in which the hole is filled by the closing head of the rivet, prevents the premature loosening or failing of the rivets at high fatigue loads (Oldersma 1992).

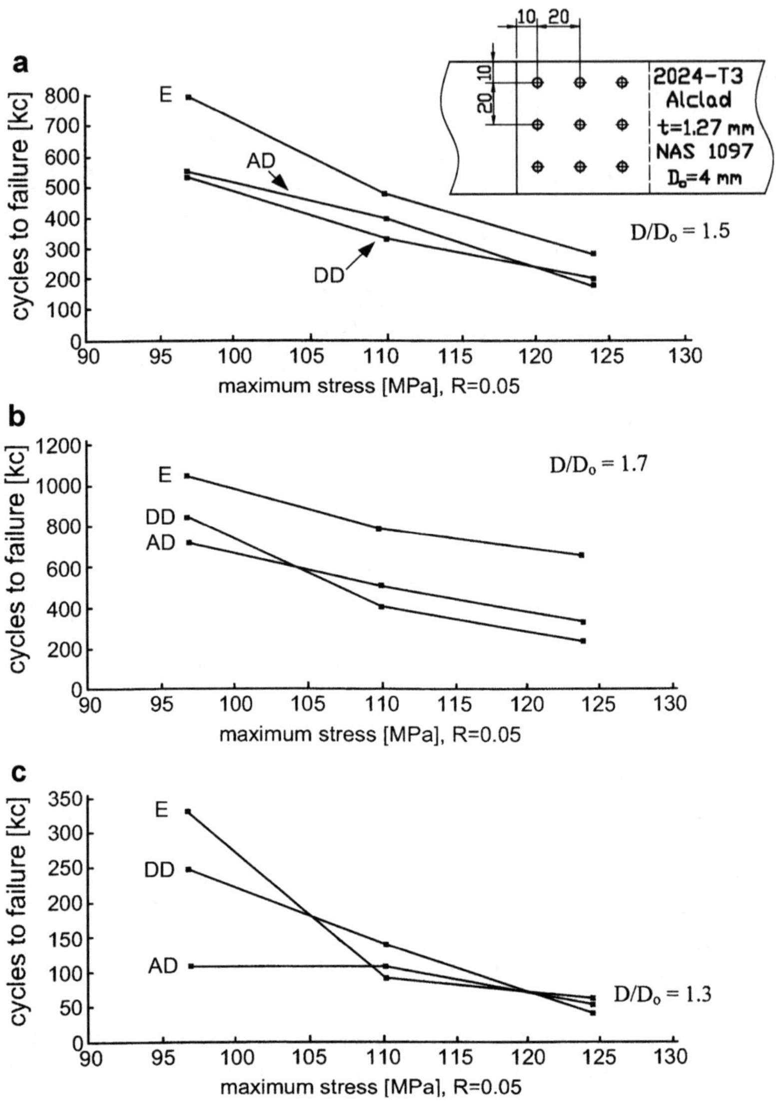

Fig. 3.9 Effect of rivet material on lap joint fatigue properties at moderate (**a**), high (**b**), and low (**c**) squeeze force (Müller 1995)

Hartman and Duyn (1952) carried out tests under monotonic and fatigue loading on lap joint specimens with two rows of rivets placed either in line or staggered. The sheet material was the 24ST aluminium alloy and the 17 S rivets were riveted cold in the solution heat-treated temper. The *S-N* curves were determined for fluctuating tension with a minimum load of 1 kN. Results of the tests on specimens with the protruding head, countersunk and dimpled hole rivets are compared in Fig. 3.11.

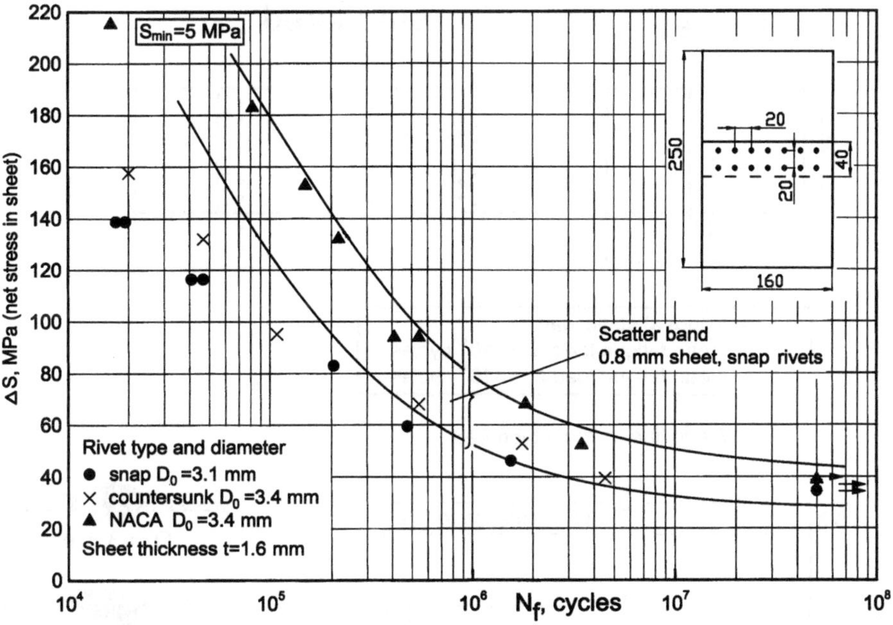

Fig. 3.10 Effect of rivet type on fatigue performance of 24 ST Alclad (2024-T3) lap joints (Hartman 1954)

There is no marked difference in the static and fatigue properties between all tested joints. On the other hand, CA tests by Schütz (1963) on lap joint specimens with two rivet rows exhibit a slight superiority of the protruding head rivet over the dimpled hole rivet, especially for a greater rivet diameter ($D_o = 4$ mm), Fig. 3.12. The sheet material used was an Al Cu Mg alloy ($S_y = 340$ MPa, $S_u = 460$ MPa, $\delta_5 = 24\%$).

Blind rivets are designed to be installed where the access to both sides of a sheet assembly or structure is not possible or impractical (Hoffer 1984; Barret 1990). These rivets are not used in longitudinal fuselage connections.

Several types of special fasteners, like Lockbolt and Hi-Lok fasteners, were also developed for aircraft applications. Nearly all such fasteners can be installed only with the aid of special tools manufactured by the producer. These fasteners are basically outside the scope of this book. However, some concise information on the Hi-Lok and the NAS Lockbolt fasteners is given below because they are used in some tests to simulate underdriven rivets or to investigate the load transmission or secondary bending in lap joints (Bakuckas et al. 2007; Starikow 2002).

The Hi-Lok fastener depicted in Fig. 3.13a consists of two parts, namely a pin and a collar, which are made of a high-strength steel to obtain the high static shear strength. A hexagonal wrench at the end of the collar is used for tightening the nut part of the collar. When the preset preload is reached, the hexagon of the collar breaks off at the torque-off groove. The Hi-Lok fastener ensures a significant clamping of the sheets and good fatigue properties of the joint.

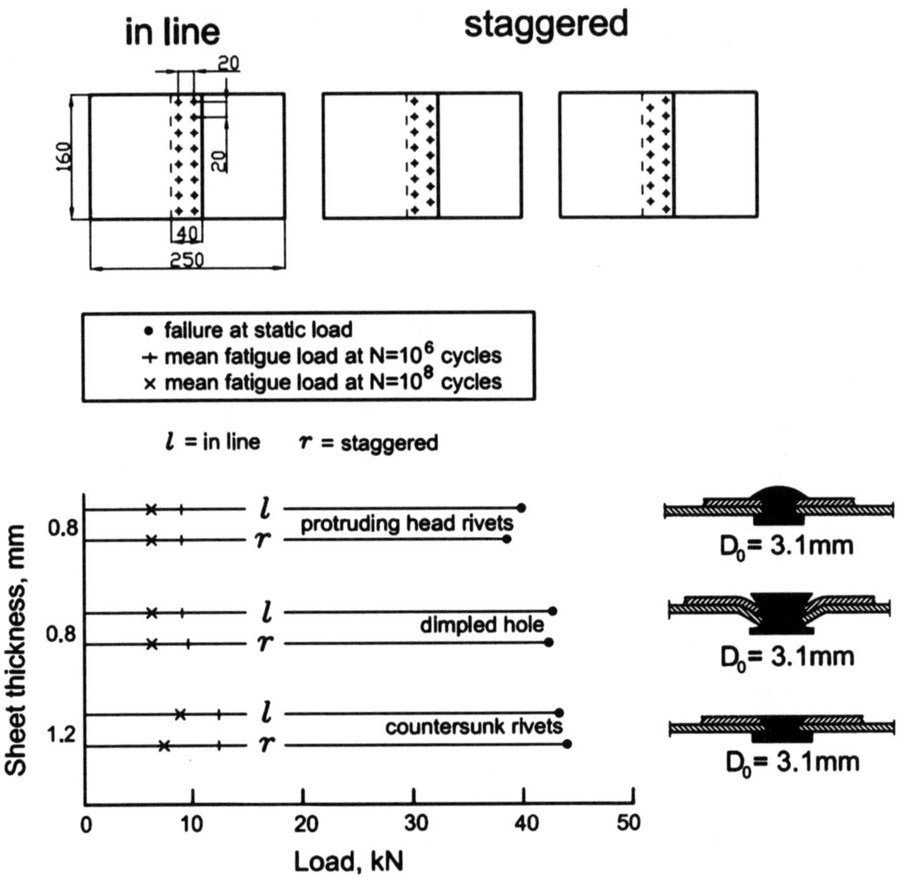

Fig. 3.11 Static and fatigue strength of lap joints with different rivet types (Hartman and Duyn 1952)

The Lockbolt fastener presented in Fig. 3.13b consists of an alloy steel pin and an aluminium alloy collar. Again a special tool is needed in order to install this type of a fastener. When the tool is turned on, the axial tension is applied to the pin stem, thus pressing the collar material into the stem grooves, which results in a permanent rigid joint. At the end of the operation, the stem is fractured at the break-neck groove due to the increasing axial tension.

CA and programme fatigue test series with DD rivets and Hi-Lok and NAS Lockbolt fasteners performed by Hartman (1968) demonstrated that, considering the scatter in the fatigue data, the fatigue strength remained unaffected by the fastener type, as depicted in Fig. 3.14.

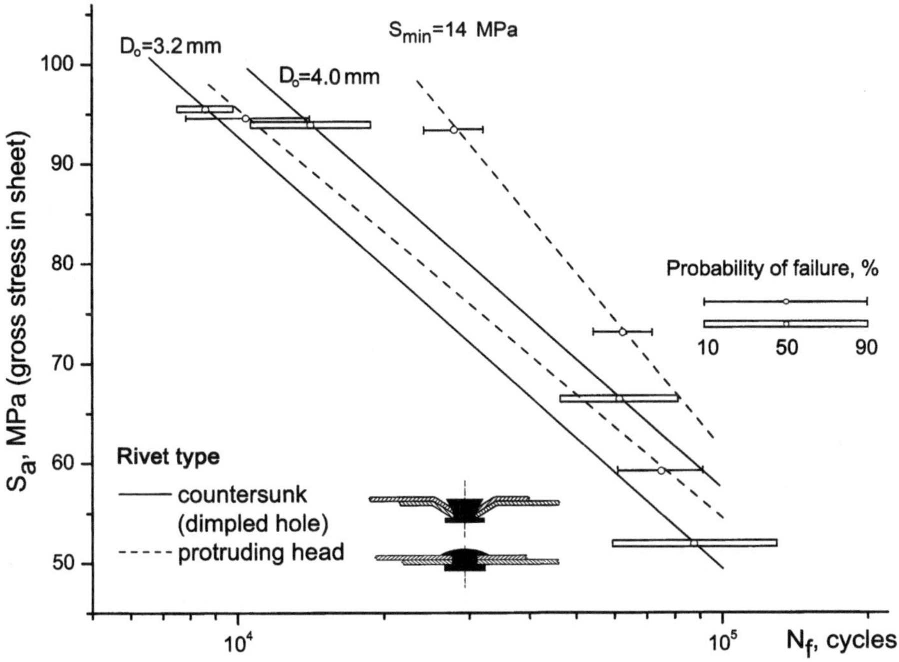

Fig. 3.12 Comparison between *S-N* curves for single lap joints with two rivet types (Schütz 1963)

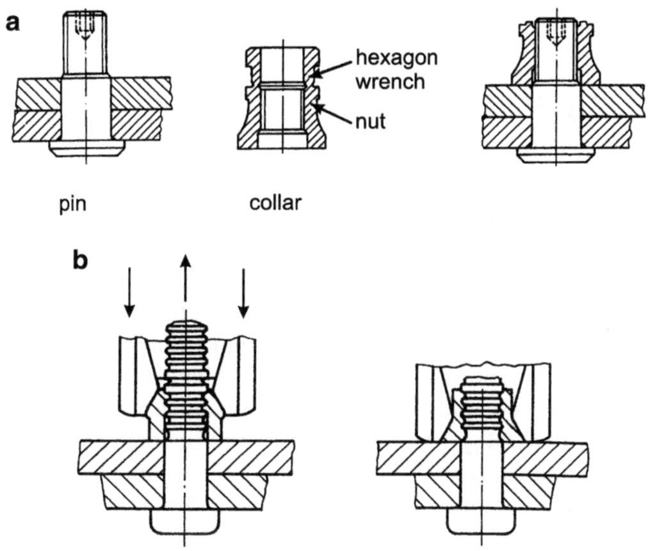

Fig. 3.13 Two types of fastener: (**a**) Hi-Lok; (**b**) NAS Lockbolt (Hoffer 1984)

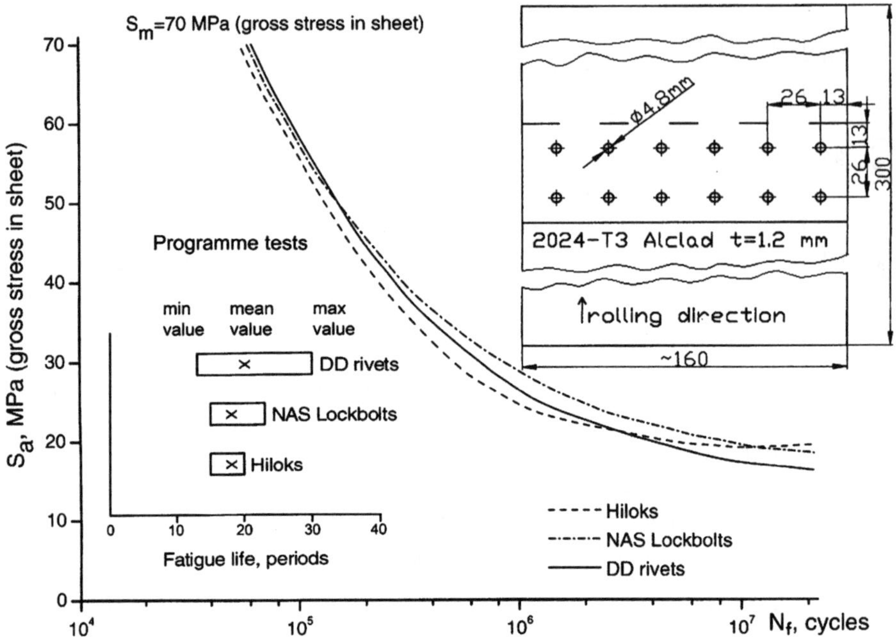

Fig. 3.14 Comparisons between fatigue test results for lap joints with several types of fastener (Hartman 1968)

3.3 Manufacturing Process

3.3.1 Riveting Method

It is well known that the fit between the rivet shank and the rivet hole, the clamping force of the rivets, and friction between the mating sheet surfaces will affect the fatigue behaviour of a joint. All these aspects depend on the method of manufacturing.

Hartman (1968) investigated the effect of the riveting method, i.e. hand, pneumatic and hydraulic riveting, on the lap joint fatigue properties. The corresponding fatigue test results are presented in Fig. 3.15. Manufacturing by using a light pneumatic rivet gun in combination with a heavy stationary rivet die was considered to be the standard riveting procedure in the aircraft industry and taken as the reference for comparisons with the other riveting techniques. With all riveting methods applied it was aimed to obtain the same driven head diameter of about 5.0 mm. As seen in Fig. 3.15, the *S-N* curves for all riveting procedures fall close together and, given the scatter in the data for the reference method, the differences between these procedures can be considered insignificant. This conclusion was

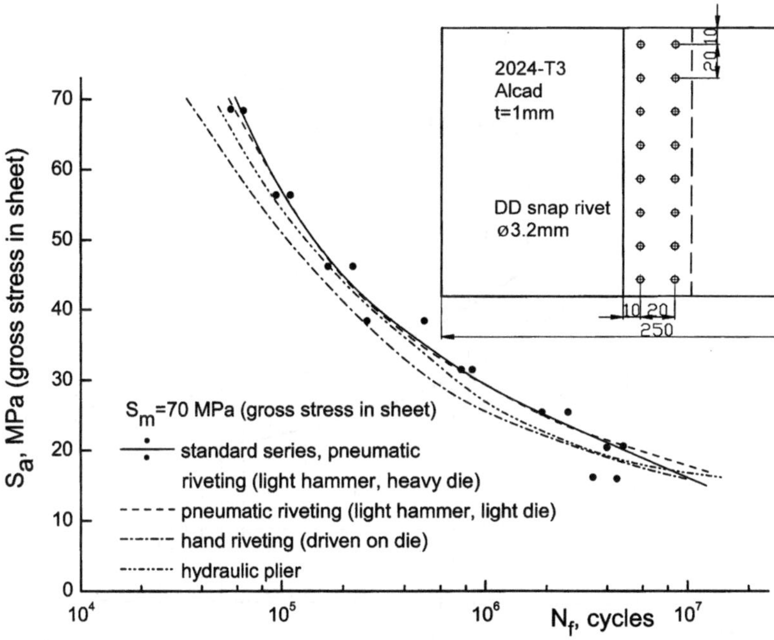

Fig. 3.15 Effect of riveting procedure on lap joint fatigue strength (Hartman 1968)

supported by the programme fatigue tests, as for all cases the fatigue life was about 20 periods. Also the stress at the static failure was nearly the same for all specimens and equalled about 250 MPa (gross stress in sheet).

On the other hand, Oldersma (1992) pointed out that the fatigue properties of riveted joints could be affected by the riveting method. He compared *S-N* curves for two specimen series which had similar dimensions and were manufactured from the same aluminium alloy. The difference in manufacturing the specimens was due to utilizing a flat or a concave rivet die. As can be seen in Fig. 3.16, the divergence between the corresponding mean *S-N* curves is significant, the flat rivet die leading to better results. An opposite trend is seen for the indirect riveting when data collected by Fokker Aerostructures (Ottens and Wanhill 2003) are analysed. The lap joint specimens were prepared from a 4.0 mm thick 7175-T7351 aluminium plate anodized and painted. 2017A and 7050 solid protruding head rivets were used for manual and automatic riveting respectively. A significant difference between manual direct and indirect riveting is observed, but there is no difference between automatic and manual direct riveting, Fig. 3.17.

A great deal of fatigue testing aiming to check the effectiveness of the automatic riveting process using the ADRM (Automatic Drilling and Riveting Machine) was carried out by the British Aerospace Ltd Manchester (Simpson 1989). In this process the work piece is positioned manually, but the hole drilling, countersinking and

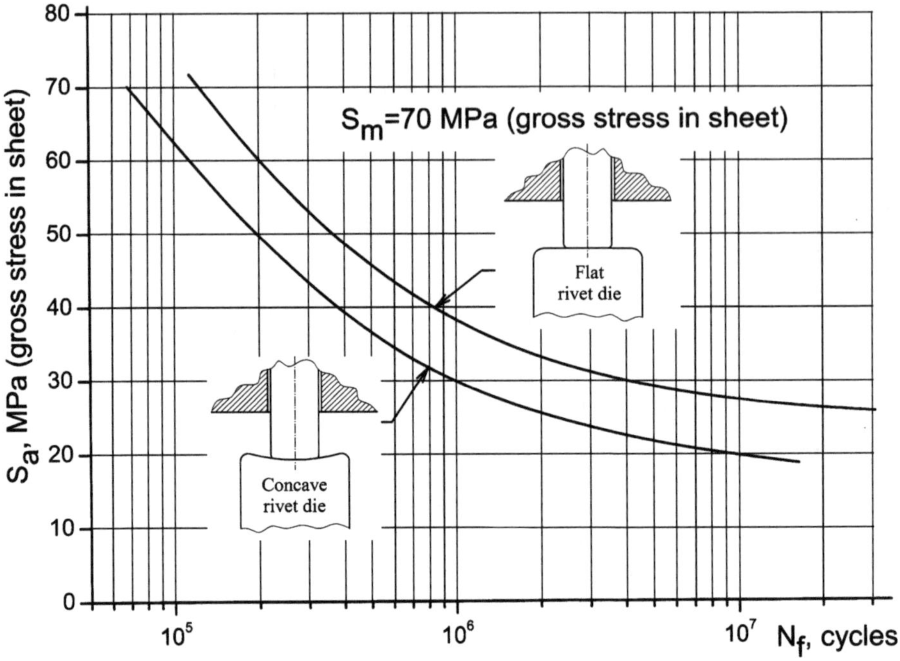

Fig. 3.16 Effect of stationary rivet die shape on lap joint fatigue strength (Oldersma 1992)

rivet installation are all functions if the machine has been set in advance for the particular task. In the ARDM, rivets are not hammered but squeezed utilizing a hydraulic system. By controlling the squeeze force applied on the rivet, the desired deformation of the driven rivet head can be obtained. The investigation covered the effect of the basic process for different rivet head forms and different rivet materials. In all cases the joints riveted by the automatic process showed the fatigue life similar to that of joints made by the traditional method, Fig. 3.18. These results confirm the trend observed in the Fokker tests referred to above (Ottens and Wanhill 2003).

Another automated riveting method is the EMR (Electro Magnetic Riveting), which is a single hit hammering method, the rivet being formed with the electromagnetic impulse process, e.g. Zieve (1986).

On the basis of the available experimental evidence addressed in this section, it can be concluded that the riveting method can be a factor affecting the fatigue properties of riveted joints.

3.3.2 *Imperfections of Rivet Holes*

Laboratory riveted specimens are usually produced under relatively ideal conditions. In practice, such conditions are not always achievable and fasteners can be installed

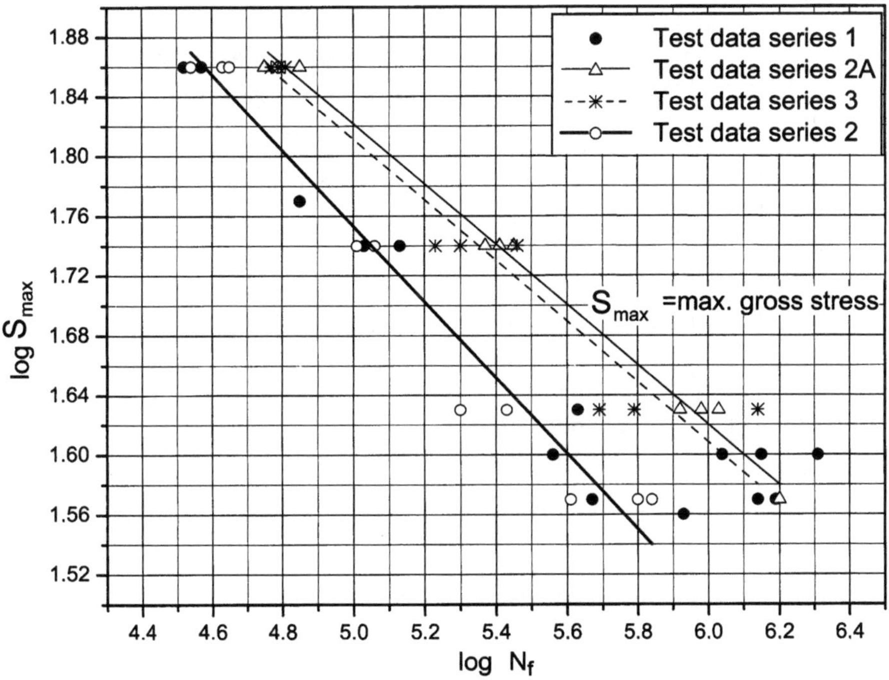

1 - manual riveting with concave die ends (indirect riveting) - reference series
3 - automatic riveting with flat die ends
2 - manual riveting with flat die ends (indirect)
2A - manual riveting with flat die ends (direct)

Fig. 3.17 Effect of riveting procedure on fatigue strength of lap joints (Ottens and Wanhill 2003)

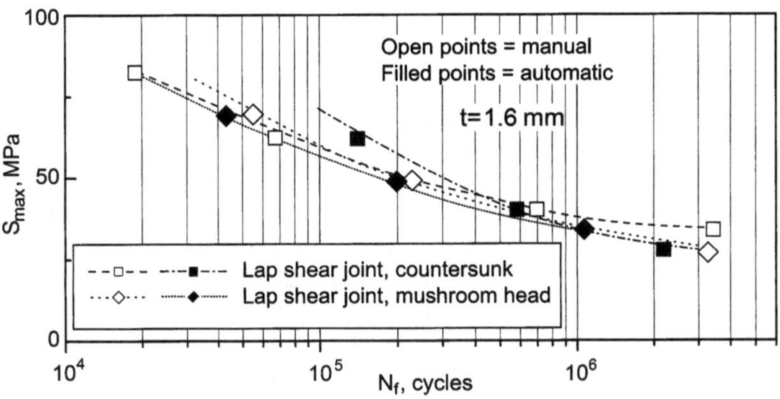

Fig. 3.18 Effect of manual and automatic riveting on lap joint fatigue strength (Simpson 1989)

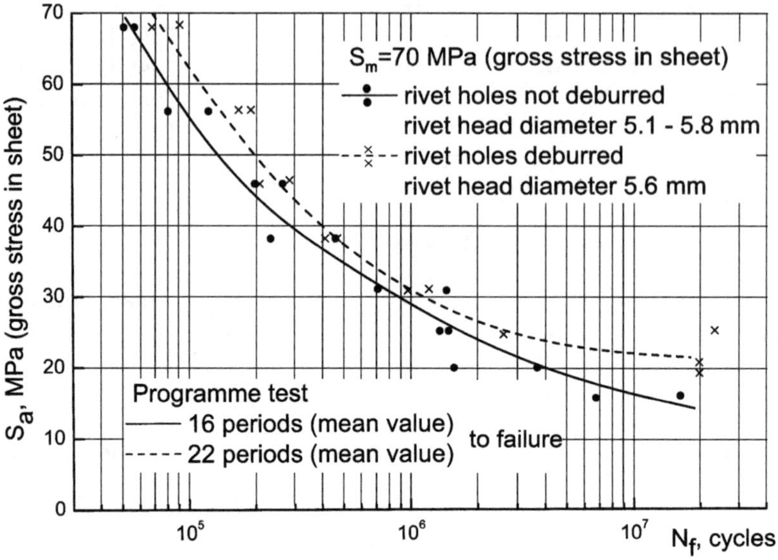

Fig. 3.19 Effect of deburring on lap joint fatigue strength (Hartman 1968)

in holes which are defective in a way. Among the most likely faults are burrs, too deep countersinking and mismatched holes. It should be emphasized that the number of experiments to elucidate the effect of imperfections is rather limited.

Hartman (1968) carried out fatigue tests under CA and programme loading on specimens with the rivet holes drilled when the two sheets were clamped. The holes were, however, not deburred in order to imitate conditions that often occur in the repair work. In Fig. 3.19, the comparison between the *S-N* curves from the CA tests on riveted specimens with the deburred and non-deburred rivet holes is presented. Both series of the lap joint, two-row specimens (2024-T3 Alclad, sheet thickness 1.0 mm, DD rivet 3.2 mm dia) were riveted by pneumatic hammering and the driven rivet head size was comparable. The results in Fig. 3.19 suggest that deburring the rivet holes only slightly improves the fatigue behaviour. It is also seen that the data for the joints with the non-deburred holes show more scatter. A somewhat more beneficial influence of deburring occurred under the programme loading, as indicated in Fig. 3.19.

Schra et al. (1995) investigated specimens cut from a longitudinal fuselage lap joint of the Fokker 100 full scale test article TA15. The countersunk and cylindrical hole misalignment causing a substantial gapping of the rivet heads was found. Fatigue cracks were observed at the rivets with extensive head gapping and inferior hole filling. Also, smaller driven rivet head diameters were found in the vicinity of frames and under the stringer head. The smaller diameters were due to poor accessibility of the rivets for riveting tools. These imperfections were considered to be the main factor responsible for poor fatigue properties of the joints compared to similar specimens produced in laboratory conditions.

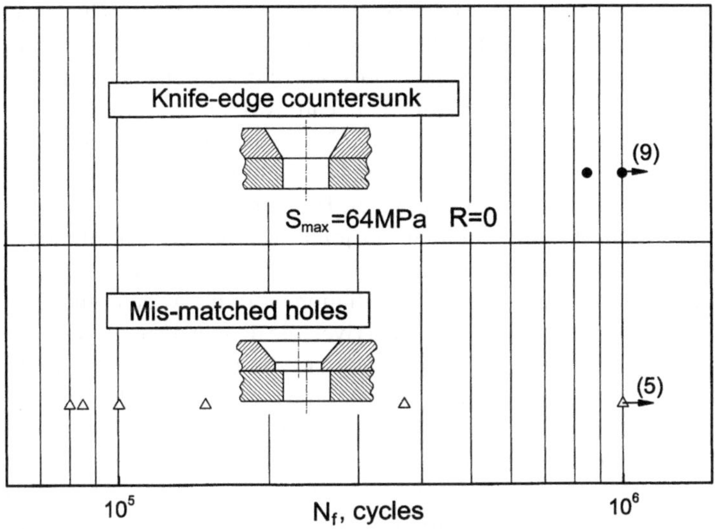

Fig. 3.20 Effect of overdeep countersinking and mismatched holes on fatigue life of lap joints with Alfast crowned rivets. In brackets number of specimens that did not fail (Simpson 1989)

Effects of the overdeep countersinking (the knife edge effect) and mismatched holes were investigated by the British Aerospace Ltd (Simpson 1989). Lap joints with Allfast crown rivets were used in the experiments. The CA test results suggested that while a degree of overdeep countersinking could be tolerated, the presence of mismatched holes introduced a possibility of fatigue life reduction by an order of magnitude, Fig. 3.20. It was concluded that all mismatched holes should be rectified before the rivet installation. It should be noticed that the overdeep countersinking combined with high stresses in the skin can cause catastrophic failure. The knife edge effect and stress concentration in the corner of the automatic direction finding (ADF) window enhanced by out-of-plane bending was recognized as an origin of the Comet I crash on January 10, 1954. Comet I was the first high-altitude jet-propeller passenger aircraft in the world (Swift 1987).

Atre (2006) and Sławiński et al. (2010) simulated using the FE method riveting in the case of mismatched rivet holes. The hole misalignment increases the unsymmetrical deformation of the rivet leading to larger hole expansion in the lower skin. This skin shows a non-uniform increase of compressive residual stresses. A small diameter hole drilled at the fastener hole location prior to drilling the final fastener hole is the so-called pilot hole. This pilot hole becomes a guide for a larger diameter hole. It was shown in an experiment performed on the 2024-T3 Al alloy (Ralph et al. 2009) that the use of the pilot hole reduces the final hole conicality, so that hand drilling with the pilot hole is roughly equivalent to machine drilling. The use of pilot holes also reduces the number of gouge marks. However, pilot hole drilling is a time-consuming operation. According to the same authors, the hole surface quality is bit length dependent. Longer bits were observed to increase surface roughness for both hand and machine drilling. The bit length also significantly affected conicality, although in opposite ways. A longer bit produced less conicality for hand drilling, and more conicality with machine drilling.

Bakuckas et al. (2007) performed a teardown inspection of some fuselage panels removed from a retired Boeing 727-232 passenger airplane. The airplane accumulated 59,497 flight cycles and 66,412 flight hours and was near its design service goal (DSG) of 60,000 flight cycles. Hole drilling and rivet installation were characterized along the longitudinal lap joint. The driven head dimensions were measured for each rivet (countersunk) along the outer row and compared to the Boeing specification for fasteners' installation in order to determine if the rivets were underdriven or overdriven. For each rivet, the diameters in two directions (hoop and forward-aft direction) were measured and the maximum and minimum head heights were determined. In general, 57% of the rivets were underdriven, 40% were within the specification, and only 3% were overdriven. Disassembly of the rivets revealed a dependence of the detected cracks on the rivet condition: 80% of the underdriven rivets showed cracks, while in the case of the overdriven rivets there were only 33% with cracks. Additionally, expansion of the rivets was not uniform through the depth of the hole. This appeared to be partly due to hole misalignment and the hole in the lower skin being of a smaller diameter than the hole in the outer skin layers. These imperfections were, according to the authors' opinion, the reason for the worse fatigue behaviour of the tested panel compared to another geometrically similar panel that showed a better overall joint quality.

3.3.3 Cold Working of Rivet Holes

Cold working (CW) of rivet holes is used for both repair and new designs in order to increase the fatigue properties of bolted and riveted joints and their damage tolerance characteristics (Leon 1988). For any CW process, the mechanism by which the gain in the fatigue life is achieved is the introduction of compressive residual stresses in the material surrounding the hole. The compressive stresses in the tangential direction are created due to the radial expansion of material caused by an oversized tapered pin which is drawn through the hole. In addition, the rubbing of the tapered pin against the hole can smooth the surface and have, therefore, a positive effect on fatigue life (Chakherlou and Vogwell 2003). The method is very effective because the residual stresses can be high, as seen in Fig. 3.21. This figure presents the results obtained by Rans (2007) who applied the FE method to simulate the formation of residual stresses generated by drawing a tapered 3.04 mm diameter pin through the 2.9 mm diameter hole drilled in 2024-T3, 1 mm thick sheet. It can be seen that hole expansion is not uniform through the sheet thickness. According to Schijve (2009a), expansion by 2–3% is sufficient to obtain a significant strength improvement. This suggestion was recently backed up with fatigue tests by Chakherlou et al. (2010) on specimens made from the 4.5 mm thick 7075-T6 Al

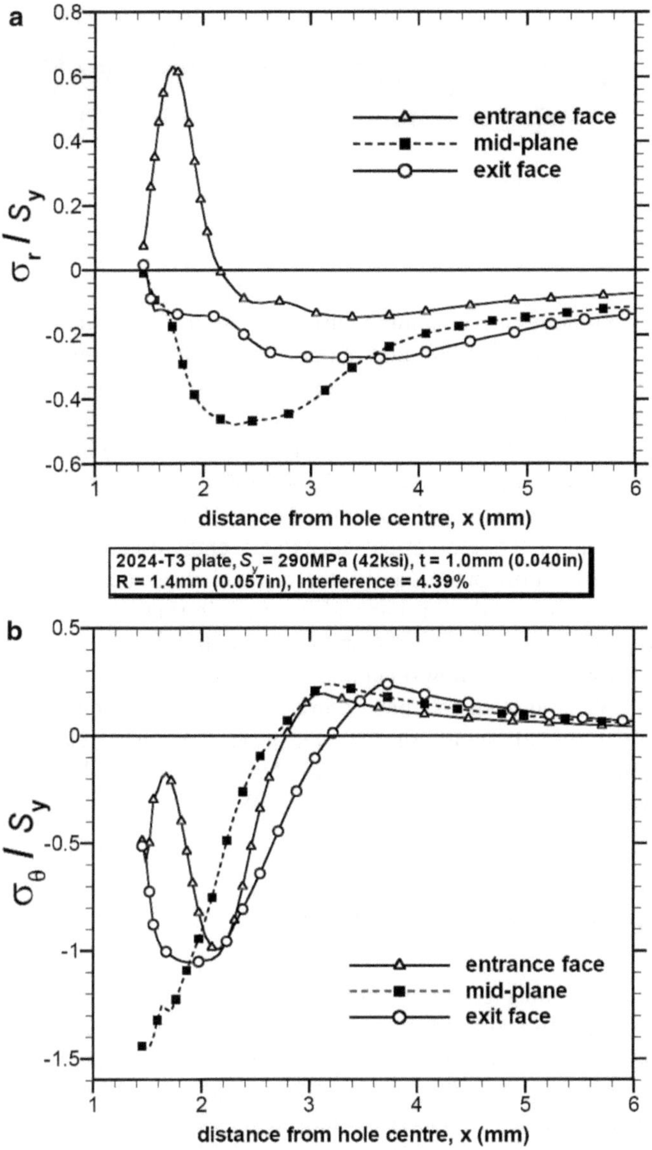

Fig. 3.21 Residual stress distribution for cold expanded hole in a 2024-T3 plate according to FE calculations: (**a**) radial stress; (**b**) tangential stress (Rans 2007)

alloy sheet with an oversized steel pin remaining inside a hole of 5 mm nominal diameter. The interference varied between 1.5% and 4%. Under CA cyclic tension with $R = 0$ an increase in fatigue life was observed up to a 2% interference. The effect of interference fit was more beneficial at high amplitude loading.

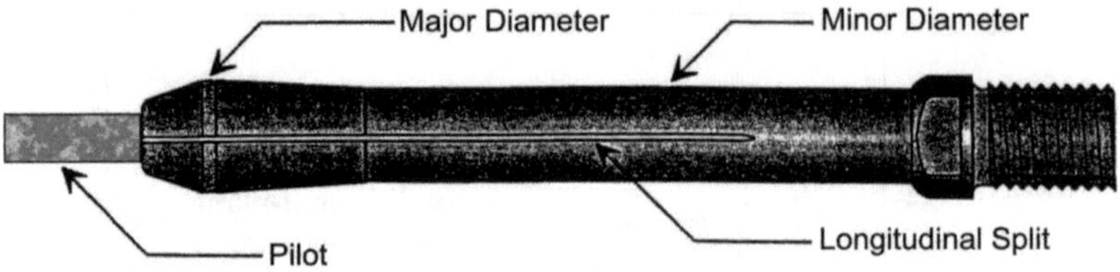

Fig. 3.22 Split mandrel (Rodman and Creager 1994)

Two CW techniques are in common use in the aircraft industry. One of these is a split-sleeve method (Leon 1988) and the other is a split-mandrel method, Fig. 3.22 (Rodman and Creager 1994). The mandrel has two mutually perpendicular slits. When the mandrel is inserted into the hole, the slits are closed by some elastic bending of its four quadrants. Next, a pilot is inserted into the mandrel centre hole. This prevents closure of the slits during the pulling out of the mandrel. Friction between the mandrel and the hole is reduced by applying a lubricant.

A significant amount of theoretical and experimental researches has been devoted to the CW process in order to quantify the residual stress field around the hole, to determine the time required to initiate and grow a fatigue crack from the cold worked hole, and to predict the effect of CW on the fatigue life of aircraft components, e.g. (Park and Atluri 1993; Krasnowski et al. 2001).

Actually, CW is applied for thicker aircraft components (e.g. lug holes) or for repair of bolted joints if small cracks occur in service (Jarfall 1986). CW of holes for riveted joints is hardly ever done. The number of holes is too large and, moreover, the solid rivet installation is already a way of CW.

3.3.4 Surface Treatment of the Sheets

Anodizing is the most common surface treatment for aircraft components. It produces an artificial oxide layer which is used as a pre-treatment for corrosion protection systems (e.g. painting) and adhesive bonding. In order to study the effect of anodizing, Hartman (1968) performed fatigue tests on three series of 2024-T3 Alclad single lap joint specimens ($t = 1$ mm) with two rows of snap rivets (DD type, 3.2 mm dia). The chromic acid, sulphuric acid anodized, and bare specimens were compared. The results obtained for CA loading are presented in Fig. 3.23. A small favourable effect of anodizing (observed also in the programme tests) compared to the bar material may follow from a higher squeezing force used for the riveted joints of the anodized sheets (larger driven head diameters). It can be seen that the chromic acid coating and the sulphuric acid coating showed a similar behaviour. Nowadays, the aerospace industry widely employs phosphoric acid anodizing. The chromic acid anodizing is eliminated in order to reduce environmental damage caused by chromates (Fredell 1994).

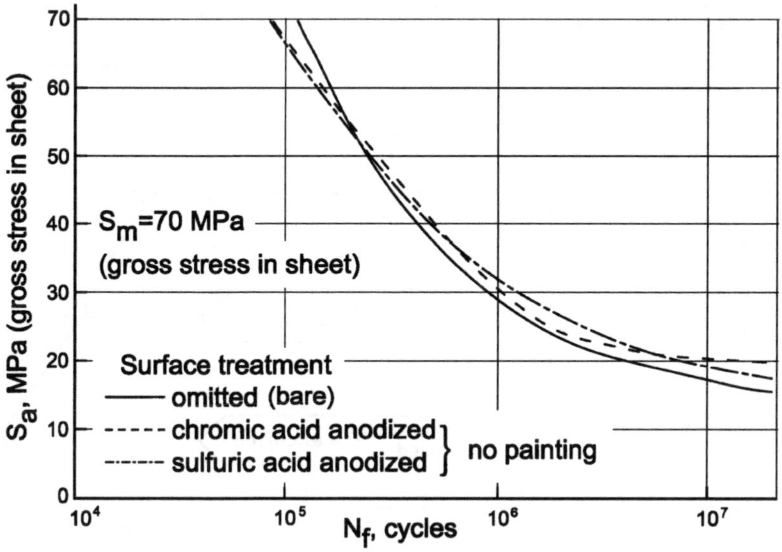

Fig. 3.23 Effect of sheet surface treatment on lap joint fatigue strength (Hartman 1968)

Wanhill's (1996) experiments on 2024-T3 lap joints indicate that anodizing is detrimental on top of cladding, but beneficial on the unclad material, as illustrated in Fig. 3.24. The specimens were tested under a gust spectrum miniTWIST that is representative for the wing root of a transport aircraft. A similar behaviour has been observed by Skorupa et al. (2009) for the 2024-T3 alloy and by Giummara and Zoker (2005) for the 7085-T7651 sheet material. In addition, Giummara and Zoker's tests back up a finding from Hartman's study (1968) referred to above, namely that the joint fatigue properties do not depend on the method of anodizing (chromic acid or sulphuric acid coating).

Wanhill (1996) also noted that for all surface conditions the lap specimens made of the 2024-T3 material exhibited better fatigue properties than the lap joints manufactured from the 7475-T761 alloy.

As seen in Fig. 3.24, the best fatigue properties are obtained when a chemically milled (non-clad) sheet is anodized. This confirms results of earlier tests by Schijve et al. (1976) on sheets tapered by chemical milling. According to Wanhill, the most likely explanation of these results is that anodizing inhibits or postpones faying surface fretting, but when cladding is present it yields cracking of the anodized layer, which then promotes cracking of the cladding. Next, the cracks progress into the metal core. Painting of the anodized sheets results in a slight decrease in fatigue strength under CA loading conditions (Hartman 1968). Schütz and Lowak (1974) tested 2024-T3 Alclad lap joint specimens with protruding head rivets applying a standard flight-load spectrum for wing components of transport aircraft TWIST. No distinct differences between the fatigue lives for the anodized sheet surfaces and the alodined and painted surfaces were exhibited.

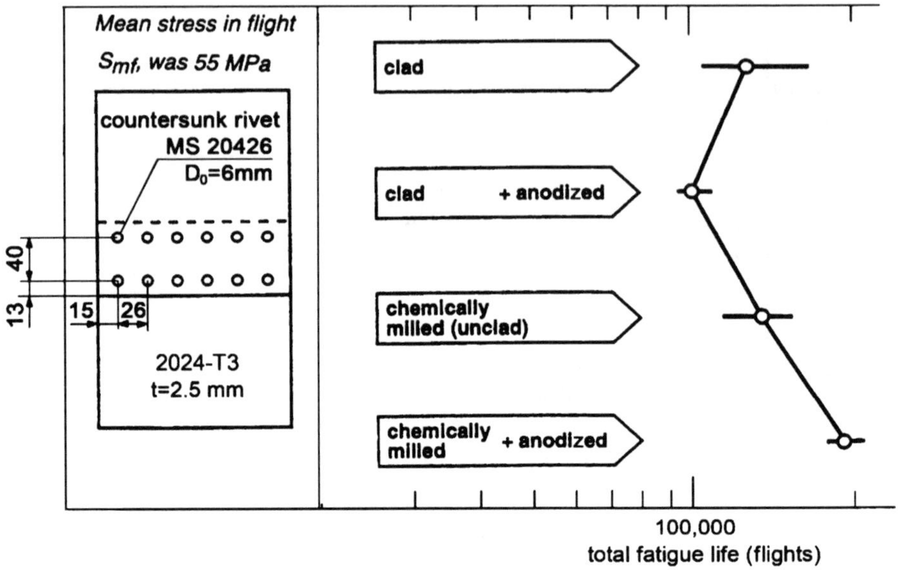

Fig. 3.24 Effect of different methods of sheet surface treatment on fatigue lives (mean value and scatter) for lap joints under the miniTWIST gust spectrum (Wanhill 1996)

Another type of surface treatment is the application of water-displacing fluids in order to avoid corrosion problems. The basic idea is that the interfaying fluid prevents the entry of water into the crevices of joints. As a result, electrolytic corrosion can no longer occur. However, the presence of this fluid between the faying sheets may change the load transmission in a riveted joint and a decrease in fatigue strength can be observed. Such a detrimental effect of water-displacing penetrant is revealed in fatigue tests by Schijve et al. (1977) on 2024-T3 Alclad lap joints with countersunk rivets and joints with fully dimpled holes under flight-simulation loading (TWIST). These results show a significant deterioration in static and fatigue strength for the joints with countersunk rivets due to the LPS-3 penetrant application. However, the static and fatigue properties of the lap joints with dimpled holes were hardly influenced by the application of penetrant. These observations can be rationalized in terms of the differences in the load transmission for both types of the joint. For a joint with dimpled holes load transmission by friction forces may be assumed to be small. Load transmission relies more on the dimpled zones around the rivets, and will not be affected by a lubricant. For a lap joint with countersunk rivets, the frictional force is decreasing and, hence, more unfavourable pin loading of rivets occurs. The absence of any effect of penetrant on the fatigue life of dimpled hole joints was also confirmed by tests under CA loading carried out at the Fokker-VFW factories referred to by Schijve et al. (1977). Application of LPS-3 before testing or during the tests after 25% or 50% of the mean life obtained with the dry specimens did not have any noticeable effect on fatigue life.

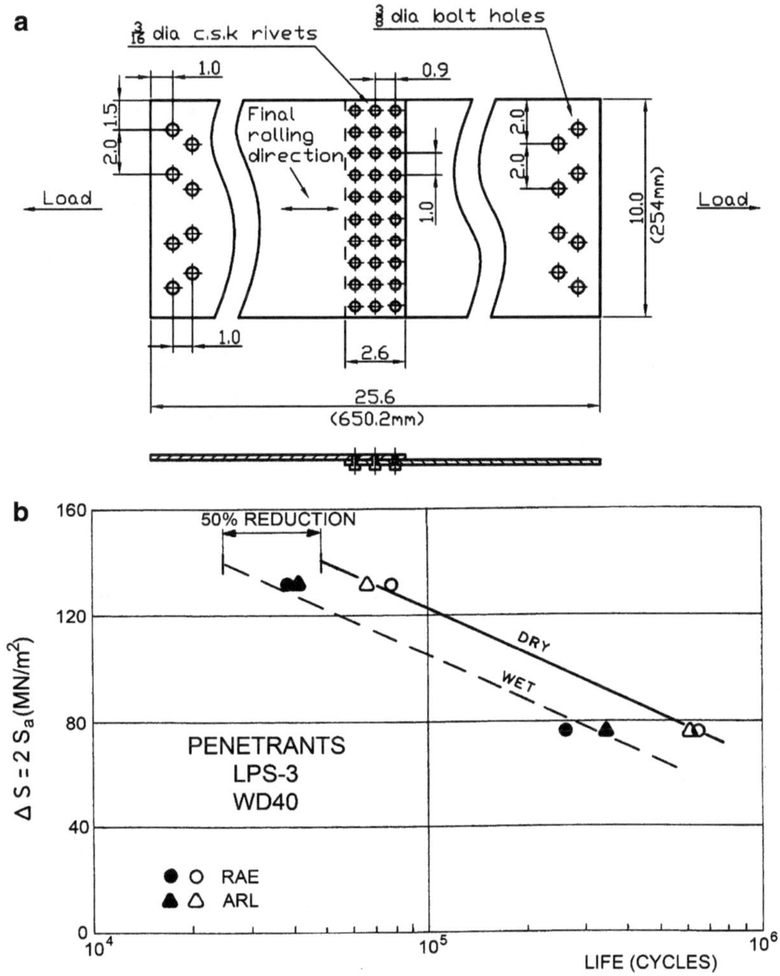

Fig. 3.25 Effect of penetrants on fatigue life of lap joints: (**a**) test piece (dimensions in inches) (**b**) test results (Schijve et al. 1977)

Results on the impact of anti-corrosion penetrants on the fatigue life of test pieces representing a fuselage skin connection (2024 Alclad sheet 1.6 mm thick, three rows of countersunk rivets 4.8 mm dia) were published by the Royal Aircraft Establishment (RAE) in England and the Aeronautical Research Laboratories (ARL) in Australia (refs 2–4 in Schijve et al. 1977). The specimens were tested in both laboratories under CA loading ($R = 0.15$ with frequency 30 Hz and $R = 0$ with frequency 7.5 Hz). As shown in Fig. 3.25, the agreement between the results from both laboratories was good. It can be seen that the application of a corrosion inhibitor approximately halves the fatigue life of the joint. Another conclusion from these tests is that the fatigue life is only slightly affected by the penetrant type and

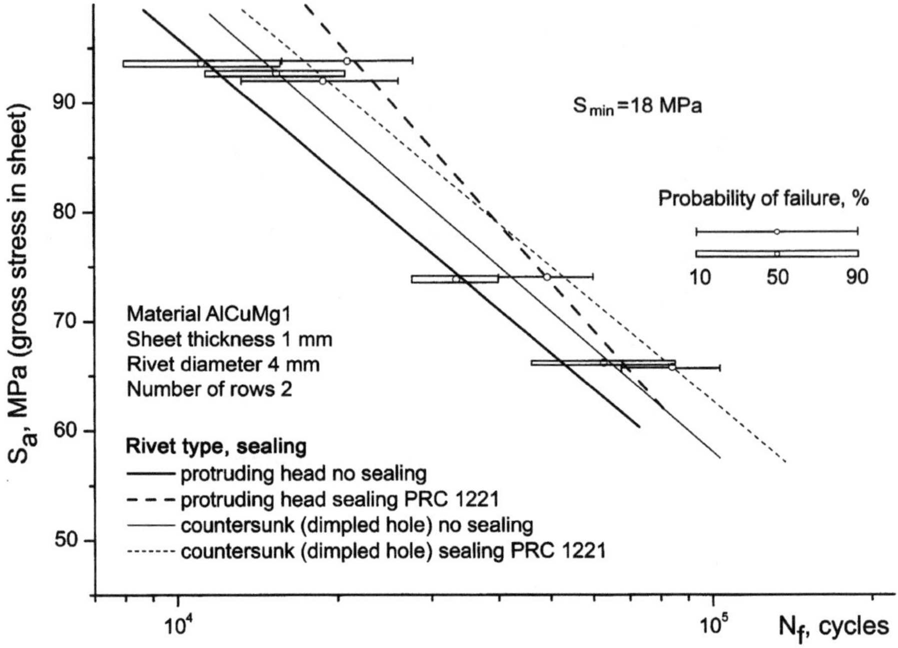

Fig. 3.26 Effect of sealing on fatigue life of lap joints (Schütz 1963)

by its application either before or after assembling the riveted joint. Recently, a detrimental influence of CICs (Corrosion-Inhibiting Chemicals) on fatigue life of riveted lap joints typical for small aircraft (2024-T3 bare Al sheets 1 mm thick, universal rivets, manual riveting technique, and 2024-T3 Alclad sheets 1.6 mm thick, countersunk rivets, force control riveting) was observed by Jaya et al. (2011). In general, the application of CICs reduced the fatigue life of these joints up to a factor of 2.

It should be emphasized that the presence of penetrant diminishes fretting damage. This could be explained by the lubrication action of the penetrant and exclusion of the aggressive environment, such as a humid air.

The information on the influence of a sealant on the fatigue life of lap joints is extremely limited. Schütz (1963) compared series of double-row lap joints riveted with protruding head rivets and countersunk dimpled hole rivets with and without the sealant PRC 1221. While a beneficial effect of the sealant was observed for the protruding head rivets, for the dimple hole specimens the application of the sealant was less favourable, Fig. 3.26. From the statistical point of view, however, the difference between the two test series was insignificant. The favourable effect of the sealant (PR1221A2) for the protruding head rivets was also indicated earlier by Hamburger Flugzeugbau factory (Hertel 1969). Double-row lap joint specimens were tested under CA loading at two maximum stress levels, i.e., 80 and 145 MPa

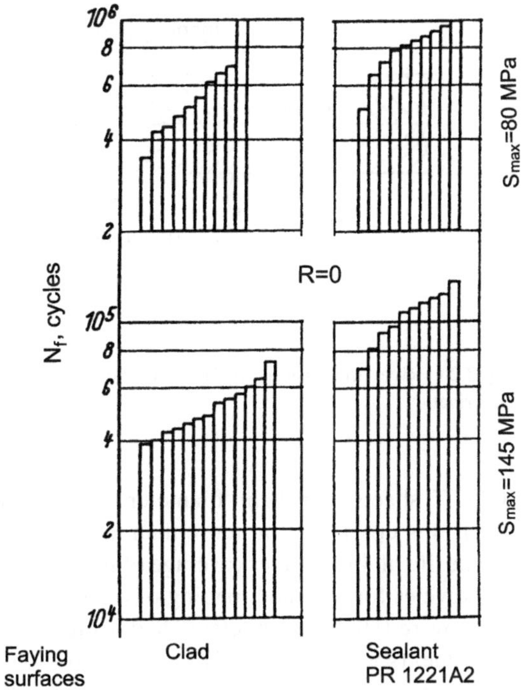

Fig. 3.27 Influence of sealant on fatigue strength of double row lap joint specimens. 2024-T3 Alclad sheet, protruding head rivet 3.2 mm dia (Hertel 1969)

($R = 0$), Fig. 3.27. It is seen again that the influence of the sealant is more pronounced at the higher loads (shorter fatigue lives). The reason for the favourable effect of the sealant on the fatigue life of riveted joints is that it acts as adhesive.

Dependence of lap joint fatigue behaviour on faying surface conditions is further considered in the context of secondary bending (Sect. 6.5) and fretting (Sect. 7.2).

3.3.5 *Squeeze Force*

The rivet installation process consists in squeezing with a large force a rivet inserted into the hole drilled in the components being joined. The resulting plastic flow of the rivet leads to the formation of a rivet driven head, Fig. 3.28.

3.3.5.1 Effect of the Squeeze Force on Fatigue Life

Numerous experimental results, some of them presented below, prove that driven head dimensions significantly influence the fatigue properties of riveted joints. Smith (1960) described an investigation made in Convair into the effect of a rivet driven head diameter on the fatigue life of riveted joints from 0.102 in. thick

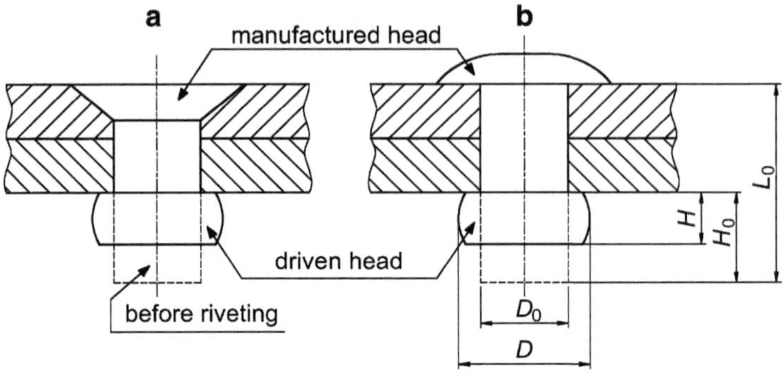

Fig. 3.28 Rivet dimensions before and after riveting: (**a**) countersunk rivet; (**b**) protruding head rivet

7075-T6 sheets with ¼ in. diameter countersunk rivets. It was found that a head which was upset 0.03 in. more than the standard 0.33 in. diameter value yielded a fatigue life of 458 kcycles (average of five specimens) compared with the standard of 317 kcycles for repeated loading of a 10 ksi gross stress. Joints wherein the driven heads were 0.015 in. undersize had a life of only 201 kcycles, Fig. 3.29. Similar experiments were done by Hartman (1968) on two series of double-row lap joint specimens with protruding head DD rivets installed using hydraulic pliers. For either specimen series a different driven head diameter was made, namely $D = 1.6D_o$ and $D = 1.7D_o$, where D_o is the rivet shank diameter. $1.7D_o$ implies increasing the driven head by about 6% compared to $1.6D_o$, considered to be a standard value. From Fig. 3.30, where the fatigue test results for CA loading are shown, an increase in the fatigue strength at 10^7 cycles with a factor of 1.4 is observed if the driven head diameter is increased from $1.6D_o$ to $1.7D_o$. A similar beneficial effect has been revealed in the programme tests ($\approx$28 periods for $D = 1.7D_o$ versus $\approx$ 14 periods for $D = 1.6D_o$).

A detrimental influence of decreasing the driven head diameter of snap rivets on the fatigue properties of double-row lap joints was reported by Schütz (1963). He observed an average fatigue life ratio of about three between series of nominally identical joints riveted by the same manufacturer in different years. Variations in the driven head dimensions were supposed to be responsible for the divergences in the fatigue lives.

For anti-symmetric riveted lap joint specimens used in Müller's fatigue tests (1995) and shown at the top of Fig. 3.31, fatigue cracks are initiated at both end rows if the rivets are equally squeezed. The increasing number of cycles to failure with increasing D, especially for the larger D_o of 4 mm, was noted. The fatigue life improvements of approximately ten times were obtained by increasing D from 1.2 to $1.75D_o$ for sheet thicknesses of 1 and 1.6 mm used in the experiments. Only for the smaller D_o-value of 3.2 mm and $t = 1.6$ mm a quadruple increase in life occurred.

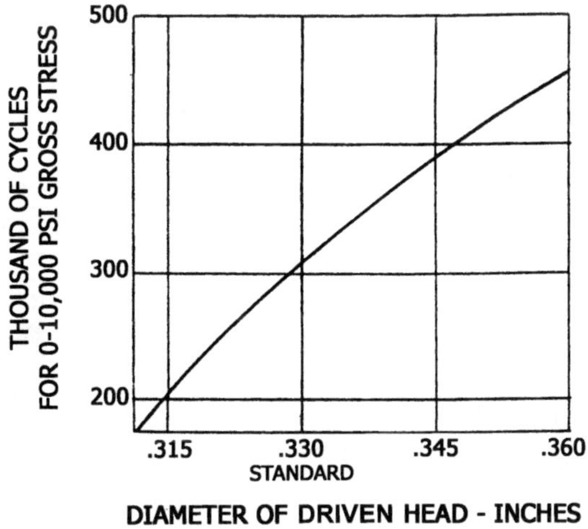

Fig. 3.29 Lap joint fatigue life versus rivet driven head diameter (Smith 1960)

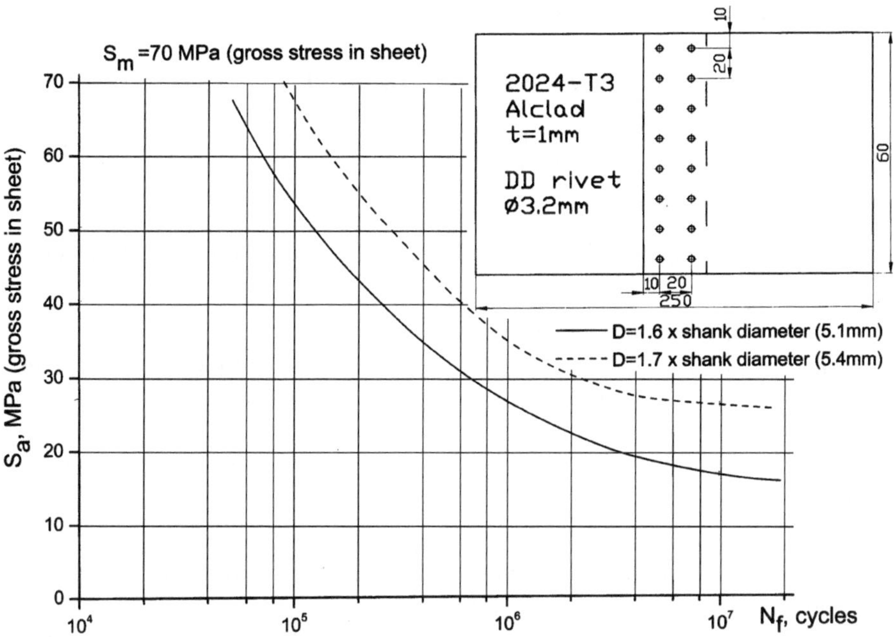

Fig. 3.30 Effect of a driven head diameter on lap joint fatigue life (Hartman 1968)

Experimental trends revealed in the works referred to above can be explained if it is recalled that the increase in the rivet head diameter is caused by a more intense riveting, i.e. by applying a higher squeeze force. This leads to a better filling of the hole and to an increased clamping between the sheets. Several consequences of

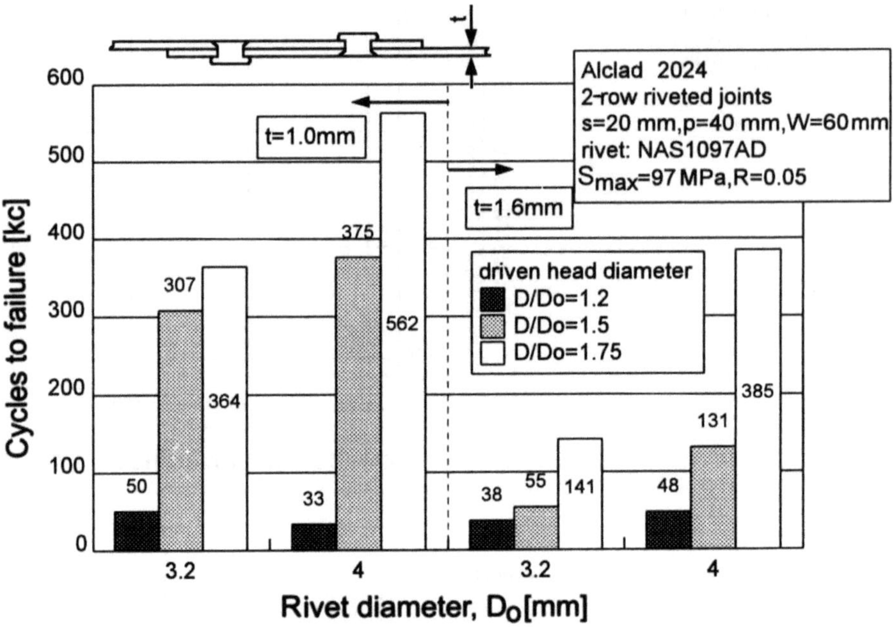

Fig. 3.31 Comparison of lap joint fatigue lives to show the effect of a driven head diameter for two sheet thicknesses and two rivet diameters (Müller 1995)

applying a higher squeeze force which beneficially affect the fatigue behaviour of riveted joints are:

- increased load transmission by friction,
- increased rivet hole expansion, which implies lowering of the local mean stress and a reduction of the local stress amplitude. Also the stiffness associated with fastener flexibility is improved,
- shift of the location of maximum secondary bending stresses away from the rivet hole.

Measurement results according to Schijve (2009a) showing the influence of increasing the riveting force on the driven head diameter and height as well as on the amount of hole expansion are presented in Fig. 3.32.

The effect of the riveting force on fatigue life can be quantified if the riveting is done under force control, as in several works addressed below. An example of the beneficial effect of increasing the squeeze force on lap joint fatigue life can be experimental results produced by Schijve (2009a) and shown in Fig. 3.33. Specimens used in the fatigue tests by Müller (1995) and presented in Fig. 3.34a were consistent with the configuration used in Airbus A340. The *S-N* curves obtained for three different squeeze forces are given in Fig. 3.34b. Increasing the squeeze force from 16.95 to 36 kN caused a threefold change in the fatigue life. A similar trend was observed by this author for lap joints of thinner sheets ($t = 1.6$ mm, rivet 4.0 mm dia).

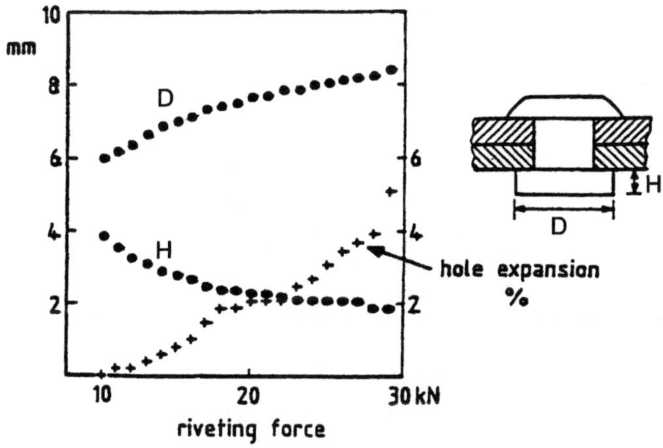

Fig. 3.32 Deformation of a rivet driven head and hole expansion as a function of rivet squeeze force (Schijve 2009a)

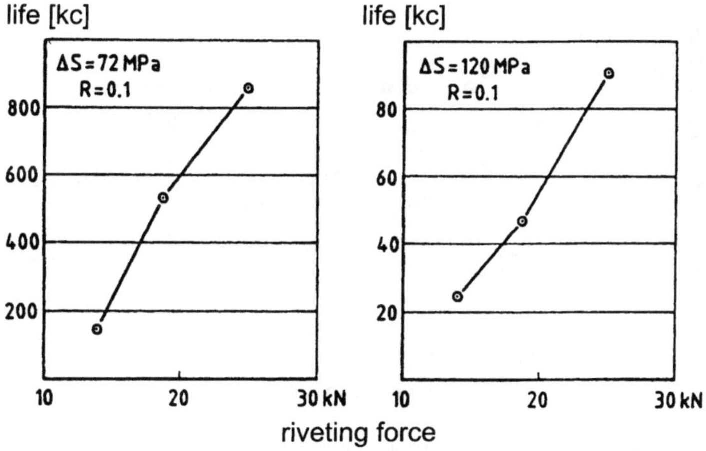

Fig. 3.33 Effect of a squeeze force on the riveted joint fatigue life at two applied stress levels. 2024-T3 sheet 2 mm thick, rivet diameter 4.8 mm, $R = 0$ (Schijve 2009a)

Shown in Fig. 3.35, fatigue lives observed in experiments by Müller and Hart-Smith (1997) for three squeeze force levels present another evidence of the beneficial influence of a high squeeze force. Here, a tenfold increase in the fatigue life is achieved by a threefold increase in the riveting force. All results in Fig. 3.35 are from CA tests with the loading frequency of 10 Hz on lap joint specimens from 1.6 mm thick 2024-T3 Alclad sheets joined by three rows of 2024-T31 DD countersunk rivets NAS1097, 4.8 mm in diameter (six rivets per row, row pitch = rivet pitch = 25.4 mm).

Similarly, Harish et al. (1999) reported that increasing the squeeze force level yielded increasingly longer fatigue lives for single lap joints with three rows of

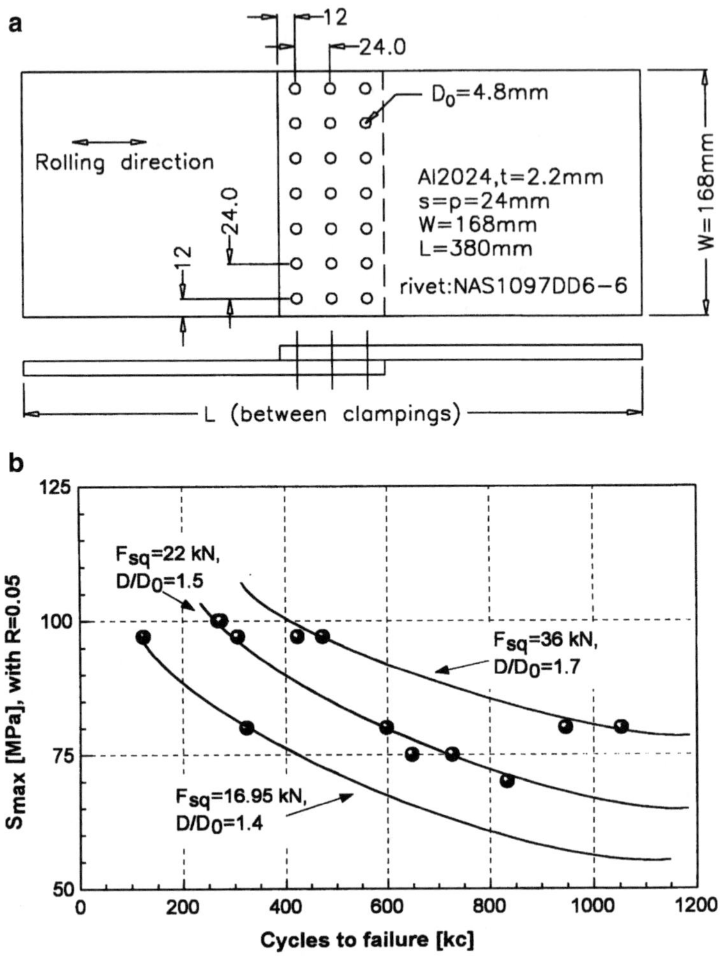

Fig. 3.34 Fatigue life of a three-row riveted lap joint (Airbus A340 configuration) as a function of the squeeze force: (**a**) specimen; (**b**) *S-N* curves for three squeeze force values (Müller 1995)

2117-T4 universal rivets. The loading frequency of 10 Hz and the stress ratio $R = 0.1$ were the same for all tests. As specified in the table in Fig. 3.36, the force controlled riveting was done at four different squeeze force levels. The duplicate tests were conducted at several applied stress values, also given in the table. A clear dependence of fatigue life on squeeze force for all stress levels is evident from the stress-life data plotted in Fig. 3.36. It can be seen that fatigue life is quadrupled due to a twofold increase in the riveting force (from 2,500 to 5,000 lb).

The ratio between the driven head diameter and the rivet diameter is usually considered as a first quality control criterion for riveted joints. Acceptable D/D_o ratios are defined in specifications of aircraft manufacturers. For example, according to Fokker, $D/D_o = 1.25–1.65$. Schra et al. (1995) established the relation between

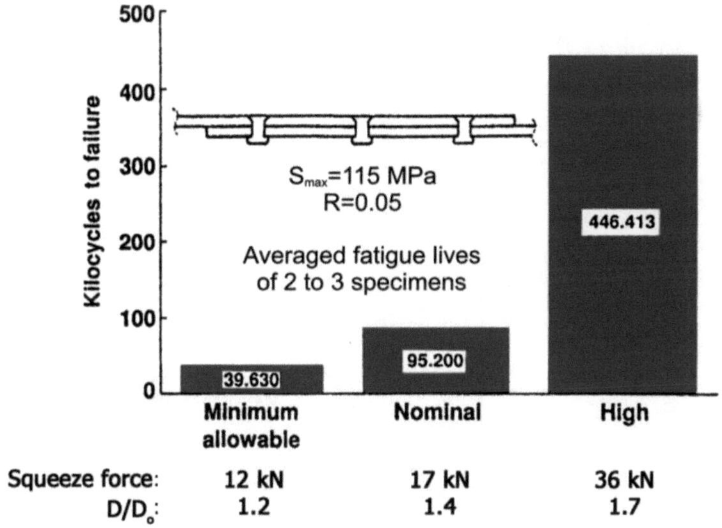

Fig. 3.35 Effect of a squeeze force on fatigue life of conventional three-row riveted lap splices (Müller and Hart-Smith 1997)

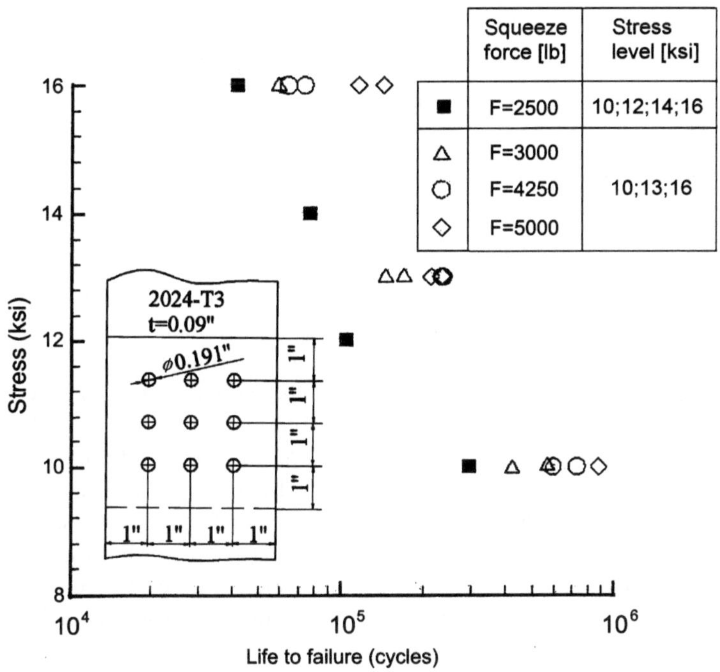

Fig. 3.36 Effect of a squeeze force on fatigue life of three-row riveted lap splices (Harish et al. 1999)

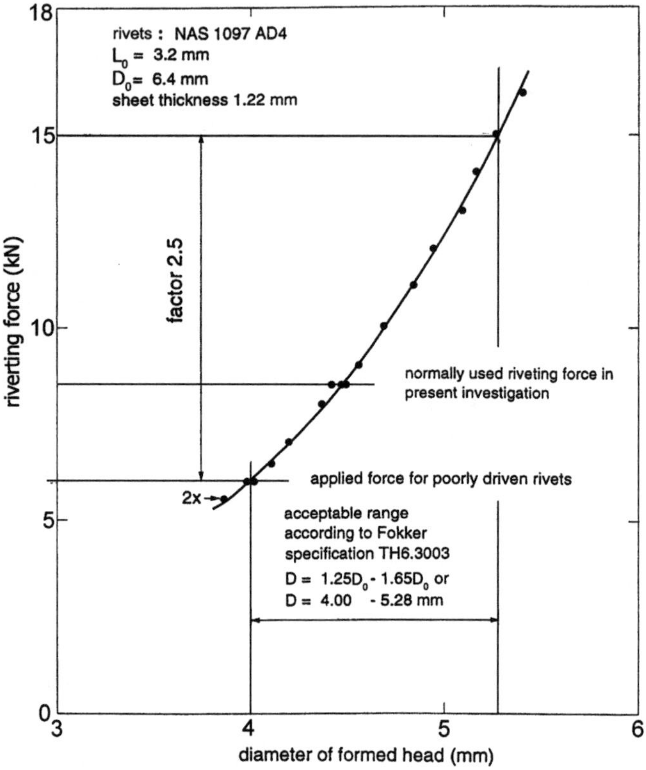

Fig. 3.37 Relation between a squeeze force and a driven head diameter for 1.22 mm thick Alclad 2024-T3 sheets assembled with NAS 1097AD4 rivets (Schra et al. 1995)

the riveting force and the driven head diameter within the range of D/D_o specified by Fokker. As can be seen in Fig. 3.37, within the acceptable D-values for $D_o = 3.2$ mm the riveting force varies from 6 to 15 kN for a sheet thickness of 1.22 mm. These authors investigated consequences of installing a certain number of countersunk rivets in both critical (outer) rows for the joint shown in Fig. 3.38a with the squeeze force of 6 kN, which was the lowest acceptable level (compare Fig. 3.37), while other rivets were riveted with a force of 8.5 kN, i.e. about 30% above the lowest acceptable value. The results on the total fatigue life observed in CA tests at $R = 0.1$ with $S_{max} = 79$ MPa are shown in Fig. 3.38b. A gradual decrease in fatigue life with an increase in the number of poorly driven rivets is observed. For the limiting case when all rivets in the critical rows were poorly riveted, a threefold decrease in fatigue life occurred. In Fig. 3.38c the total fatigue life of the joint from Fig. 3.38a observed under the same loading conditions is presented as a function of the riveting force in the instance when all rivets are installed with the same force. In agreement with other studies referred to above, an increasingly longer fatigue life with an increase in the riveting force level was noted. Interestingly, the riveting force of 2.5 kN resulted

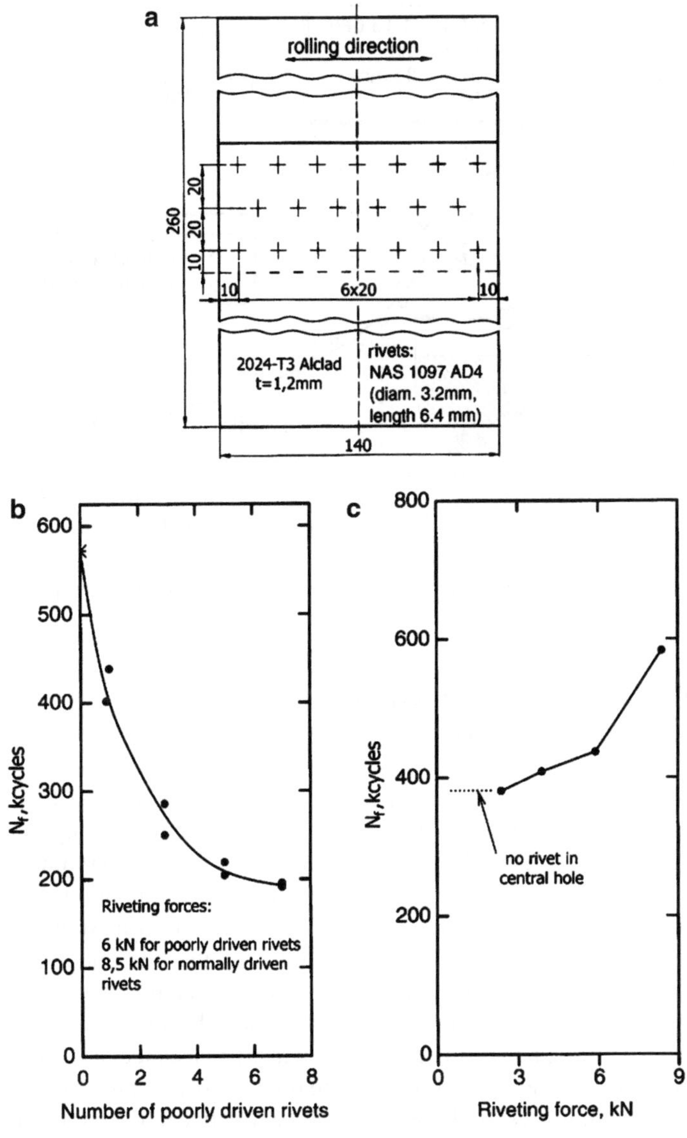

Fig. 3.38 Effect of a squeeze force on fatigue life of three-row riveted lap splices: (**a**) specimen; (**b**) fatigue life as a function of the number of poorly driven rivets; (**c**) fatigue life as a function of the squeeze force (Schra et al. 1995)

in the same fatigue life as observed for the specimen without rivets in the central hole of both outer rows. This fatigue life was about 35% lower compared to a reference specimen riveted with the force of 8.5 kN.

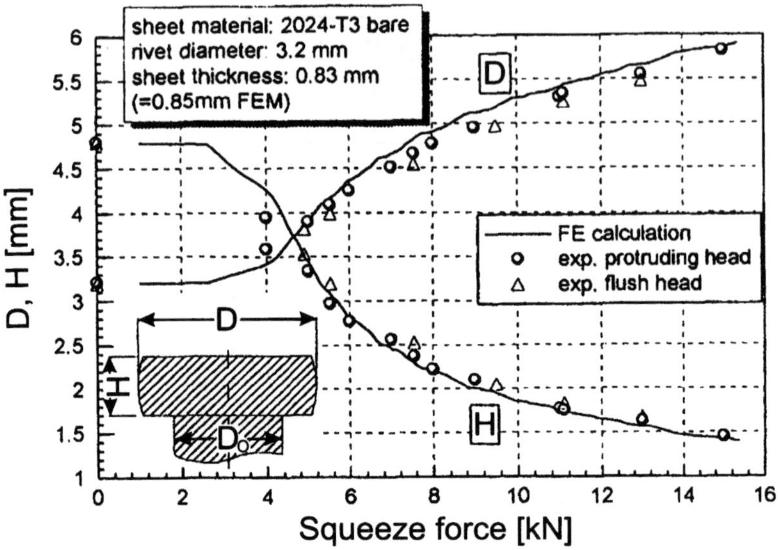

Fig. 3.39 Effect of a squeeze force on driven head dimensions according to experiments and FE computations (Müller 1995)

3.3.5.2 Dependence of Rivet Driven Head Dimensions on the Squeeze Force

The dependence of the rivet driven head dimensions on the squeeze force level was extensively studied by Müller (1995). His measurement and FE analysis results corresponding to the AD material rivet and shown in Fig. 3.39 are similar to Schijve's results presented in Fig. 3.32. The hole diameter was 0.1 mm larger than the initial rivet diameter D_0. The protruding length of the rivet was consistent with the rule of thumb formula $H_0 = 1.5D_0$. Figure 3.39 indicates that the relation between the squeeze force and the driven head dimensions is nonlinear, the variations in D and H diminishing at higher squeeze force levels. The latter behaviour is associated with the rivet material strain hardening and the increasing resistance of the sheet material to hole expansion. The results in Fig. 3.39 are practically the same for the protruding head and countersunk rivet and a good agreement between the experiment and FE computations is shown. The FE model considered by Müller (1995) for a rivet with the driven head at both sides is shown in Fig. 3.40. The rivet and sheet materials were assumed to be elastic-ideally plastic.

The same trends as those revealed in Figs. 3.33 and 3.39 have been observed by de Rijck (2005) and de Rijck et al. (2007). His measurement data shown in Fig. 3.41 come from eight specimen series with protruding head and countersunk AD rivets, 4.0 and 4.8 mm in diameter. The rivet holes were drilled 0.1 mm larger than the rivet shank diameter, according to the Fokker specification. The total sheet thickness was

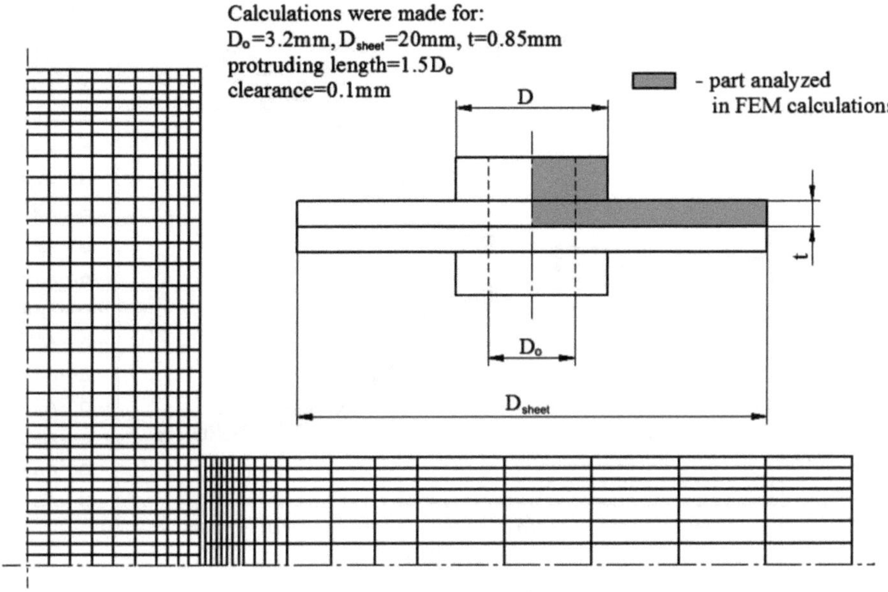

Fig. 3.40 FE model for a rivet with driven head at both sides (Müller 1995)

4 mm. In Fig. 3.41, the rivet head dimensions are presented in terms of the squeeze stress which is defined as the true stress computed for the actual deformed head area,

$$\sigma_{sq} = 4F_{sq}/\pi D^2. \tag{3.1}$$

Experiments similar to de Rijck's tests were also performed by Skorupa et al. (2009, 2010b) on round head rivets of the PA24 alloy, protruding head rivets with the compensator (a small protrusion on the mushroom head) of the PA24 and PA25 materials, and universal rivets of 2117 (MS20470). All rivet types used are shown in Fig. 3.42a. D16 sheets 1.9 mm thick were joined with the PA24 and PA25 rivets, while the MS rivets were used for joining 2024-T3 sheets, 2.0 mm in thickness. In Fig. 3.42b the normalized rivet head dimensions are presented against the squeeze stress defined according to Eq. 3.1. The notation "c", "r", and "MS" corresponds to the rivet with a compensator, round head rivet and universal rivet respectively. Also plotted in Fig. 3.42b is the scatter band of data for the AD (2117) rivets reported by de Rijck (2005) and shown in Fig. 3.41. It is seen that the results for the MS and PA24r rivets made of the same material fall within the de Rijck's data scatter band. For the PA24c rivet, the D/D_0 and H/H_0 data falling above and below respectively the scatter band of de Rijck's results confirm a higher ductility of this rivet material revealed in compression tests of samples machined out of the PA24c rivets. The opposite trend is shown in Fig. 3.42 by the results for the PA25c rivet made of the harder material.

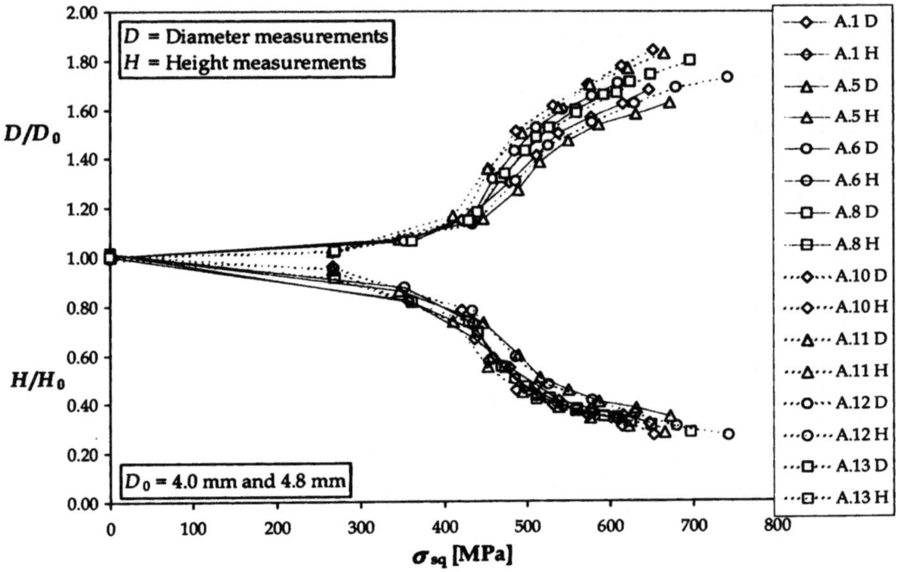

Sheet material	t [mm][1]	Rivets					
		D_0 [mm]	$L_0^{(2)}$ [mm]	H_0/D_0	Material	Rivets, type[3]	Code
2024-T3 clad	4.00	4.0	9.5	1.38	AD	NAS 1097 c	A.1
			8.0	1.00		EN 6101 c	A.5
			9.0	1.25			A.6
			9.0	1.25		MS20470 p	A.8
		4.8	11.0	1.46		EN 6101 c	A.10
			11.1	1.48		NAS 1097 c	A.11
			9.0	1.04		EN 6101 c	A.12
			10.0	1.25		MS20470 p	A.13

(1) Total thickness of combined sheets
(2) $L_0 = H_0 + t$
(3) c - countersunk, p - protruding head

Fig. 3.41 Relation between normalized dimensions of a rivet driven head and squeeze stress (de Rijck 2005)

Unlike in the experiments by de Rijck and Skorupa et al. addressed above, Li and Shi (2003) obtained a very good correlation between the driven head dimensions of the countersunk rivet and the engineering stress (squeeze pressure)

$$S_{sq} = 4F_{sq}/\pi D_o{}^2, \tag{3.2}$$

$$D/D_o = 1.4399 S_{sq}{}^{0.3322}, \tag{3.3}$$

$$H/H_o = 0.5037 S_{sq}{}^{-0.7072}. \tag{3.4}$$

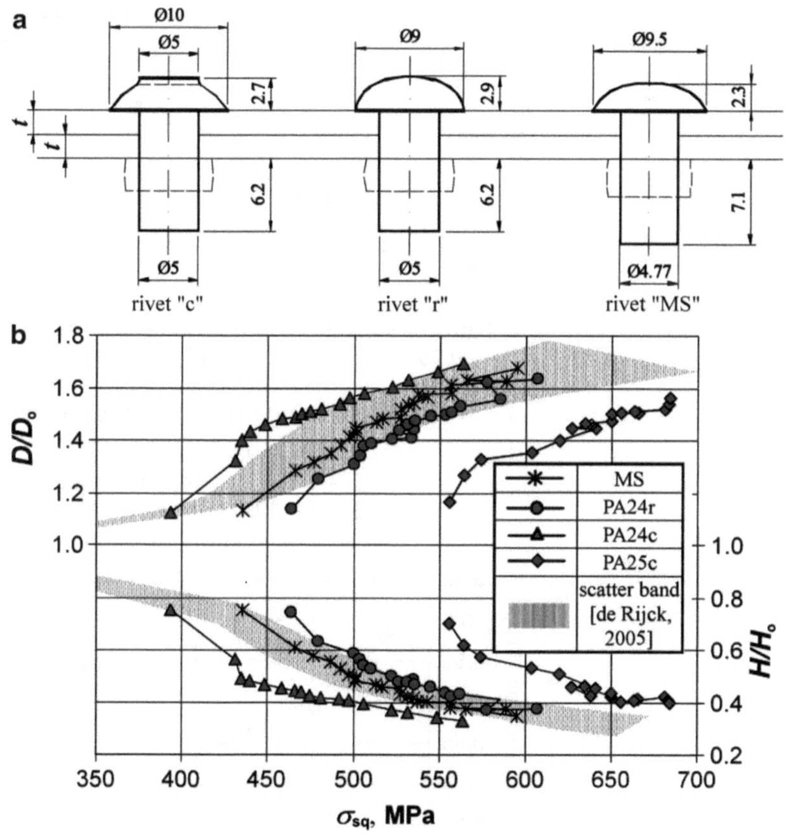

Fig. 3.42 Observed dependence of rivet driven head dimensions on the squeeze force: (**a**) rivet types; (**b**) normalized driven head dimensions as a function of the squeeze stress together with the scatter band of de Rijck's (2005) results for 2117-T4 rivets (Skorupa et al. 2010b)

Figure 3.43 presents the results derived by Li and Shi for two countersunk rivet dimensions and three sheet thicknesses. The rivets were made of the 2117-T4 Al alloy and the sheet material was the 2024-T3 Al alloy.

It should be noted that the rivet head does not remain perfectly cylindrical during its formation and adopts a barrel type shape, as shown in Fig. 3.44.

As already indicated in Fig. 3.42, the squeeze force required to form the proper driven head depends on the rivet material type and its heat treatment. If the yield stress is low, the driven head can be easily formed with a low squeeze force. It leads to a large head diameter and a relatively low head height. In Fig. 3.45, the driven head diameter and height are presented as functions of the squeeze force for the 2117 AD rivet ($S_y = 165$ MPa), 2024 DD rivet ($S_y = 324$ MPa), and 2219 E rivet ($S_y = 434$ MPa). It is seen that for the hard E rivet the largest squeeze force is required to form the desired driven head shape. On the other hand, the

Fig. 3.43 Variations of rivet driven head dimensions versus engineering squeeze stress (Li and Shi 2003)

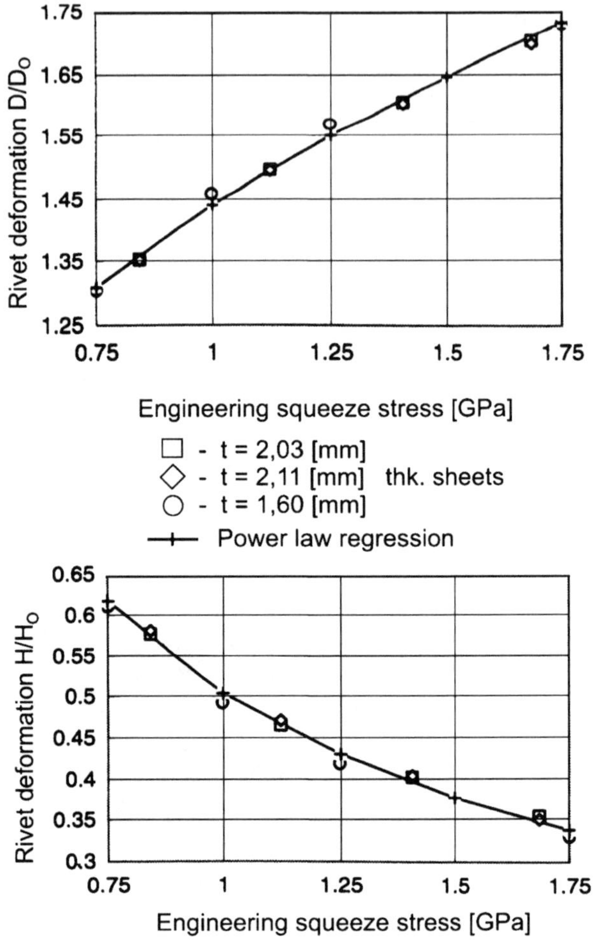

soft AD rivet shows the largest deformation with respect to the squeeze force. The above observations by Müller (1995) are fully confirmed by de Rijck's experiments (2005).

Obviously, the driven head shape for a given squeeze force level depends on the rivet length in that the driven head will be higher for a longer rivet. The protruding length of the rivet before squeezing (H_o) is typically chosen as $1.5D_o$ for protruding head rivets, as already said above, while for countersunk rivets $H_o = 1.0D_o$ (see Fig. 3.28). If the combined thickness of the sheets is larger than 7.0 mm, H_o is increased by an extra 1.0 mm. In Fig. 3.46, the results of tests by Müller (1995) for three H_o/D_o ratios covering the range frequently used in production are presented (Schijve 2005). It can be concluded that the effect of H_o/D_o is rather small, although not fully absent.

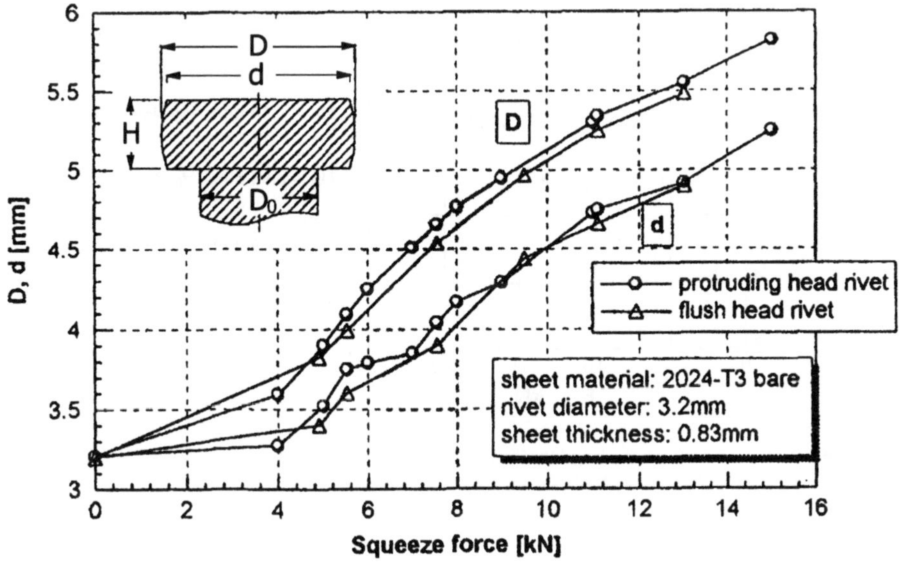

Fig. 3.44 Dependence of the largest (*D*) and smallest (*d*) driven head diameter on the squeeze force (Müller 1995)

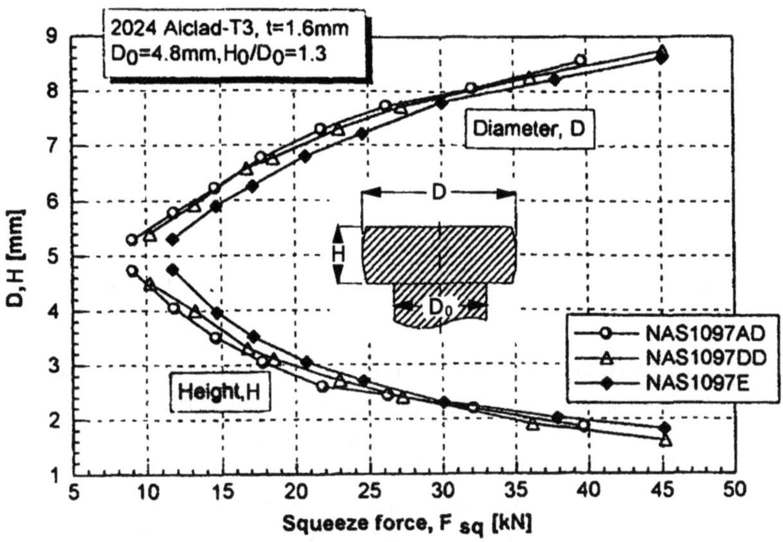

Fig. 3.45 Driven head diameter (*D*) and height (*H*) as a function of the squeeze force for countersunk 2117 AD, 2024 DD and 2219 E rivets (Müller 1995)

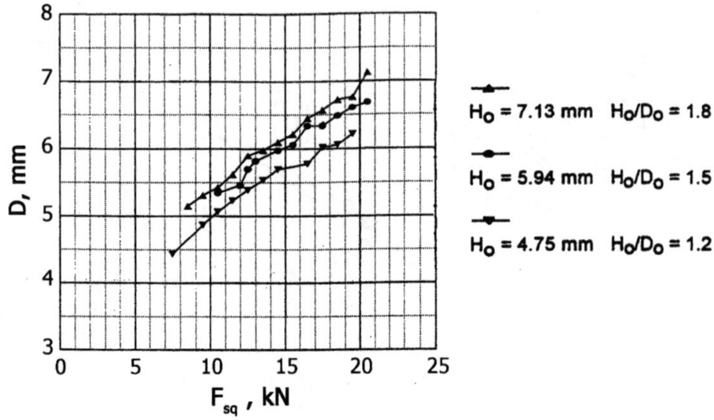

Fig. 3.46 Effect of rivet protruding length (H_o) on driven head diameter (D). 2024-T3 Alclad sheet 1.27 mm thick, rivet NAS1097AD5, $D_o = 3.96$ mm (Schijve 2005)

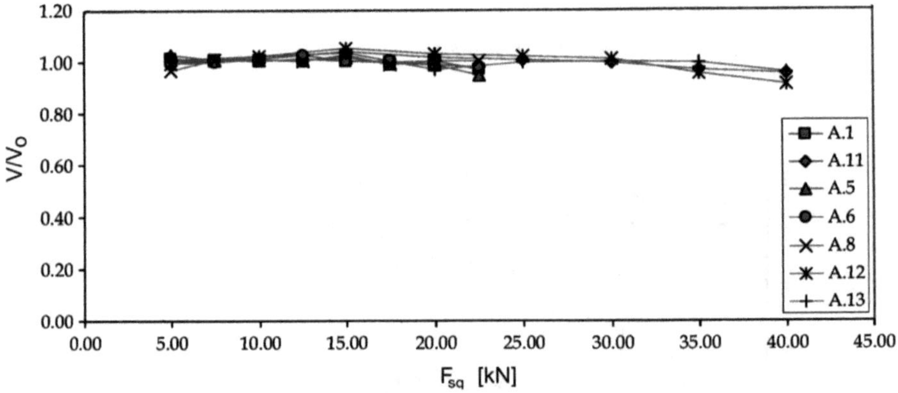

Fig. 3.47 Volume changes in forming driven-rivet heads, for legend see Fig. 3.41 (de Rijck 2005)

De Rijck's (2005) measurement results on the ratio of the rivet driven head volume to the rivet protruding part volume before squeezing V/V_o ($V = \pi HD^2/4$ and $V_o = \pi H_o D_o{}^2/4$, Fig. 3.28), for a range of the squeeze force values are shown in Fig. 3.47. These data reveal that V/V_o remains constant and close to unity. Only a small amount of the rivet material is squeezed into the rivet hole, as also backed up by Müller's (1995) experimental results presented in Fig. 3.48. The data on the driven head diameter plotted against the squeeze force, Fig. 3.49, indicate that the D-value is practically independent of the sheet thickness (Schijve 2005). For rivets of the same shank diameter of 3.96 mm, the same D diameter is obtained at a given F_{sq} level for different sheet thicknesses of 1.02 and 1.27 mm. Other experimental results (Müller 1995; Schijve 2005) indicate that also the type of the sheet material does not influence the driven head diameter.

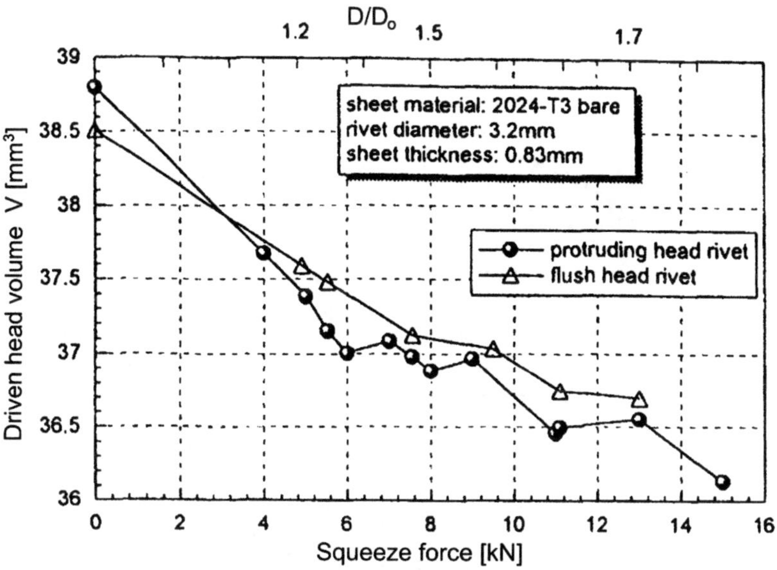

Fig. 3.48 Relation between the squeeze force and the driven head volume for a protruding-head rivet and countersunk rivet (Müller 1995)

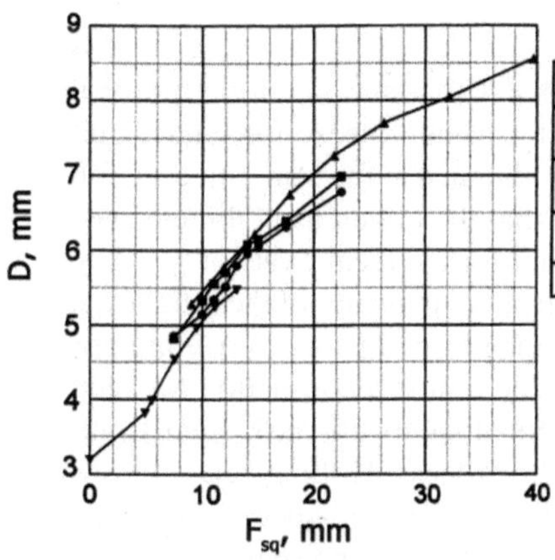

t, mm	0.83	1.02	1.27	1.61
D_0, mm	3.196	3.96	3.96	4.75
H_0, mm	4.8	5.93	5.36	6.2
H_0/D_0	1.5	1.5	1.36	1.3
Symbol	▼	■	●	▲

Fig. 3.49 Effect of sheet thickness, rivet diameter and rivet length on driven head diameter depending on the squeeze force. 2024 Alclad sheet, NAS1097AD rivet (Schijve 2005)

In general, a larger sheet thickness requires a larger rivet diameter. Not only a higher squeeze force, but also a higher squeeze pressure (S_{sq}) must be applied to obtain the desired D/D_0 ratio when the rivet diameter increases, as can be easily deduced from the data shown in Fig. 3.50a (Schijve 2005). Here, S_{sq} is defined according to Eq. 3.2. Surprisingly enough, the effect of D_0 practically disappears if the driven head shape is represented by the D/H ratio, as demonstrated by the D/H vs. S_{sq} plots in Fig. 3.50b.

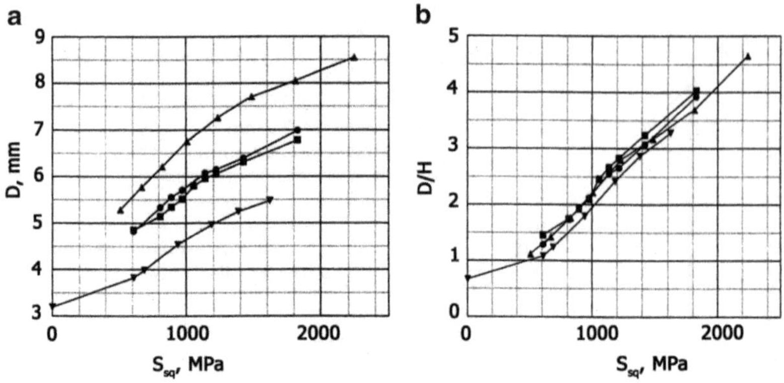

t, mm	0.83	1.02	1.27	1.61
D_0, mm	3.196	3.96	3.96	4.75
H_0, mm	4.8	5.93	5.36	6.2
H_0/D_0	1.5	1.5	1.36	1.3
Symbol	—▼—	—■—	—●—	—▲—

Fig. 3.50 Correlation between squeeze pressure and: (**a**) driven head diameter *D*; (**b**) *D/H* ratio for several combinations of sheet thickness, rivet diameter and rivet length. 2024 Alclad sheet, NAS1097AD rivet (Schijve 2005)

Müller's (1995) fatigue tests on 2024-T3 riveted lap joints prove that though rivets of different lengths riveted with the same squeeze force have quite different driven head diameters, the fatigue lives obtained at the same squeeze force are very similar, Fig. 3.51. These results imply that the driven head dimensions can reflect the fatigue performance of a riveted joint for identical rivets only, while in a more general case the squeeze force is a much better indicator of the fatigue quality. Since displacement controlled riveting rather than force controlled riveting prevails in the aircraft industry, attempts have been made to find a correlation between rivet head dimensions and the squeeze force. Difficulties of this task stem from the highly non-linear character of the rivet forming process. Hilling (ref. 2 in Schijve 2005) suggested that D^2/H should be an indicator for the squeeze force. A correlation between this parameter and the squeeze stress is presented in Fig. 3.52a. The latter graph includes experimental data by Müller (1995) and some Airbus test series for specimens with different *D* and *t* values riveted with AD, DD and E rivets. According to Schijve (2005), the *D*/*H* parameter enables a better overall consolidation of the results because the D^2-value is more than *D*-value sensitive to scatter, Fig. 3.52b. It cannot be overlooked that exact measurements of the driven head diameter in production conditions can be difficult due to its non-cylindrical shape (compare Fig. 3.44).

Based on the assumption of the constant volume of the driven head proved in his experiments (compare Fig. 3.47), de Rijck (2005) derived a simple relationship between the squeeze force and the diameter of the deformed rivet head

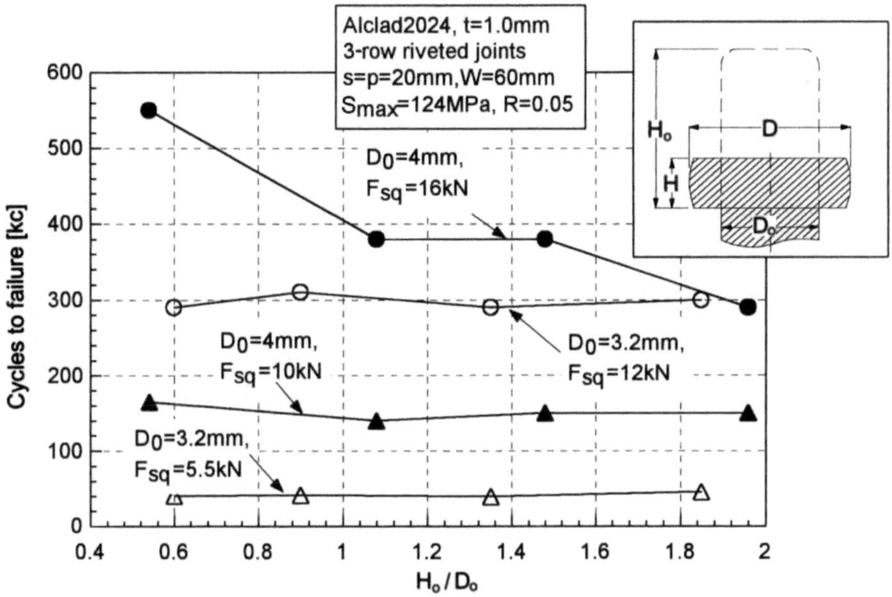

Fig. 3.51 Fatigue lives of lap joints riveted with a constant squeeze force for different initial rivet lengths. Countersunk AD rivets (Müller 1995)

$$F_{sq} = \frac{1}{4}\pi D^2 K[2\ln(D_o/D)]^n, \tag{3.5}$$

where K is the strength coefficient and n is the strain hardening exponent. The values of K and n for several rivet materials estimated from the true stress-true strain data obtained from the measurements of F_{sq}, D and H are given in Table 3.3. Equation 3.5 can be used to estimate the squeeze force from measurements of the driven head diameter. The agreement between Eq. 3.5 and the measured D values can be assessed from Fig. 3.53.

3.3.5.3 Dependence of Rivet Hole Expansion on the Squeeze Force

Additional to the driven head formation consequences of the riveting process are the radial expansion of the rivet shank and deformation of sheet material near the rivet hole. Considering the condition of sheet material adjacent to the hole, three stages can be distinguished during the rivet installation:

(i) Filling the hole. The hole diameter is initially made approximately 0.1 mm larger than the rivet diameter, Fig. 3.54a. The axial compression of the rivet results in its radial expansion. When the rivet shank touches the wall of the

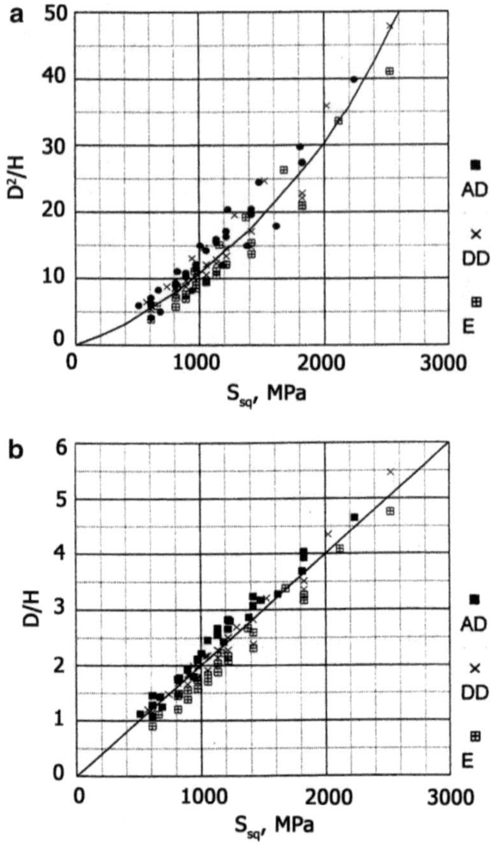

Fig. 3.52 Comparison of the correlation between squeeze pressure and: (**a**) D^2/H parameter; (**b**) D/H parameter. Rivet materials: *AD*, *DD*, *E* (Schijve 2005)

Table 3.3 Empirical constants *K* and *n* in Eq. 3.5

Rivet material	Rivet type	*K* (MPa)	*n*
2117-T4	AD	600	0.30
2017-T4	D	600	0.45
2024-T3	DD	550	0.30
7050-T73	E	350	0.30

hole, the rivet material is already plastically deformed because the transverse strain $\varepsilon \geq 2\%$ for $D_o \leq 5$ mm, Fig. 3.54b.

(ii) Expanding the hole. This phase is more complex as it involves plastic deformation of both the rivet and sheet material. The rivet begins to exert radial pressure on the hole. Consequently, deformation of the sheet material around the hole, at first elastic and then plastic, occurs. Due to the resistance to rivet deformation provided by the sheet material along the rivet shank, the free end of the rivet expands forming the driven rivet head. At the same time, a portion of the squeeze force begins to be transferred directly through the mating sheets in the shadow of the driven and manufactured rivet head and contributes to

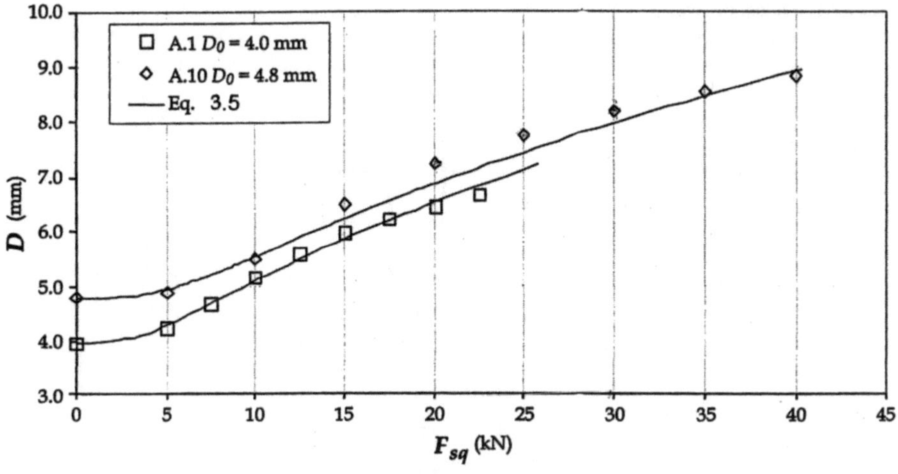

Fig. 3.53 Comparison between the measured and calculated driven head diameter. AD countersunk rivets (2117-T4), for legend see Fig. 3.41 (de Rijck 2005)

further plastic deformation of the sheet material. The process proceeds until a preset value of the squeeze force/driven head dimension is reached, Fig. 3.54c.

(iii) Springback. Due to removal of the squeeze force the rivet and sheet material relax elastically in the radial and thickness direction, Fig. 3.54d. Consequently, a residual stress state is left within the rivet-sheet system. If the sheet thickness springback exceeds the axial springback of the rivet, the clamping force of the rivet is generated.

Müller (1995) measured hole expansion by comparing dimensions of the squeezed rivet shank and the original hole. The rivets were cut vertically close to their centre line. The cutting surfaces were polished and the expansion measurements were performed. The polishing and measuring were repeated until maximum hole expansion was found. The relative motion between the sheets and the rivet popped out from the rivet hole must be prevented in this procedure. The exemplary data on the dependence of the hole expansion (measured at the mating surface) on the squeeze force for the protruding head and countersunk rivet, 3.2 mm in diameter, are shown in Fig. 3.55. Both type results match up to F_{sq} of 11 kN which corresponds to $D/D_o = 1.6$. For a larger squeeze force, hole expansion for the countersunk rivet becomes smaller than for the protruding head rivet. In the low range of F_{sq} the experimental data are consistent with the FE results.

Hole expansion varies in the thickness direction. As indicated by Müller's (1995) results shown in Fig. 3.56a, b, it is more homogenous through-the-thickness for the protruding head rivet compared to the countersunk rivet. Maximum hole expansion occurs at the outer surface under the driven rivet head. At the manufactured head, hole expansion is significantly smaller, especially for the countersunk rivet.

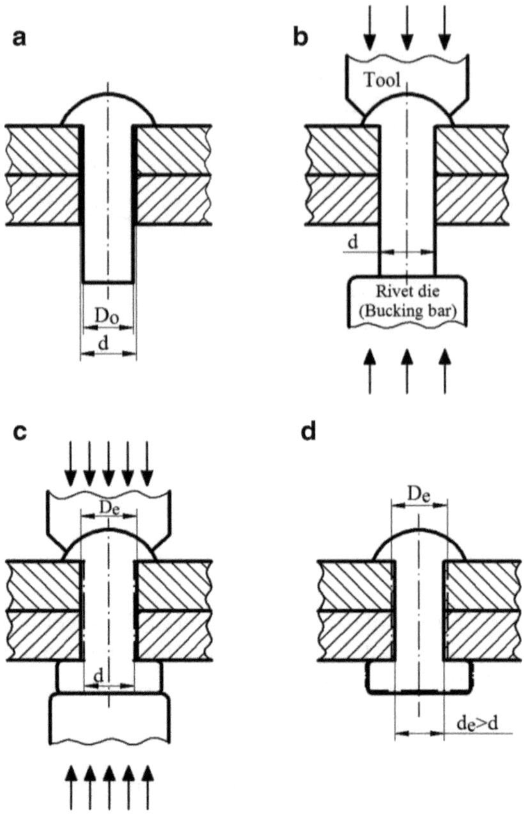

Fig. 3.54 Stages of rivet squeezing, explanation in text: D_o – rivet shank diameter, d – initial hole diameter, D_e – expanded hole diameter during squeezing, d_e – hole diameter after springback

The same conclusion can be derived from experiments performed by Wang (1998), though his results were obtained for rather low D/D_o ratios. Wang's data on hole expansion calculated based on the average value of the actual hole diameters (d_{ave}) for different D/D_o ratios are presented in Fig. 3.57. The d_{ave}-value was obtained from 54 measurements and equalled 0.1918 in. (standard deviation for Gaussian distribution = 0.00048 in.) when the nominal hole diameter was 0.191 in. It should be mentioned that actual diameters corresponding to the measured expansion values are not reported in Wang's work.

At a given squeeze force level, hole expansion is smaller for the larger thickness than for the smaller thickness, as shown in Fig. 3.58 (Müller 1995). This trend is not surprising because the same hole expansion in a thicker sheet requires a larger squeeze force. It is also evident that hole expansion must depend on the rivet material. The hard E rivet requires a larger squeeze force than the milder DD rivet and the AD rivet, Fig. 3.59a. If, however, hole expansion is presented against the driven head diameter, Fig. 3.59b, the data for three materials become consolidated. This is explained by the riveting process: forming of the driven head starts when rivet expansion in the hole becomes constrained by the sheet.

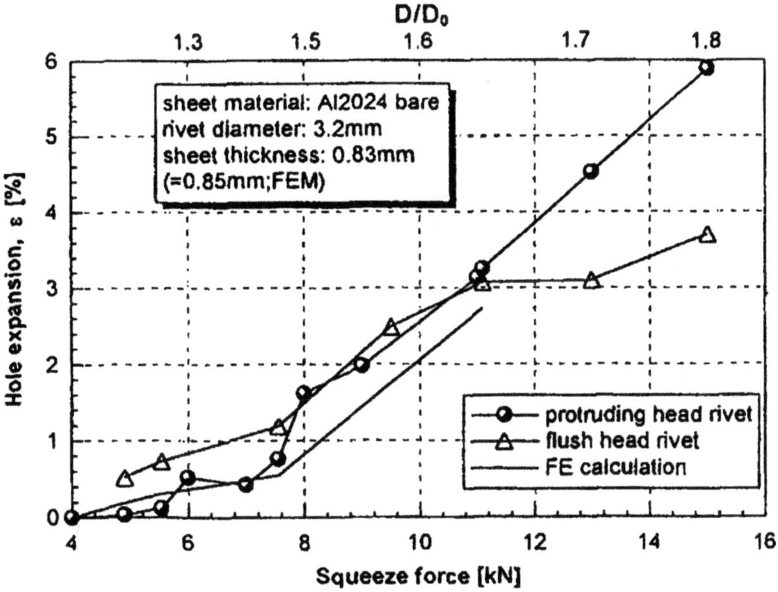

Fig. 3.55 Hole expansion at the mating surface according to experiments and FE calculations as a function of the squeeze force (Müller 1995)

Some of Müller's (1995) observations are confirmed by 3-D FE computations according to Rans (2007) and Rans et al. (2007) on hole expansion derived for the universal rivet and presented in Fig. 3.60a. A nonlinear power-law plasticity model with isotropic hardening was used and a coefficient of friction of 0.18 was assigned to each of the contact surfaces. In agreement with Fig. 3.56a, maximum hole expansion occurs at the driven head. The same trend is shown for the reduced-depth countersunk rivet (NAS1097AD4-4), Fig. 3.60b, again in accordance with Müller's data from Fig. 3.56b. In Fig. 3.60b, two dash-dot-dotted lines illustrate the divergence in the level of hole expansion at high squeeze forces between countersunk rivets that are initially perfectly flush and rivets that protrude 0.07 mm above the outer skin. It is well known from riveting practice that a small protrusion of the rivet head above the sheet surface (0.01–0.2 mm) prior to the rivet's installation has a beneficial effect on the fatigue performance of riveted joints due to the improved hole filling of the countersunk recess in the countersunk sheet. It should be noted that the hole expansion distribution over the sheet thickness can be influenced by the riveting process. For example, the difference in the fatigue behaviour between specimens made using a flat and concave die (see Fig. 3.16) is explained by the different distributions of expansion caused by the different riveting processes.

Also Li and Shi (2004a) obtained a non-uniform through-the-thickness distribution of hole expansion for the countersunk type rivet MS20426AD8-9 ($D_o = 6.35$ mm) joining two 2.03 mm thick, bare 2024T-3 Al alloy sheets. In their

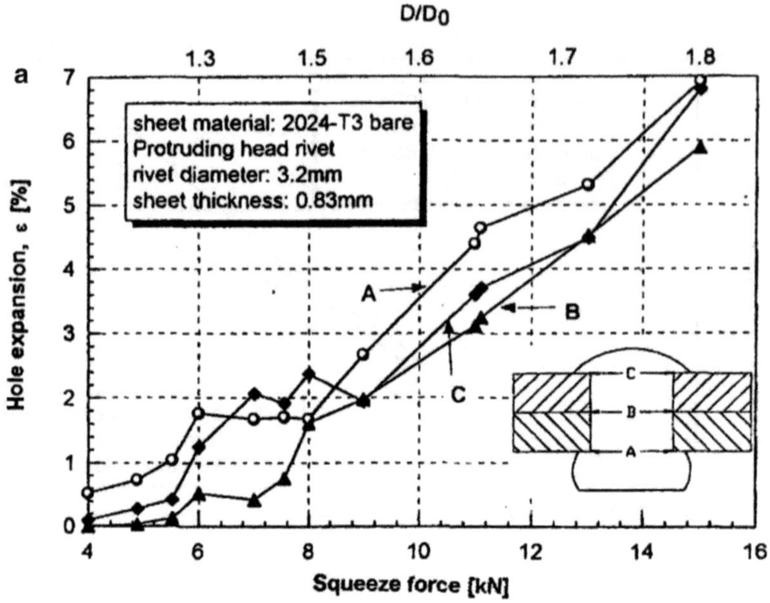

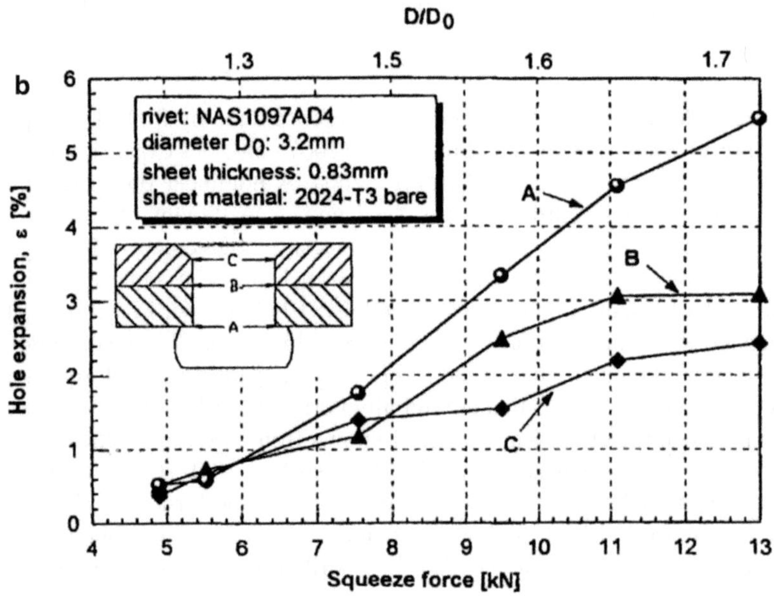

Fig. 3.56 Hole expansion measured at different locations as a function of the squeeze force for: (**a**) protruding head rivets; (**b**) countersunk rivets (Müller 1995)

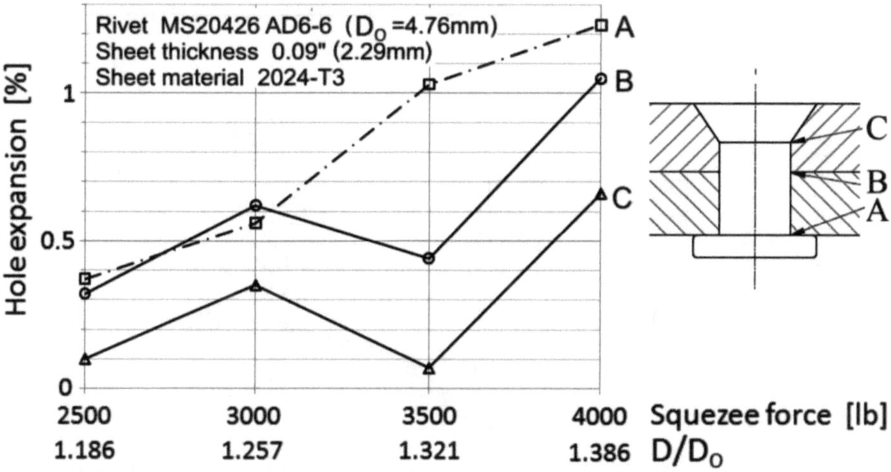

Fig. 3.57 Hole expansion measured at different locations as a function of D/D_0 ratio for the countersunk rivet (Wang 1998)

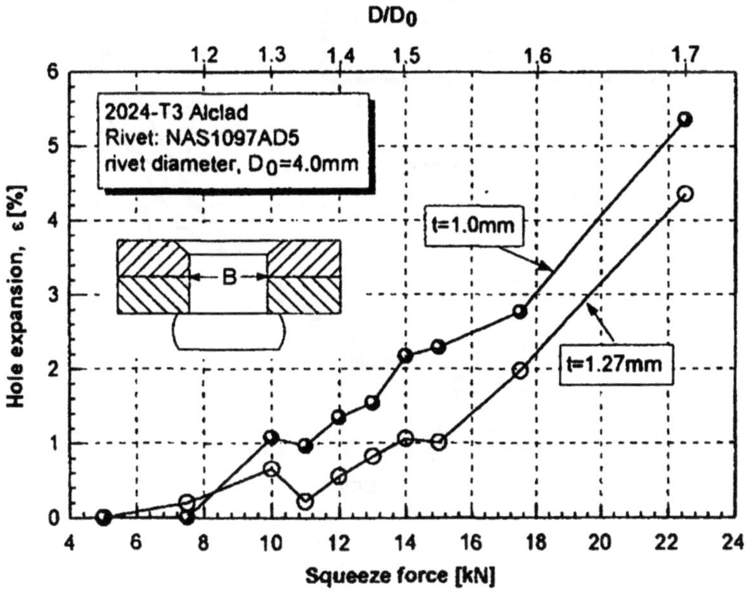

Fig. 3.58 Hole expansion as a function of the squeeze force for different sheet thicknesses, $t/D_0 = 0.25$ and 0.32 (Müller 1995)

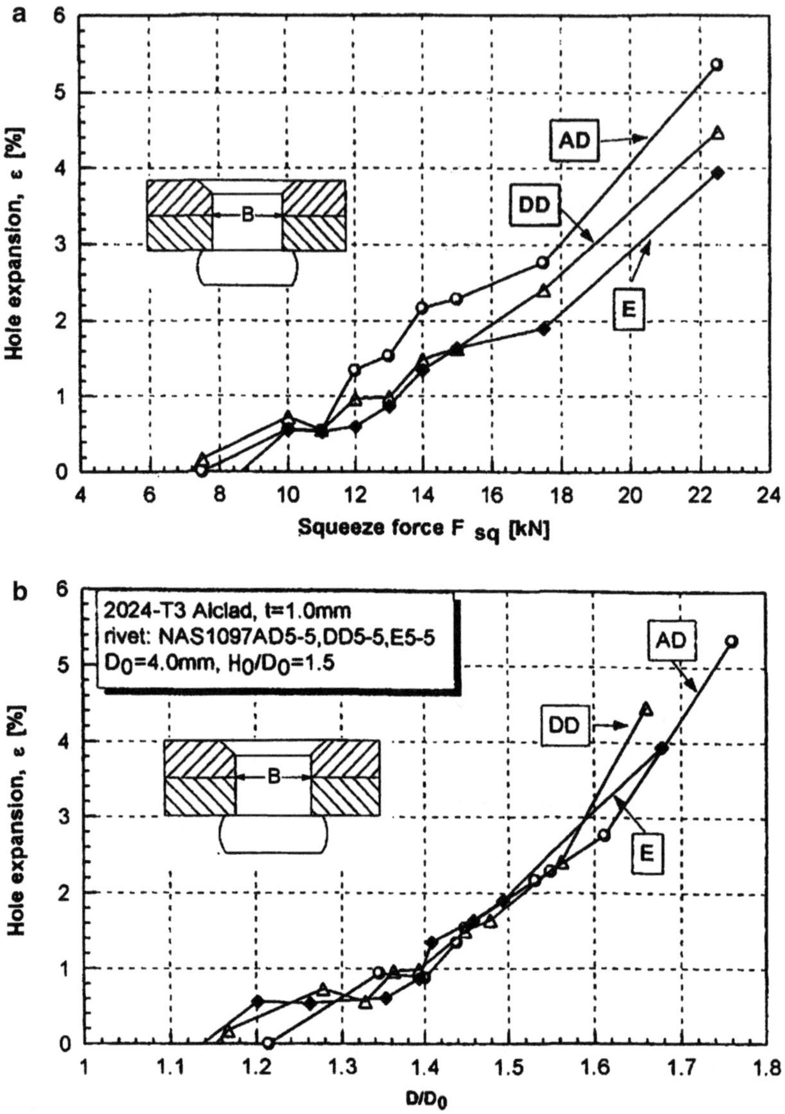

Fig. 3.59 Hole expansion at the mating surface (location B) for countersunk *AD*, *DD*, *E* rivets with $t/D_0 = 0.25$ as a function of: (**a**) squeeze force; (**b**) driven head diameter (Müller 1995)

2-D axisymmetric FE model isotropic hardening behaviour was assumed for both the rivet and sheet material, and the friction coefficient of 0.2 was chosen for all contact surfaces. The initial clearance between the rivet shank and the hole equalled 0.1 mm. A large expansion level was achieved only in the inner sheet. Outer sheet hole expansion was not sensitive to the applied squeeze force and was still very

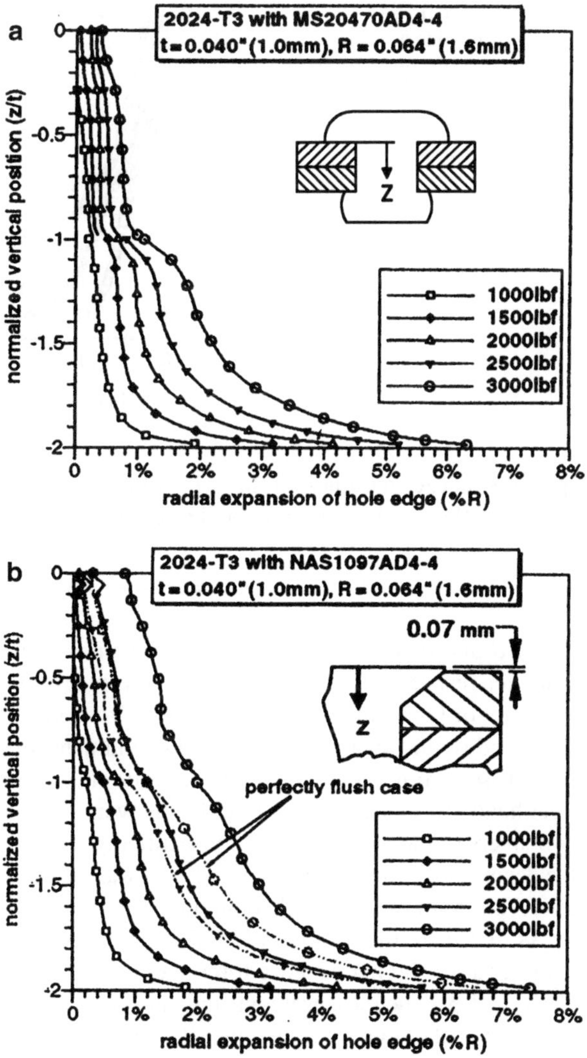

Fig. 3.60 Variations in rivet hole expansion with the squeeze force for: (**a**) universal rivet; (**b**) countersunk rivet. FE results (Rans et al. 2007)

small for the highest squeeze force (F) of 53.38 kN ($D_0/D = 1.71$), Fig. 3.61. The same conclusions can be drawn from Wang's (1998) results, but the D_0/D ratios considered in his FE analyses were relatively small (1.25–1.4).

Hole expansion measurements were also performed by Skorupa et al. (2009, 2010b). The PA24 Al alloy round head rivets (notation PA24r) and protruding head rivets with the compensator (notation PA24c) were installed in two 1.9 mm thick D16 sheets in the Alclad condition (cf. Fig. 3.42a). The rivet holes were drilled according to the process specification by the Polish aircraft manufacturer PZL Mielec, which implies that the hole diameter was nominally 0.05 mm larger than the rivet shank diameter. The rivets were installed under force control using a

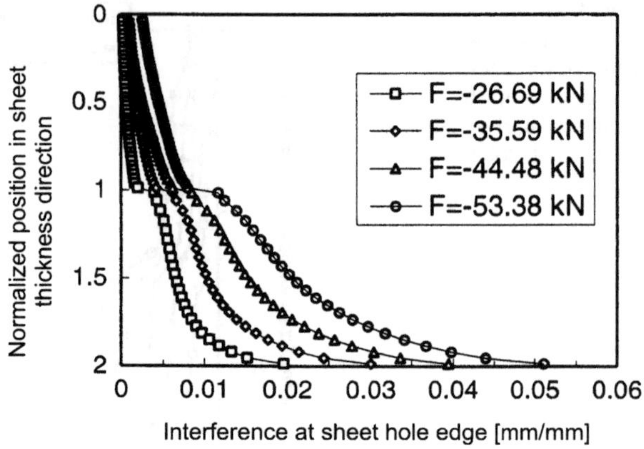

Fig. 3.61 Variations of interference along a sheet hole edge. FE results (Li and Shi 2003)

special squeezer mounted in the grips of a fatigue machine. The device effectively prevented the overlap curvature and rivet tilting. Hole expansion (*he*) was defined as

$$he = \frac{d_e - d}{d} \cdot 100\%, \tag{3.6}$$

where d and d_e is the rivet hole diameter prior to riveting and the expanded rivet diameter respectively. Both d and d_e were measured in several directions with an accuracy of 0.01 mm using an optical microscope. The results shown in Fig. 3.62 are based on the average values. The d_e diameter measurements were taken in the rivet cross sections at two locations for each sheet, namely 0.2 and 1 mm below each rivet head. To this end, both rivet heads were first cut away. Then for either sheet two layers of the material were consecutively removed by milling in order to obtain the desired locations. The machined area was subsequently polished to facilitate the observations of the rivet/hole boundary.

As expected, the results in Fig. 3.62 demonstrate that for either type of rivet *he* is larger at the higher squeeze force, i.e. for $D/D_o = 1.5$, than at the lower F_{sq} level ($D/D_o = 1.3$). The measurements indicate considerable differences between the hole expansion behaviour for both types of rivet. For rivet "c", *he* below the manufactured head is much larger than below the driven head. The reverse is true for rivet "r", but the differences between *he* in either location are much less significant. Generally, due to the presence of the compensator, *he* for rivet "c" is larger than for rivet "r", the advantage of the rivet with the compensator being particularly pronounced within the sheet adjacent to the manufactured head for $D/D_o = 1.5$. The above observation is confirmed by FE simulations of the riveting process for a rivet with and without the compensator (Szymczyk et al. 2009).

It is seen in Fig. 3.62 that for both D/D_o-values the average results for rivet "r" fall between measurement data reported by Müller (1995) for the countersunk DD rivet of $D_o = 4.8$ mm joining 1.61 and 2.19 mm thick 2024-T3 sheets. It is further seen that a reasonable agreement is revealed between the measurements for rivet "r" and FE results according to Rans (2007) derived for the universal $D_o = 3.2$ mm AD rivet installed in two 1 mm thick 2024-T3 sheets.

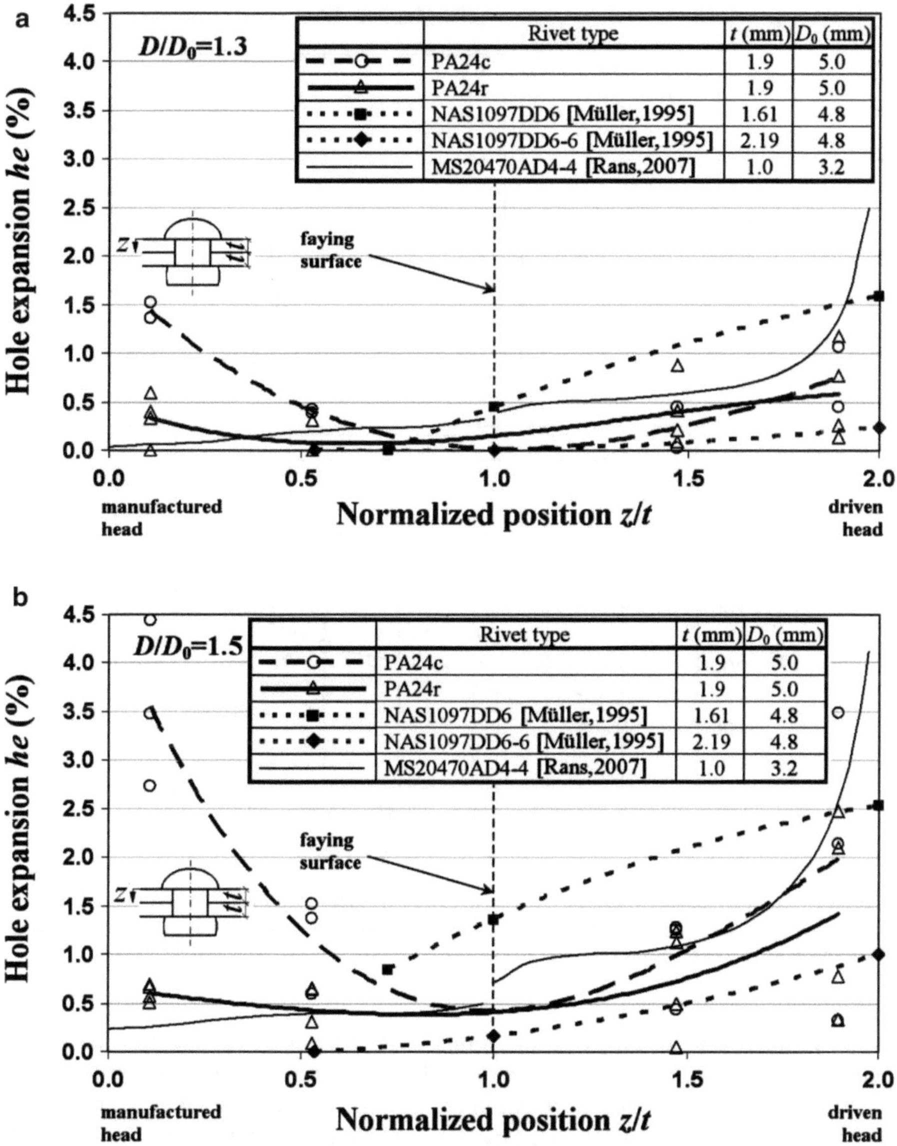

Fig. 3.62 Comparison between hole expansion for round head rivets (PA24r) and rivets with the compensator (PA24c) in D16 Al alloy sheets: (**a**) $D/D_o = 1.3$; (**b**) $D/D_o = 1.5$ (Skorupa et al. 2010b)

A beneficial effect of the compensator on the fatigue behaviour of lap joint specimens shown in Fig. 3.63a, especially for higher D/D_o values, is revealed by fatigue life values plotted in Fig. 3.63b. As indicated in this figure by the data-point denoted as PA24c-rem, removing the compensator before riveting by machining caused the fatigue life drop to the level observed for the PA24r rivet.

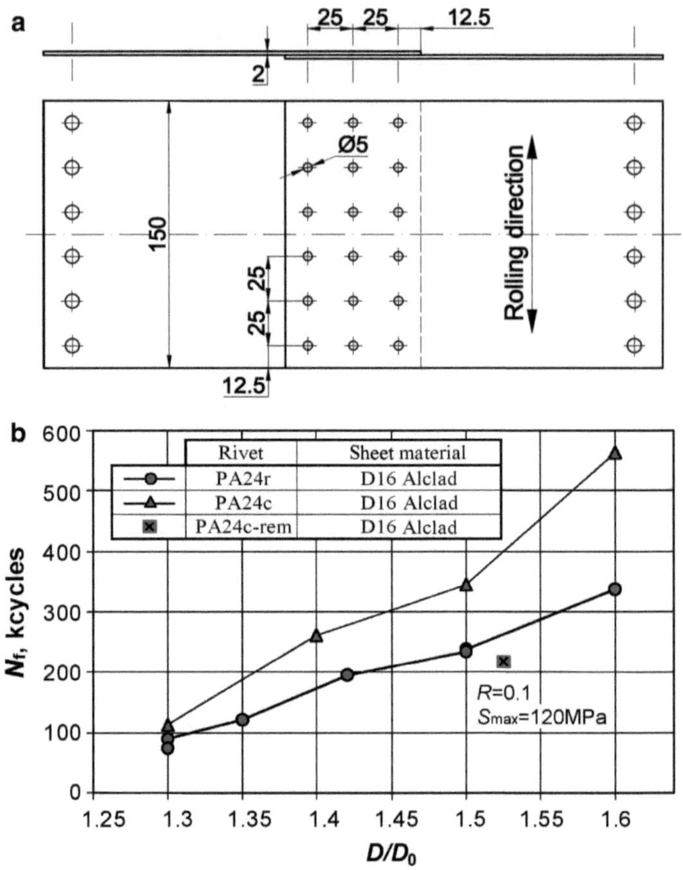

Fig. 3.63 Comparison between the fatigue lives of lap joint specimens with round-head rivets (*PA24r*) and protruding head rivets with the compensator (*PA24c*): (**a**) test sample; (**b**) fatigue lives (Skorupa et al. 2010b)

As pointed out in Chap. 1, for joints with eccentricities like the lap joints, bending of the sheets, referred to as secondary bending, occurs. In order to reduce the resulting bending stresses, local decreasing of the sheet thickness around the outer rivet rows has been proposed by Schijve (2006), for details see Sect. 6.2.2.2. Hole expansion measurements for the staggered thickness configuration of a lap joint specimen were carried out by Skorupa et al. (2009). Thickness reduction of the D16 sheets from 1.9 to 0.95 mm was made by face milling. In Figs. 3.64 and 3.65 measurement results on *he* (see Eq. 3.6) for two possible sheet configurations relative to the rivet are compared with the data corresponding to equal thicknesses (1.9 mm) of both sheets shown previously in Fig. 3.62.

When joining the 0.95 mm thick and the 1.9 mm thick sheets using the rivet 5 mm in diameter, the same driven rivet head dimensions were obtained as in

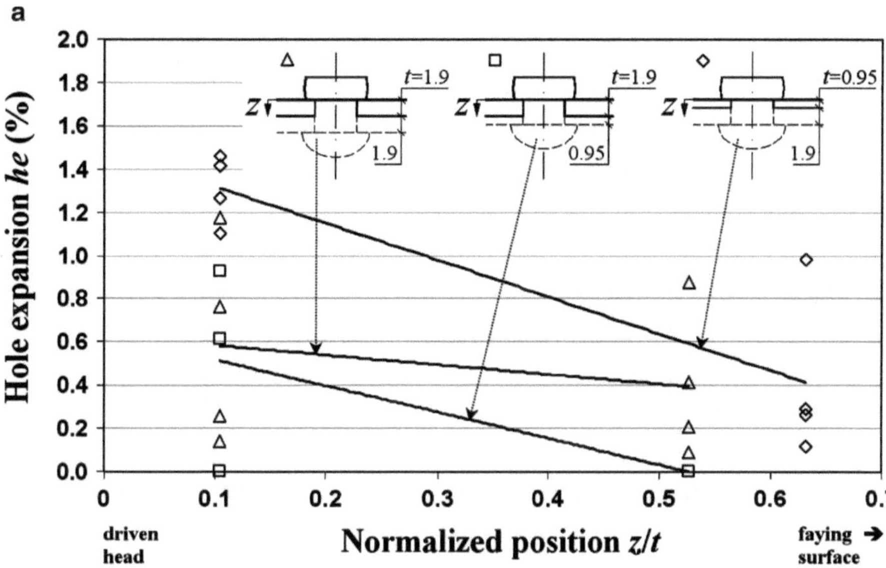

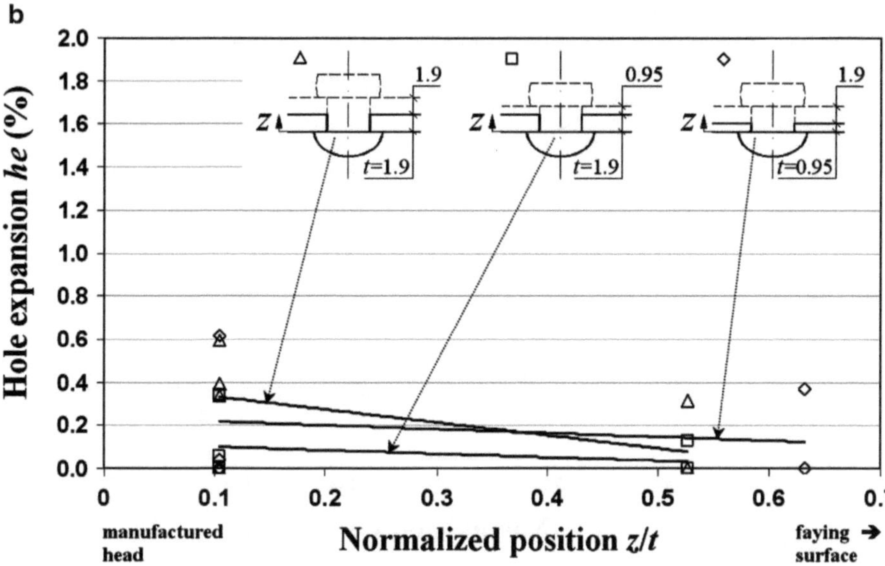

Fig. 3.64 Comparisons between hole expansion for three-sheet thickness configurations. Sheet adjacent to: (**a**) rivet driven head; (**b**) rivet manufactured head. D16 Al alloy, PA24 round head rivet, $D/D_o = 1.3$ (Skorupa et al. 2009)

the case of joining two 1.9 mm thick sheets with a rivet of the same diameter if the squeeze force and protruding rivet length ($H_o = 6.2$ mm) were the same for all configurations. The d_e measurements in the 0.95 mm thick sheet were taken

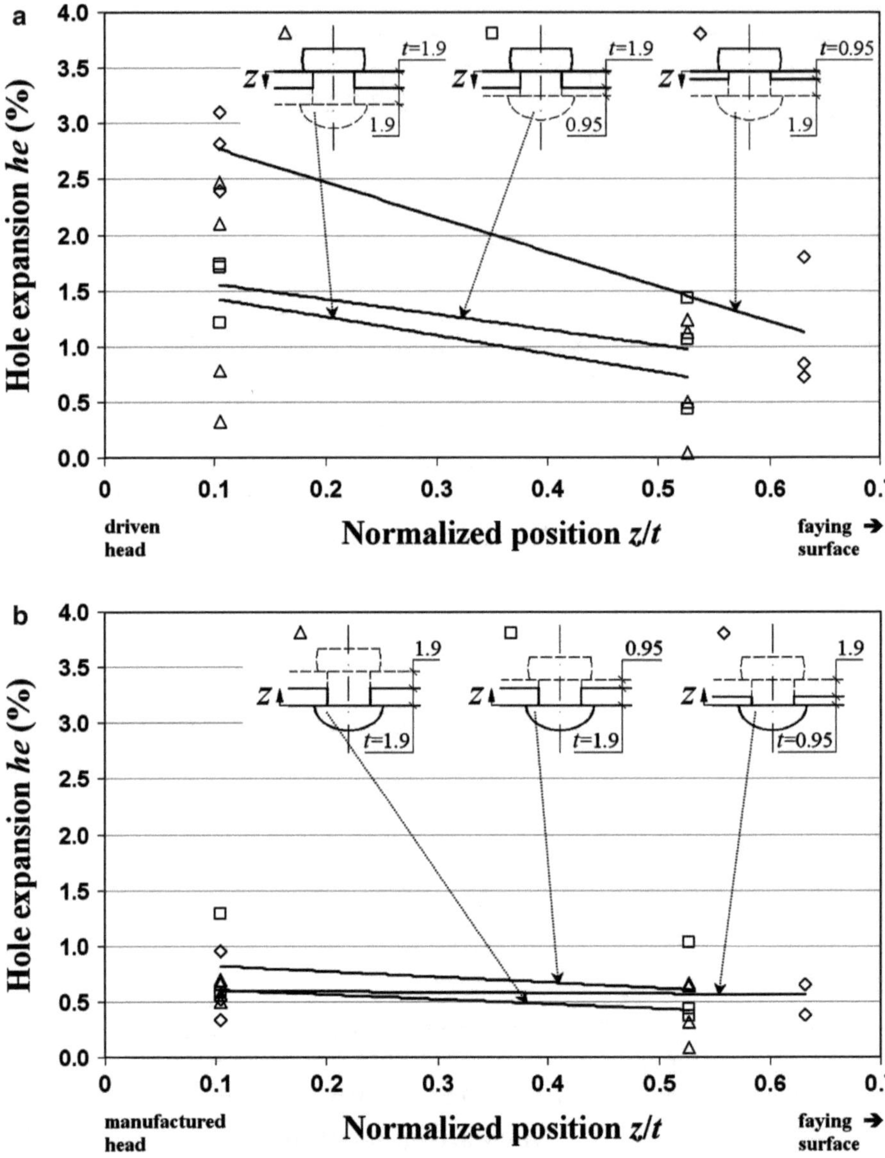

Fig. 3.65 Comparison between hole expansion for three sheet thickness configurations. Sheet adjacent to: (**a**) rivet driven head; (**b**) rivet manufactured head. D16 Al alloy, PA24 round head rivet, $D/D_o = 1.5$ (Skorupa et al. 2009)

for $z = 0.1$ and 0.6 mm. In the 1.9 mm thick sheet, d_e was measured at the same locations, as shown in Fig. 3.62. Again, hole expansion for the lower D/D_o ratio of 1.3, Fig. 3.64, is smaller than for the larger driven head diameter ($D/D_o = 1.5$),

Fig. 3.65. Other trends revealed in Figs. 3.64 and 3.65 are common for both D/D_o ratios. For the sheet adjacent to the rivet driven head, *he* in the thinner sheet is always larger than in the thicker sheet. Within the sheet adjacent to the manufactured head, hole expansion remains not affected by the sheet thickness configuration.

The scatter of the results revealed in Figs. 3.62, 3.64 and 3.65 is due to the scatter in the d-values rather than in the d_e-values. It was observed that for a given squeeze force level and type of rivet the expanded rivet diameters showed only minute differences. At the same time, due to the relatively large tolerance of the rivet hole diameter the actual diameters could differ by even 0.1 mm.

3.3.5.4 Residual Stresses Due to the Riveting Process

The nature of the residual stress field created due to the riveting and its impact on the subsequent loading of the joint plays an important role in the nucleation and growth of fatigue cracks in the vicinity of the rivet hole.

As said earlier, squeezing of the rivet leads to plastic deformation of the sheet material around the rivet. This process is to some extent similar to the earlier considered cold working process (see Sect. 3.3.3) when the fastener hole is expanded to a desired value before inserting a mechanical fastener. The main differences between cold working and riveting are the following:

- after cold working the hole is open. Then the radial residual stress on the hole must be zero. After riveting a rivet shank remains in the hole resulting in a radial pressure,
- the rivet heads exert a clamping pressure through the thickness of the sheets, which results in a 3-dimensional residual stress state.

Because no analytical models exist for predicting riveting induced residual stresses, only numerical calculations can provide realistic results. A FE analysis of residual stresses in the riveted lap joint (AD rivet) has been performed by Müller (1995), his symmetric model being shown in Fig. 3.40. Müller's results on the residual tangential (σ_θ) and radial (σ_r) stress distribution in the radial direction along the mating surface are presented in Fig. 3.66a, b. It can be seen in Fig. 3.66a that for small driven heads ($D/D_o = 1.2$ and 1.3) the tensile tangential stresses are present close to the hole edge, which implies that the hole expansion is not sufficient. Compressive tangential stresses occur for larger driven heads ($D/D_o = 1.5$ and 1.7). The area with compressive σ_θ stresses becomes larger with increasing squeeze force level. Under cyclic loading conditions, compressive tangential stresses reduce the local mean stress at the hole boundary. The distance from the hole at which the tensile σ_θ stress occurs increases with the squeeze force value. This explains the observations on the location of the fatigue crack nuclei. For a low squeeze force the cracks initiate at the edge of the hole, whereas for a higher squeeze force the location of the crack nuclei shifts to a small distance away from the hole.

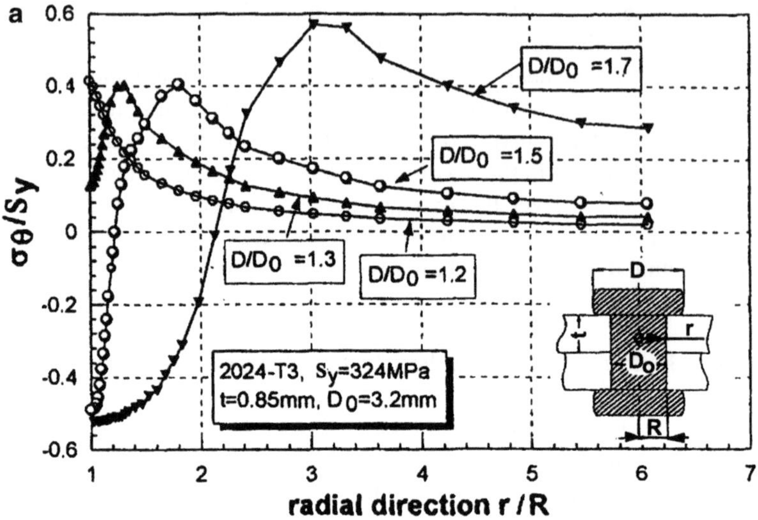

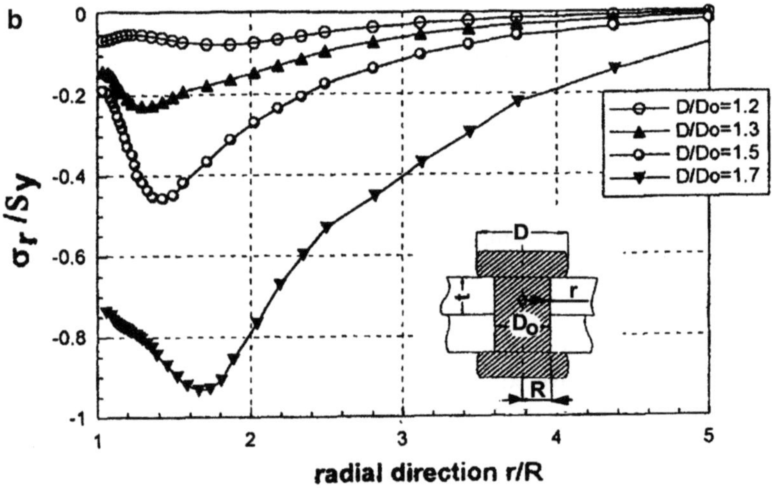

Fig. 3.66 Residual stress distribution at the mating surface according to FE calculations: (**a**) tangential stresses; (**b**) radial stresses (Müller 1995)

The residual radial stresses are highly dependent on the squeeze force, Fig. 3.66b. For larger squeeze forces significant compressive σ_r stresses at the hole edge are induced. Like the hole expansion (cf. Fig. 3.56), the radial stresses are inhomogeneous through the thickness of the sheet, as shown in Fig. 3.67.

A strong dependence of the residual stress distribution on the squeeze force was observed by Li and Shi (2004a) for the countersunk rivet MS20426AD8-9 ($D_o = 6.35$ mm, rivet hole diameter 6.45 mm) joining two 2024T-3 bare alloy, 2.03 mm thick sheets ($S_y = 310$ MPa). A mean clearance between the hole and

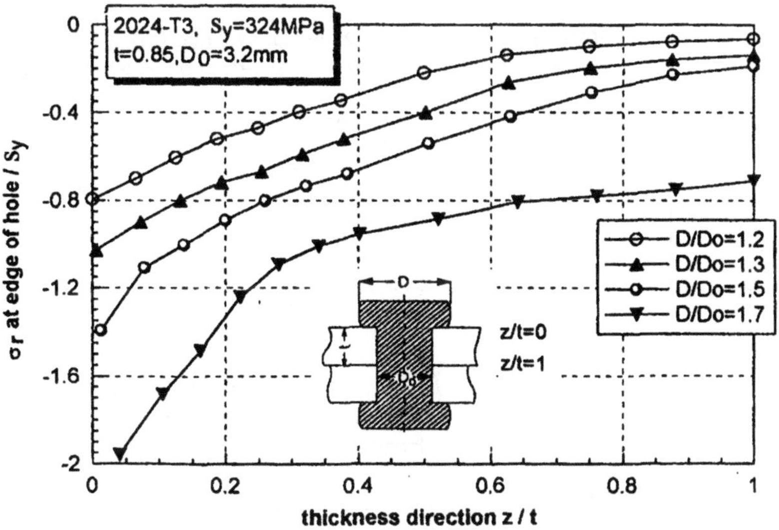

Fig. 3.67 Through-the-thickness variations in residual radial stresses on the edge of the hole for different driven head sizes according to FE calculations (Müller 1995)

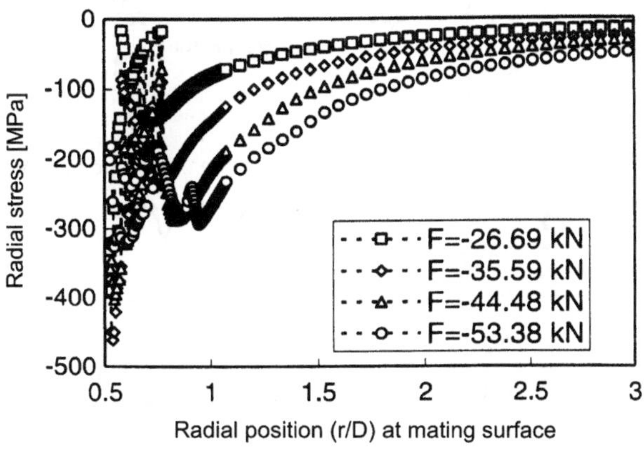

Fig. 3.68 Residual stress distribution along the mating surface according to FE calculations for a countersunk rivet (Li and Shi 2004a)

the rivet shank was 0.08 mm. The radial stress distribution at the mating surface along the radial direction obtained from the FE analysis for the squeeze force (*F*) varying from 26.69 to 53.38 kN ($D/D_o = 1.35$ and 1.71 respectively) is presented in Fig. 3.68. The results in Fig. 3.68 are in a close correspondence with the hole expansion distributions shown in Fig. 3.61. It can be deduced that a tight connection between the inner sheet and the rivet can be achieved when a large squeeze force is applied.

Müller's (1995) FE simulation of the countersunk rivet installation indicates that residual stress distributions at the hole edge shown in Fig. 3.69a and b are more favourable for the non-countersunk sheet compared to the countersunk sheet due to more compressive tangential stresses in the former sheet.

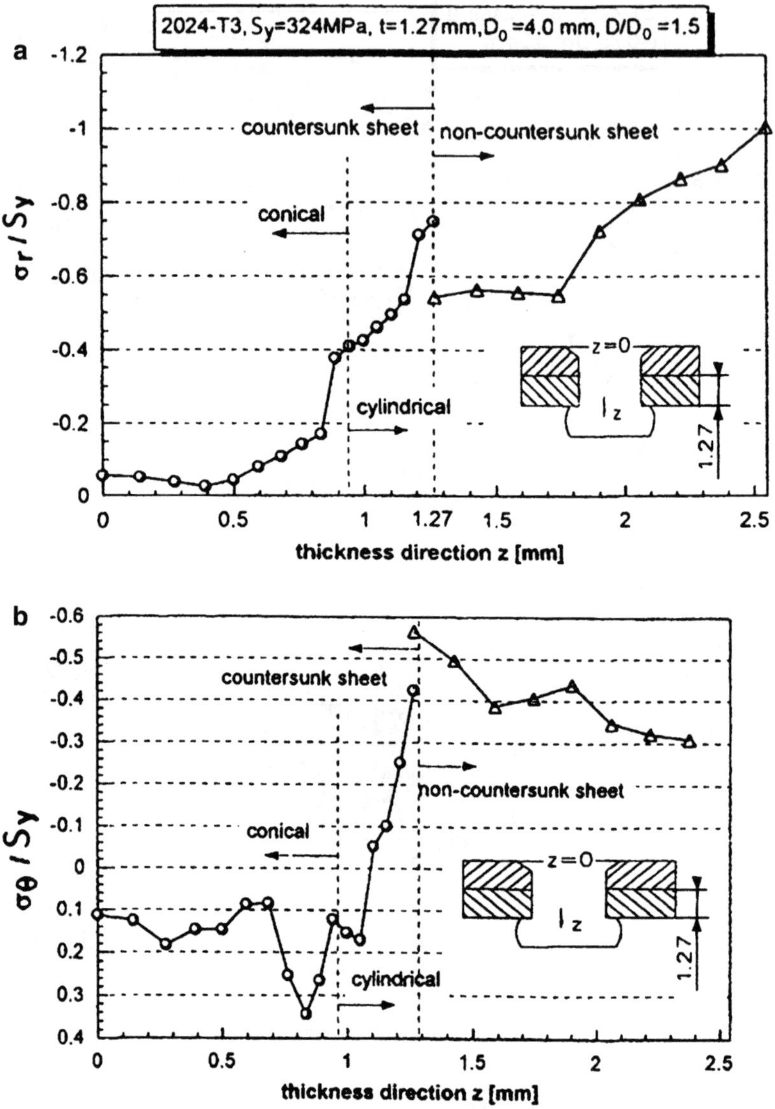

Fig. 3.69 Through-the-thickness variations in residual stresses on the edge of the hole for the countersunk rivet according to FE calculations: (**a**) radial stresses; (**b**) tangential stresses (Müller 1995)

Residual stresses around the protruding head and countersunk rivet were also modelled by Rans et al. (2007), the information on the joint being given earlier in this section in the context of Fig. 3.60. The residual tangential stress distributions along the radial direction on the faying surface of the outer and inner sheet for

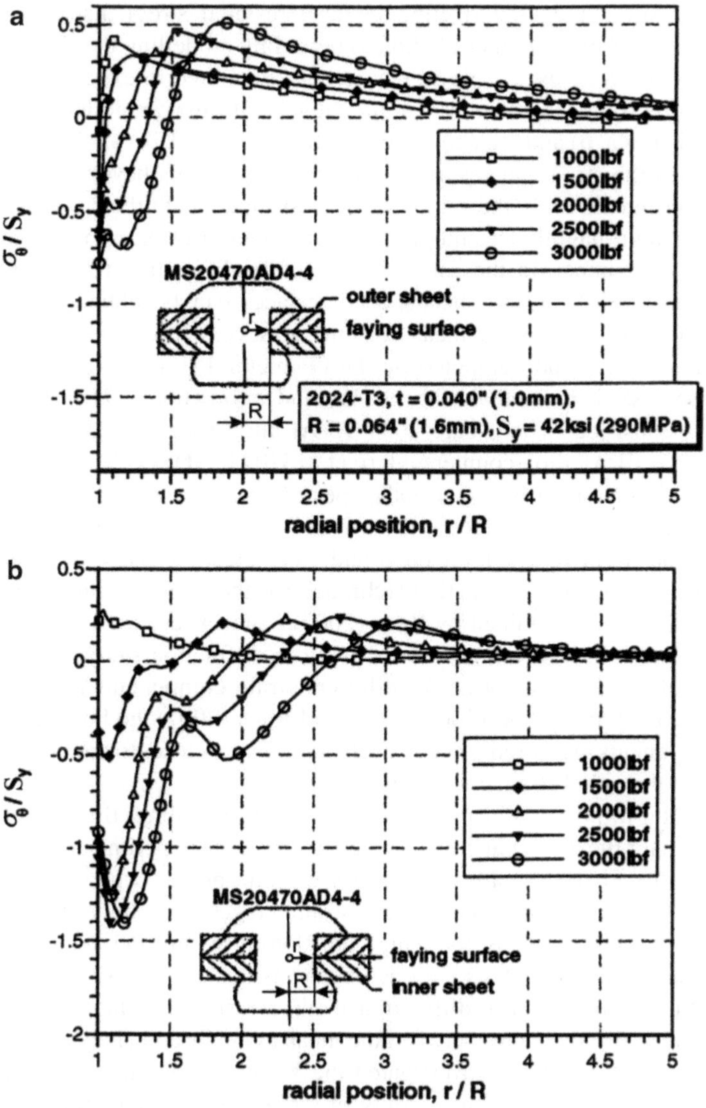

Fig. 3.70 Influence of the squeeze force on faying surface residual tangential stresses for the universal rivet: (**a**) outer sheet; (**b**) inner sheet (Rans et al. 2007)

the universal rivet are presented in Fig. 3.70a, b, as an example. In agreement with Müller's (1995) results referred to above (compare Fig. 3.66a), Rans et al. (2007) observe the region of the compressive σ_θ stress to increase when the squeeze force becomes higher. Again in accordance with Müller's results from Fig. 3.69b, Fig. 3.70 implies that the tangential stresses are more compressive on the faying surface of the inner sheet (under the driven head).

The FE solution on the riveting induced residual stress distribution can be critically affected by a number of choices regarding modelling the rivet installation and also by the code used. Several researchers attempted to experimentally verify the FE results by means of micro-strain gauges with the gauge length of 0.8 mm (Li and Shi 2004a, b) and the neutron diffraction method (Li et al. 2004, 2006; Rans 2007). When the strain gauges are employed, only strains beyond the rivet heads can be determined, while the region of interest underneath the rivet heads, where stress and strain magnitudes and gradients are much greater, remains inaccessible for the measurements. In the case of the neutron diffraction technique, the data close to the rivet/sheet and sheet/sheet interfaces can be obtained. Figure 3.71 shows comparisons between measurements at several positions by micro-strain gauges (Fig. 3.71a) and computed by the FE method strain variations during the riveting process in the radial (Fig. 3.71b) and hoop (Fig. 3.71c) direction. The specimens consisted of two bare 2024-T3 Al alloy sheets, 2.03 mm thick each, and the 2117-T4 Al alloy countersunk rivet MS20426AD8-9 ($D_o = 6.35$ mm). The overall good agreement between both types of data is revealed, somewhat larger discrepancies occurring for the radial strains. Exemplary comparisons between the residual hoop strain values in the outer and inner sheet measured by Li et al. 2004, 2006) using the neutron diffraction technique for the joint considered in Fig. 3.71 and the distributions obtained by Rans (2007) from FE calculations are presented in Fig. 3.72. The gauge volume for the neutron diffraction study was 1 mm^3, but the shape of the volume varied for different strain components and locations. In the studies by both Li et al. (2004, 2006) and Rans (2007) the hoop and clamping strain components measured by neutron diffraction are closer to the FE computation results than the radial strain component.

In addition to possible inadequacies of the FE modelling, there can be several other causes for the discrepancies observed between the experimental and computed data. First, the FE results are resolved for discrete points, whereas the measured results are averaged over a specific area (strain gauges) or volume (neutron diffraction method). Differences in the gauge positioning within the neutron diffraction method and the FE model, as well as variations in the grain size and orientation which influence the neutron diffraction method results can be additional factors contributing to discrepancies between the measurements and analyses.

Gadalinska et al. (2009) investigated the X-ray diffractometry applicability to measure residual stresses in the vicinity of a rivet joining aluminium alloy sheets. This method enables measurements only on the outer surfaces of a joint outside the rivet heads. A special surface preparation is required which includes removing the clad or anodized layer and subsequent polishing in a few steps. The measured stress values were in a satisfactory agreement with the FE computation results.

The effect of the riveting process induced residual stresses on the stress response of a lap joint to the applied tensile loading was investigated numerically using the 3-D FE modelling and experimentally (micro-strain gauge measurements) by Li and Shi (2004b) and Li et al. (2006) for a single row configuration and by Li et al. (2007) for a three-row configuration shown in Fig. 3.73a. The joint consisted of 1.6 mm thick Al 2024-T3 bare sheets connected using Al 2117-T4, countersunk

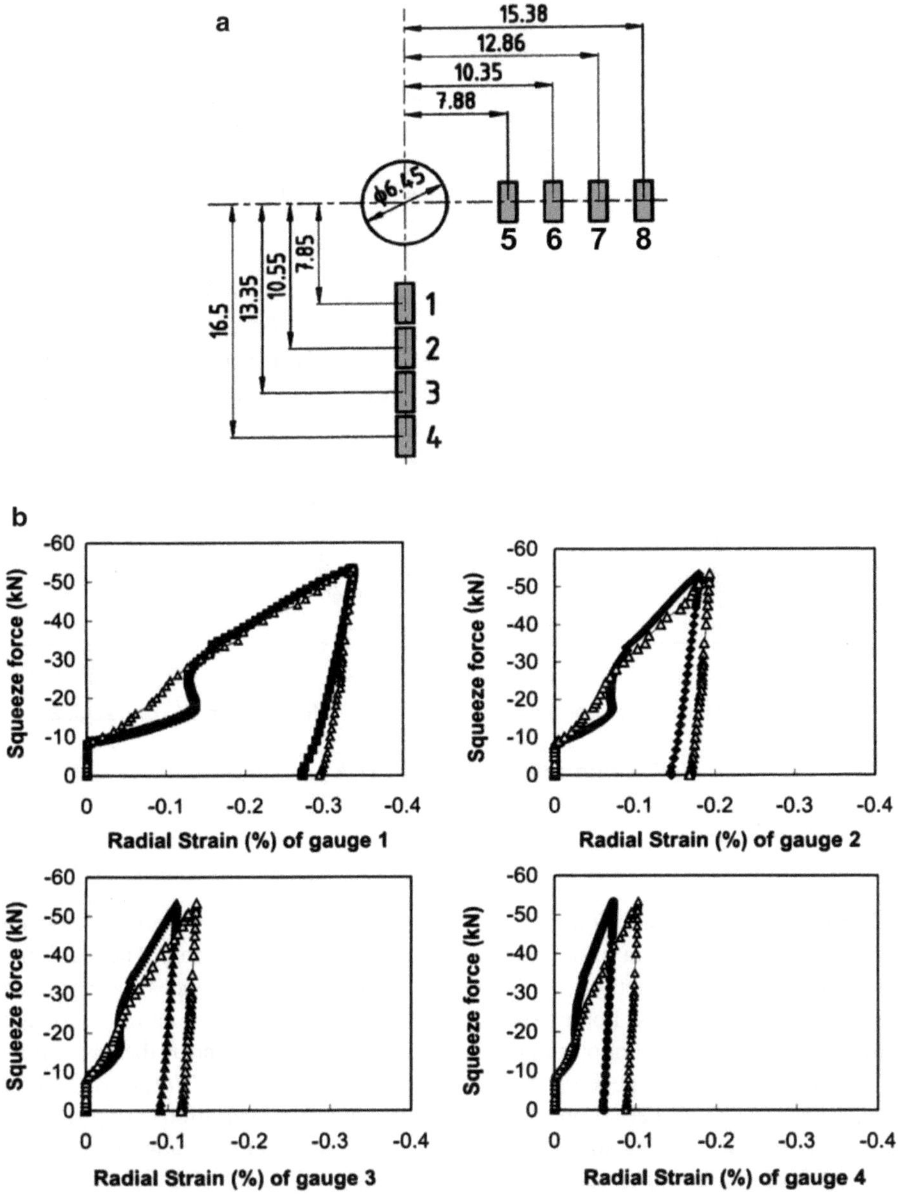

Fig. 3.71 Comparison of strain variations during the whole riveting process under a 53.38 kN squeeze force obtained from strain gauge measurements and FE predictions: (**a**) locations of micro-strain gauges on the joint inner sheet surface; (**b**) radial strains; (**c**) hoop strains (Li and Shi 2004a)

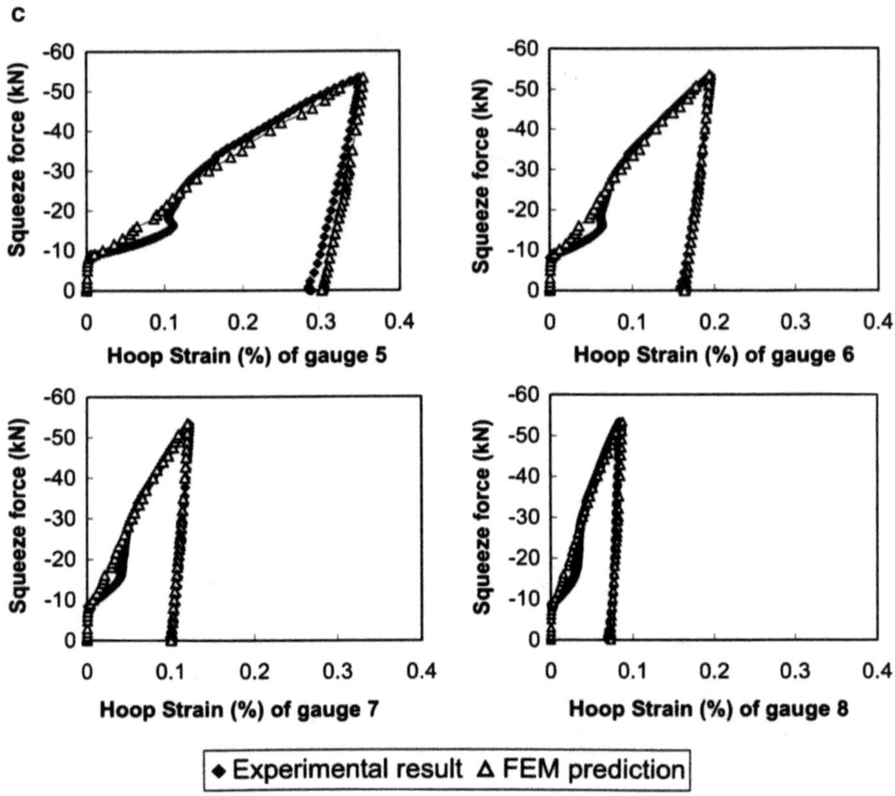

Fig. 3.71 (continued)

MS20426AD5-6 ($D_0 = 3.97$ mm) rivets. The rivet hole mean diameter was 4.09 mm which implied a 0.06 mm initial clearance between the rivet and hole. The Coulomb friction model with a friction coefficient of 0.2 for all contact surfaces was assumed in the FE computations. Force controlled riveting was simulated numerically and the resulting residual hoop (tangential) stress distributions near the rivet hole corresponding to squeeze force values of 10, 14 and 18 kN ($D/D_0 = 1.32$, 1.52 and 1.66 respectively) are plotted in Fig. 3.73b. The stress variations are presented for the outer sheet faying surface along the transverse direction. In agreement with results from other works reviewed earlier in this section, the residual stresses in the hole vicinity become more compressive when the squeeze force is increased. Compared to the squeeze force of 10 kN, the squeeze forces of 14 and 18 kN yield the hoop stress at the hole edge lower by 16% and 49% respectively. Figure 3.73c gives the hoop stress distributions under an applied stress of 98.6 MPa. The local response to the applied cyclic stress ranging between 0 and 98.6 MPa, derived from Fig. 3.73b, c for two extreme squeeze forces, is presented in Fig. 3.74 in terms of the cyclic mean stress and the stress range. It is seen that within the compressive

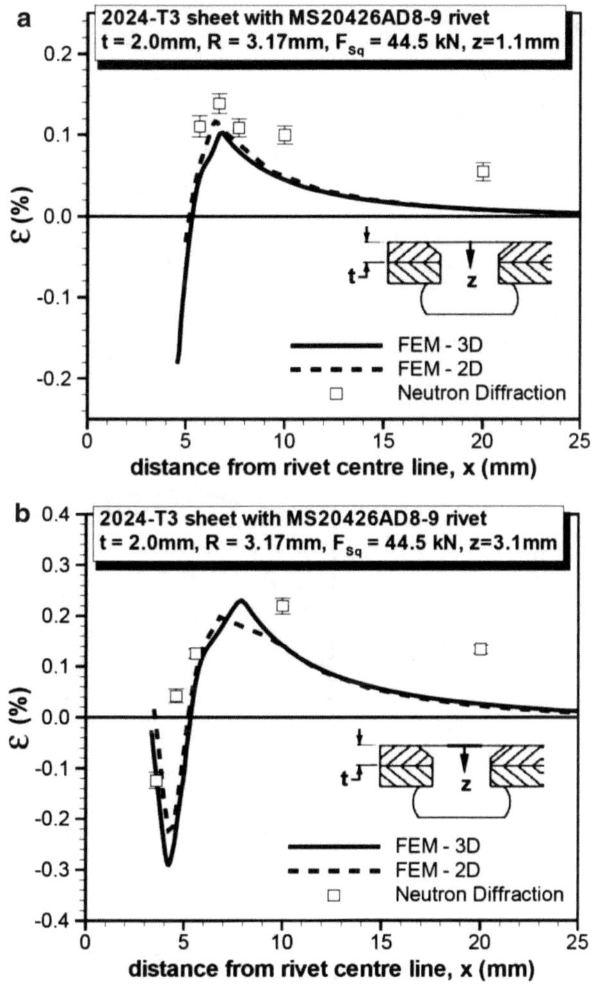

Fig. 3.72 Comparison between FE computations and neutron diffraction measurement on residual tangential strains: (**a**) outer sheet; (**b**) inner sheet (Rans 2007)

residual stress region the higher squeeze force causes a more negative local cyclic mean stress and a slightly lower local stress amplitude than observed for the lower squeeze force. Note also in Figs. 3.73c and 3.74 that for $F_{sq} = 18$ kN the region adjacent to the hole edge experiences a fully compressive cyclic loading, whereas for $F_{sq} = 10$ kN the maximum cyclic stress at the hole edge is positive.

The above results suggest that lowering the local cyclic mean stress coupled with a slight reduction of the local stress amplitude can be among factors contributing to improved fatigue properties of a lap joint riveted with a higher squeeze force. At the same time, Figs. 3.73c and 3.74 reveal that the larger squeeze force moves the high-stressed region away from the hole vicinity. This behaviour can lead to a shift of the crack initiation location away from the rivet hole observed at higher squeeze forces, as detailed in Chap. 7.

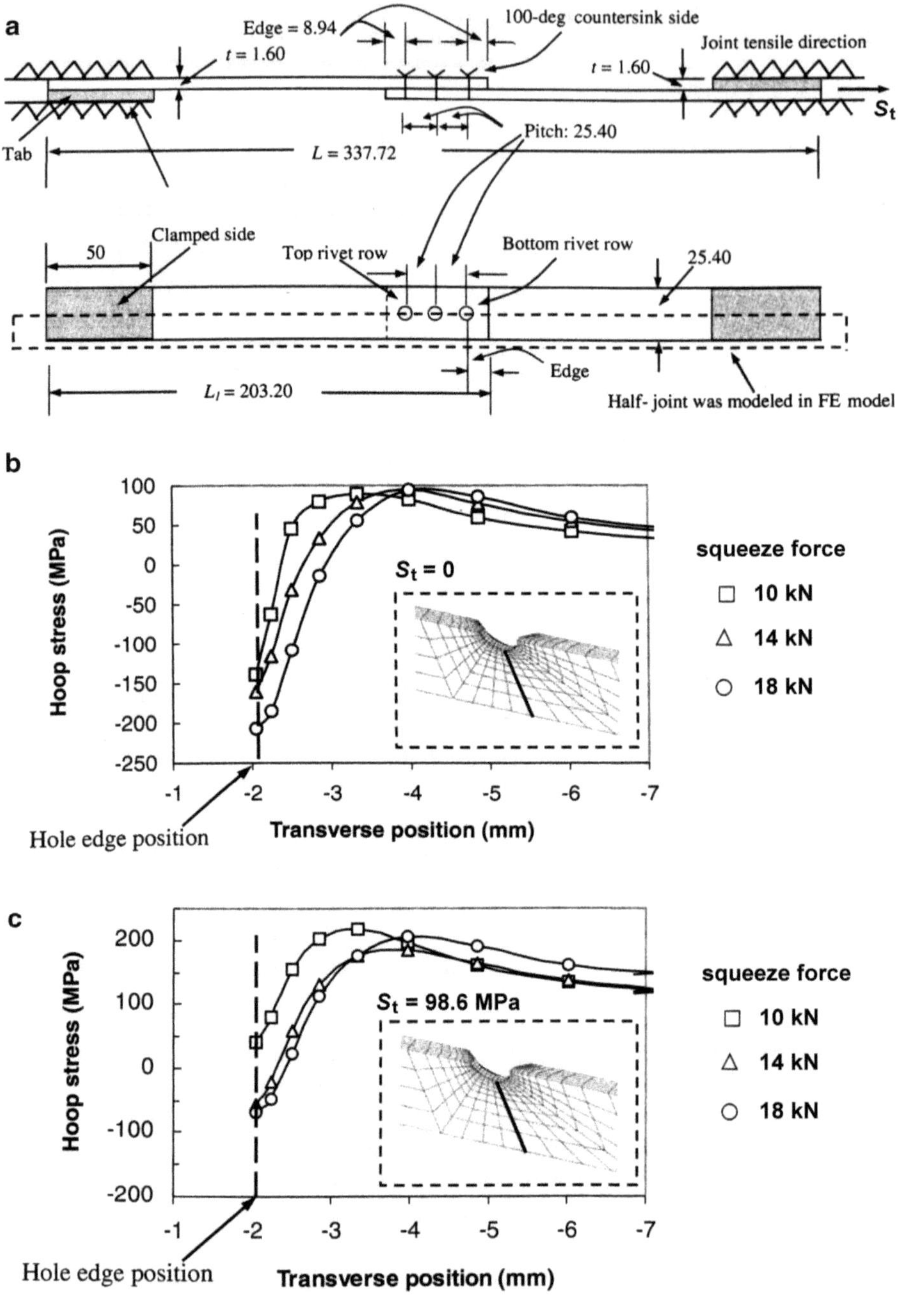

Fig. 3.73 FE results on hoop stress distribution under tensile loading on the lap joint for different squeeze forces: (**a**) FE model; (**b**) residual stress distributions; (**c**) stress distributions at applied stress of 98.6 MPa (Li et al. (2007))

Fig. 3.74 Cyclic local mean hoop stress and hoop stress range variations for the joint from Fig. 3.73a under applied stress ranging from 0 to 98.6 MPa (based on results from Fig. 3.73b, c)

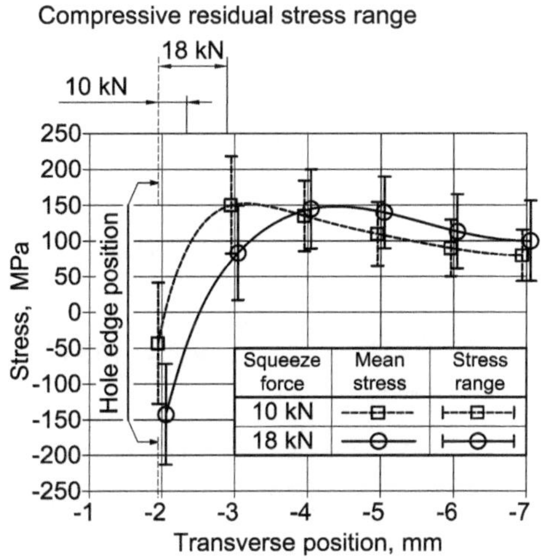

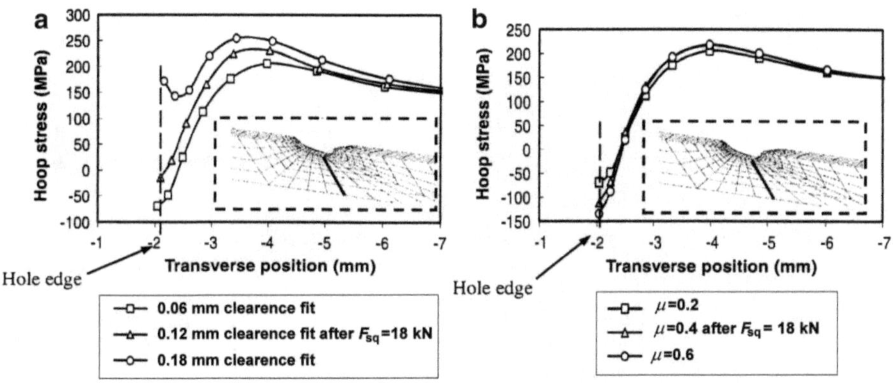

Fig. 3.75 FE results on hoop stress distribution under tensile loading on the joint from Fig. 3.73a for: (**a**) different initial clearance fits; (**b**) different friction coefficients (Li et al. 2011)

In a subsequent work Li et al. (2011) studied numerically the influence of the rivet/sheet initial clearance fit and of the friction coefficient on the stress state in a three-row lap joint shown in Fig. 3.73a. Two additional clearance values of 0.12 and 0.18 mm and two additional friction coefficients of 0.4 and 0.6 were considered. The squeeze force of 18 kN corresponding to the D/D_0 ratio of 1.66 was applied. Variations in the hoop stress on the outer faying surface along the transverse path at the top-row fastener for all three clearance values under the remote tensile stress of 98.6 MPa are presented in Fig. 3.75a. It is seen that for the clearance fit of 0.18 mm high tensile hoop stresses at the hole edge occur. As the clearance increases, the location of the maximum tensile stress moves towards the hole edge. These results suggest that fatigue strength of lap joints with a large clearance fit

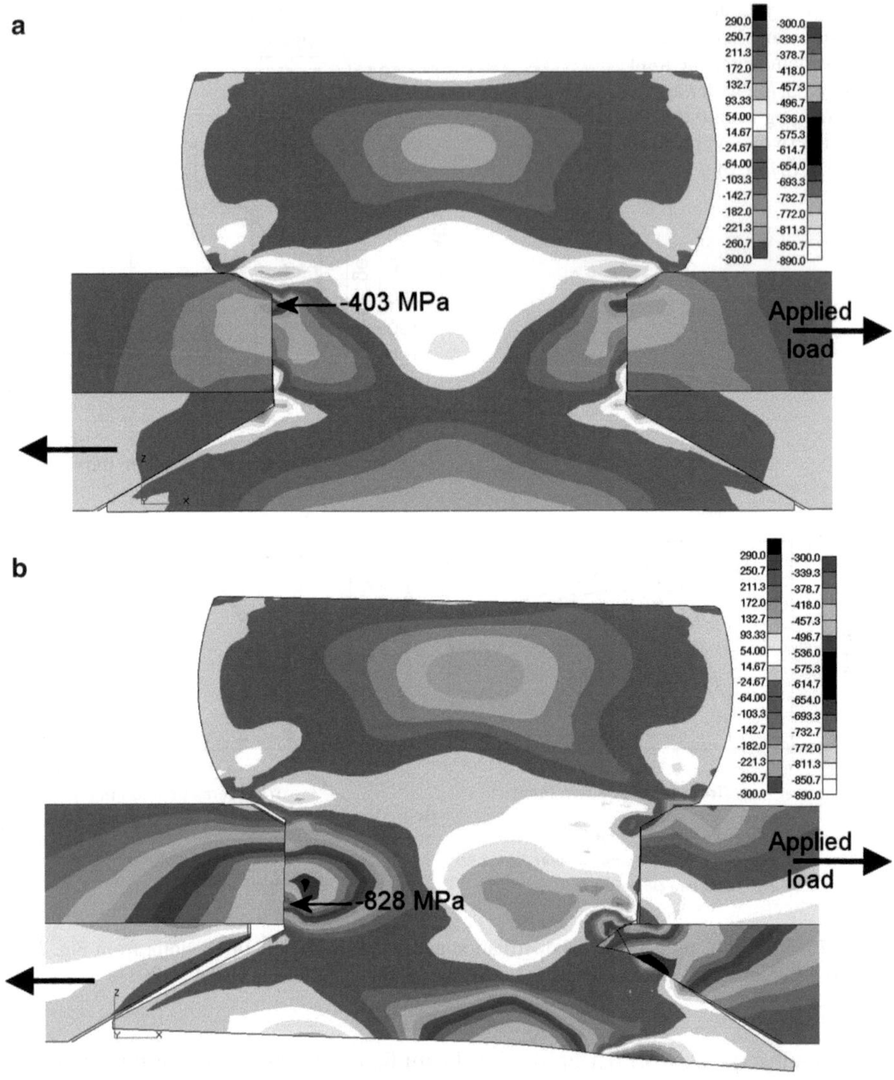

Fig. 3.76 Stresses in the loading direction and joint deformation near the rivet in the plane of symmetry for a double-row lap joint at applied stress of: (**a**) 1.8 MPa; (**b**) 61.5 MPa (Wronicz and Kaniowski 2010)

should be significantly reduced, which is consistent with experiments (Vlieger and Ottens 1998). The influence of the friction coefficient was studied by Li et al. (2011) for the clearance fit of 0.06 mm. Figure 3.75b indicates that the friction coefficient has a certain impact only on the stress at the hole edge, while the magnitude and location of the peak tensile hoop stress remain not affected.

FE analyses of Wronicz and Kaniowski (2010) aimed at the determination of the stress distribution and deformation due to the riveting process and the subsequently applied tensile load for a double-row lap joint. The joint consisted of 1.2 mm thick 2024-T3 alloy sheets and PA25 alloy countersunk rivets 3.5 mm in diameter. The rivets were squeezed to obtain the D/D_0 ratio of 1.5. The simulations have shown that at higher applied stresses (50–60 MPa) a small gap between the rivet head and countersink is formed, Fig. 3.76.

3.4 Summary of This Chapter

Two basic types of sheet material are used for the fuselage skin, namely the 2024-T3 (and its D16Cz Russian equivalent) and the 7075-T6 aluminium alloys. Fatigue properties of the 2024-T3 alloy are slightly better then those of the 7075-T6 alloy, but its static strength is lower. In contrast to 7075-T6, fatigue crack growth in 2024-T3 is relatively insensitive to environmental effects. Although static properties of both materials show a noticeable dependence on the rolling direction, the directionality effect can be neglected considering the fatigue behaviour of riveted joints. The joint fatigue behaviour is also not very sensitive to variations in the properties of nominally the same materials delivered by different producers.

Solid rivets most frequently used in fuselage skin joints are the protruding head rivet and the countersunk rivet. The dimple hole rivet and the NACA rivet are a kind of the countersunk rivet. The countersunk rivet is applied when a flat outer side of the joint is required. Aluminium alloys used for the solid rivets are designated as AD (East-European equivalent PA24), D and DD, and E. The D and DD rivets must be installed in a quenched condition. For joints manufactured according to typical riveting specifications fatigue properties of the AD and DD materials are similar. The E rivet surpasses the AD and DD rivets, but at a larger squeeze force cracks in the driven head can occur. The effect of the rivet type on the joint fatigue behaviour is revealed at relatively high loads, then the NACA rivet proves to be superior. Also, differences between the various types of rivet become more significant for thicker sheets.

Anodizing is the most common surface treatment used for corrosion protection of aircraft components. The sulphuric acid coating and the chromic acid coating show a similar behaviour. Nowadays, the aerospace industry widely employs phosphoric acid anodizing in order to reduce environmental damage caused by chromates. Anodizing is detrimental on top of cladding but beneficial on the unclad material. Also, water-displacing fluids are applied in order to avoid corrosion problems. A significant deterioration in fatigue strength due to the application of these penetrants is observed for joints with countersunk rivets. Sealants have a favourable effect on fatigue life for protruding head rivets.

A method of rivet hole preparation (hand or machine drilling, drilling with or without a pilot hole) and hole imperfections, like overdeep countersinking (the so-called knife edge effect) and mismatched holes, can be among factors affecting the fatigue behaviour of riveted lap joints.

The riveting method and, still to more extent, the squeeze force are primary production variables influencing fatigue properties of riveted lap joints. Increasing the squeeze force leads to increased load transmission by friction and creates residual stresses around the rivet hole, both phenomena being beneficial for the joint fatigue behaviour. There is a strong dependence between the squeeze force and the rivet driven head diameter which is a first control criterion for the riveting process. For that reason, many experimental works and FE analyses were performed to obtain a relationship between these quantities.

Rivet hole expansion due to the squeeze force varies in the thickness direction. It is more homogenous for the protruding head rivet compared to the countersunk rivet. Due to hole expansion tangential and radial residual stresses around the rivet hole are generated. Compressive tangential stresses occur for a larger squeeze force and the area of these stresses becomes wider if the squeeze force level is increasing. This explains why for a higher squeeze force the location of fatigue crack nuclei shifts a small distance away from the hole, whereas for a low squeeze force the crack initiates at the hole edge. Also, decreasing the initial clearance between the sheet and rivet shank has a beneficial influence on the fatigue behaviour of riveted lap joints.

Chapter 4
Design Parameters Influencing the Fatigue Behaviour of Riveted Lap Joints

Geometric variables related to the design of a riveted lap joint are the number and spacing of rivet rows, the rivet pitch in a row, the distance of the rivet from the sheet edge, the sheet thickness, the joint size, and the rivet pattern, Fig. 4.1. Because each of these parameters can exert influence on stress concentration at the rivet hole or modify the secondary bending, they also affect joint fatigue behaviour. The corresponding experimental observations reported in the literature are reviewed in the present chapter.

4.1 Number of Rivet Rows

The number of rivet rows is increased when the joint has to sustain higher loads. The effect of the number of rows cannot be regarded as free from the influence of thickness because the increasing load requires a thicker sheet. Results of fatigue tests by Bürnheim (ref. XVII/4 in Hertel 1969) and by Russel et al. (ref. 28 in Schijve 1956), performed as early as the 1940s, clearly show that lap joint fatigue strength significantly increases with an increase in the number of rows. As follows, however, from the *S-N* data by Bürnheim presented in Fig. 4.2, this beneficial effect is revealed up to three rows only because the data for the four-row and five-row joint fall within the scatter band for the three-row configuration. On the other hand, experiments by Smith and Lindenau (1957) indicate a beneficial effect of increasing the number of rivet rows from three to four (clad aluminium alloy, sheet thickness 0.051 in., 3/16 in. 100° countersunk rivet, row pitch 1 in., rivet pitch 0.5 in., 4 rivets in row, $R = 0.1$, $S_{max} = 10$ ksi). The mean fatigue life was about 331 kcycles for the three-row specimens and 500 kcycles for the joints with four rows. However, from the statistical point of view, the sample in these experiments was too small to draw valid conclusions (altogether five specimens). The results addressed above

A. Skorupa and M. Skorupa, *Riveted Lap Joints in Aircraft Fuselage*, Solid Mechanics and Its Applications 189, DOI 10.1007/978-94-007-4282-6_4,

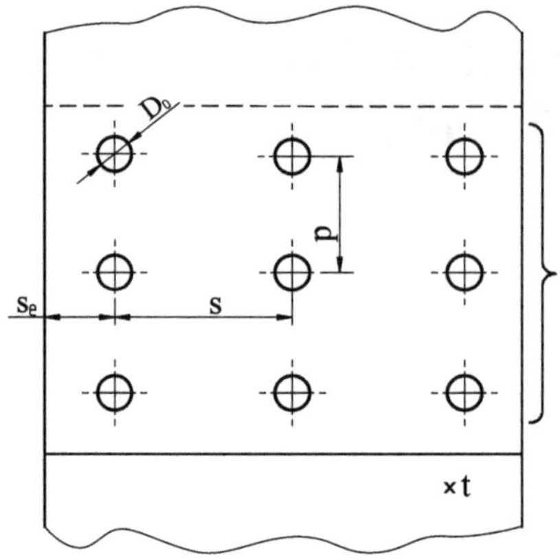

Fig. 4.1 Nomenclature for dimensions of a longitudinal riveted lap splice joint

imply that the lap joint fatigue properties may not be improved in a meaningful way if more than three rivet rows will be applied. However, a four-row riveted lap splice design is applied in some cases (for example Boeing 747-400).

Figure 4.3 gives the graphs based on results obtained by Russel et al. on lap joints in which the number of rivet rows was the only variable because the number of rivets (protruding head) per row and the sheet thickness was the same for all specimens. As can be seen, the joint fatigue strength increases with an increase in the number of rows. It is confirmed again that, under static loading, joints made of the 7075 Al alloy are stronger than those made of 2,024.

Another test series by Russel et al. (ref. XVII/5 in Hertel 1969) on lap joint specimens with one, two and three rows of countersunk rivets showed the same trend as observed in Fig. 4.3 for the protruding head rivet. In line with these results are the *S-N* curves from more recent fatigue tests by Skorupa et al. (2011) on three-row and double-row riveted lap joints, both with the same row spacing $p = 25$ mm, shown in Fig. 4.4.

The influence of the number of rivet rows on joint fatigue performance observed in the experiments addressed above can be explained by lowering the load transfer through the critical outer rows when the number of rows is increasing (see Sect. 5.1). In addition, when the row spacing is kept unchanged, a joint with more rivet rows implies a reduction of secondary bending due to the longer overlap. A pure effect of changing the number of rows can be demonstrated when the same distance between the end rows is maintained, as for the double-row joint with $p = 50$ mm and the three-row configuration with $p = 25$ mm, Fig. 4.4. It is seen that in that case the benefits from increasing the number of rows are less pronounced than when, for the three-row joint, the same row pitch as for the double-row joint is adopted.

Hartman's (1954) experiments involved lap joint specimens with two and three rivet rows, but with the same total number of rivets for either configuration because

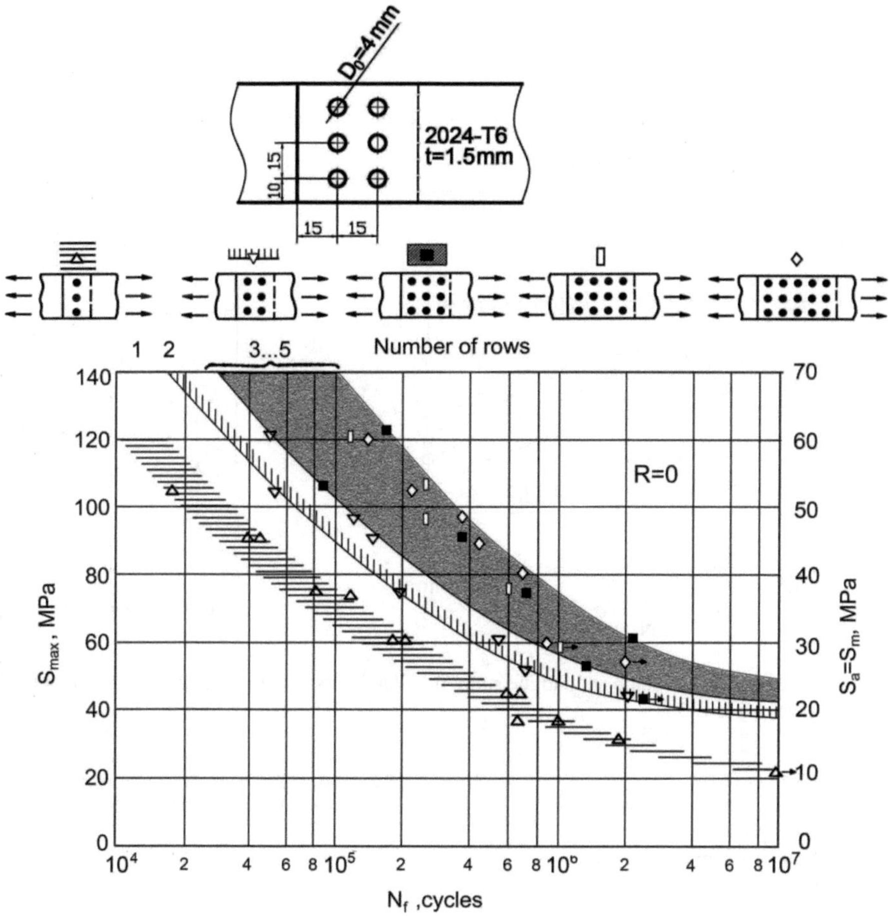

Fig. 4.2 Influence of the number of protruding head rivet rows on fatigue strength of single lap joints (Hertel 1969)

for the three-row joints the rivet pitch for the outer rows was doubled, as shown in Fig. 4.5. *S-N* data for the double-row and three-row joints plotted in this figure are combined within a single scatter band. This could suggest that if fatigue joints have the same total number of rivets, and hence their static strength (failure by shearing of the rivets) is the same, the number of rivet rows has no influence on fatigue properties. However, such a conclusion is only valid for a specific case when the reduced amount of secondary bending in a three-row specimen (larger overlap) exactly compensates the increased stress concentration and load transmission due to reducing the number of rivets in the outer rows. Generally, the multi-row lap joint with a reduced number of rivets in the outer rows is a poor design.

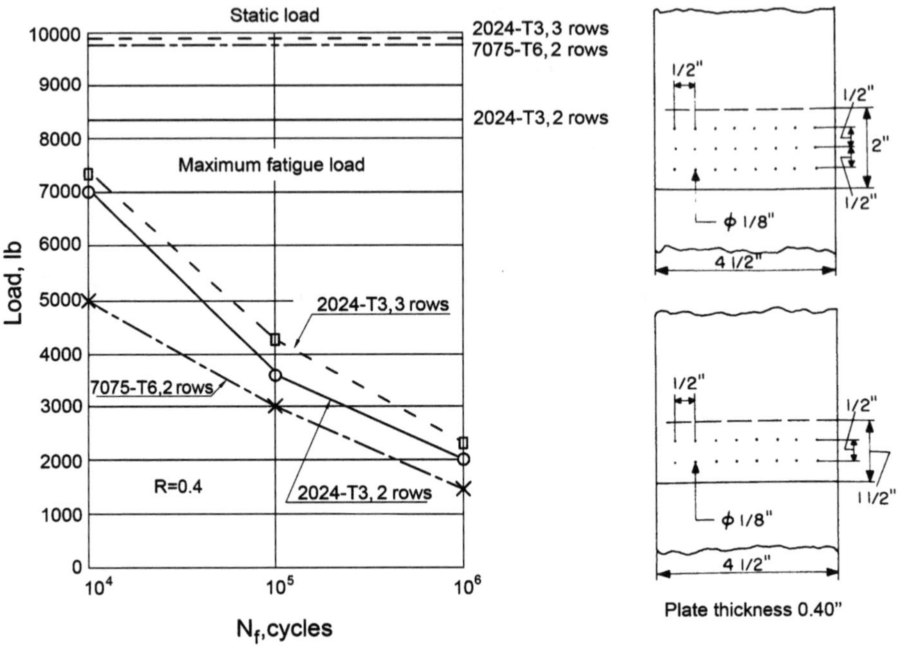

Fig. 4.3 Comparison of static and fatigue strength of lap joints with two and three protruding head rivet rows (Schijve 1956)

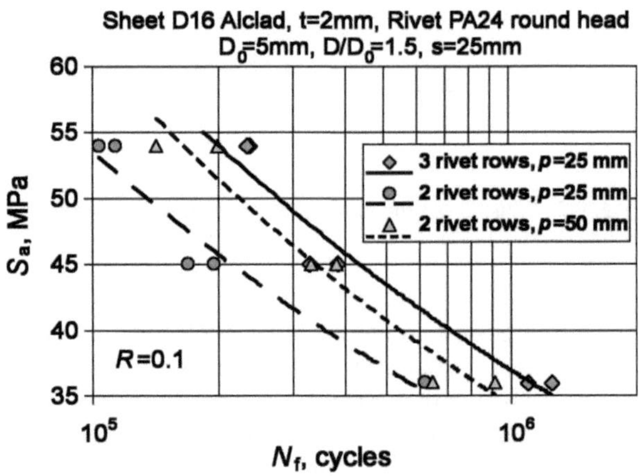

Fig. 4.4 Effect of the number of rivet rows and row spacing on lap joint fatigue lives at $R = 0.1$ (Skorupa et al. 2011)

4.2 Rivet Row Spacing

Perhaps the earliest results on the effect of rivet row spacing (p) in the lap joints come again from Bürnheim (ref. XVII/4 in Hertel 1969) and Russel et al. (ref. 28 in Schijve 1956). Bürnheim's results for specimens with protruding head rivets shown

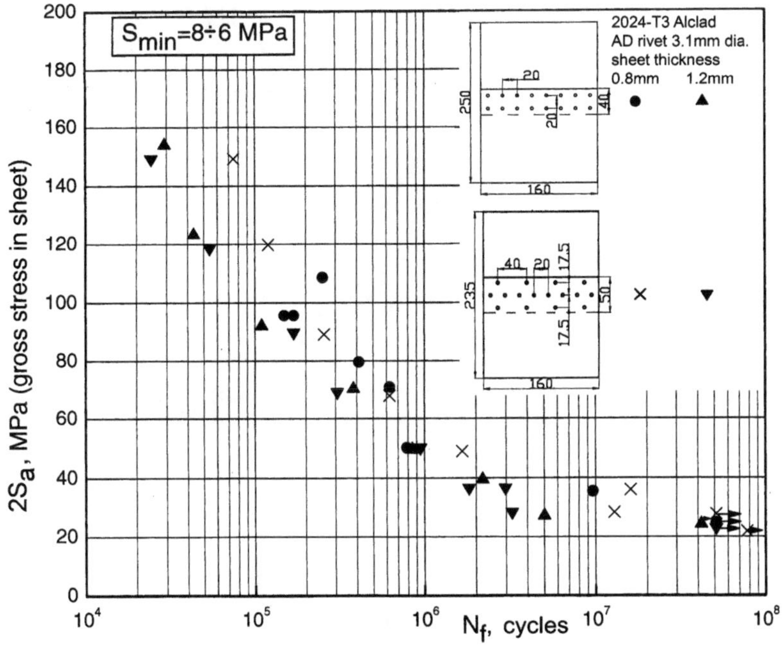

Fig. 4.5 Results of fatigue tests on single lap joints with two and three rows of protruding head rivets; same total number of rivets for both type joints (Hartman 1954)

in Fig. 4.6 demonstrate that increasing the rivet row spacing improves joint fatigue properties. The trend of bettering fatigue performance with increases in rivet spacing has also been observed by Russel et al. for countersunk rivets, Fig. 4.7.

Fatigue tests under CA loading in order to investigate the effect of the rivet row pitch were performed by Hartman and Schijve (1969) on two series of double-row riveted lap joints. The only difference between the joints was the rivet row spacing ($p = 13$ and 30 mm). As seen in Fig. 4.8, the larger row spacing again yielded a higher fatigue strength, especially at lower fatigue lives. These authors also reported a significant influence of rivet row spacing under programme loading. In the latter case the ratio between endurances of joints with large spacing and with small spacing was about 3:1. Another example of the beneficial effect of longer overlap for joints with two rivet rows can be found in Fig. 4.4 where longer fatigue lives are observed for the larger spacing of 50 mm compared to $p = 25$ mm.

As seen in Fig. 4.9, a similar trend is exhibited by the fatigue test results for three-row lap joints reported by Skorupa et al. (2011). For either of the two sheet thicknesses considered, the *S-N* curve corresponding to the larger rivet row spacing of $8D_0$ is shifted right compared to the smaller spacing of $5D_0$, the effect being much stronger for the thinner sheet of 0.8 mm than for the 2 mm thick sheet. Also Müller and Hart-Smith (1997) observed a beneficial influence of increasing the overlap on

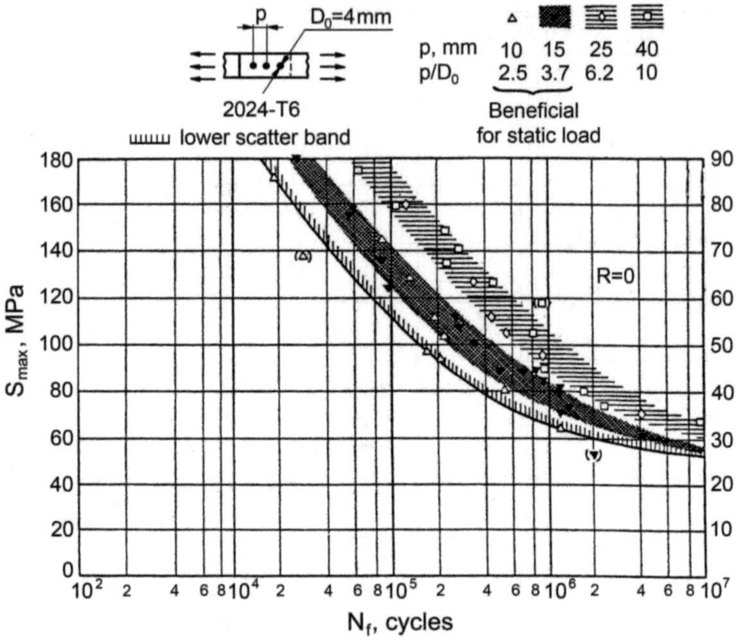

Fig. 4.6 Influence of rivet row spacing on fatigue strength of lap joints (Hertel 1969)

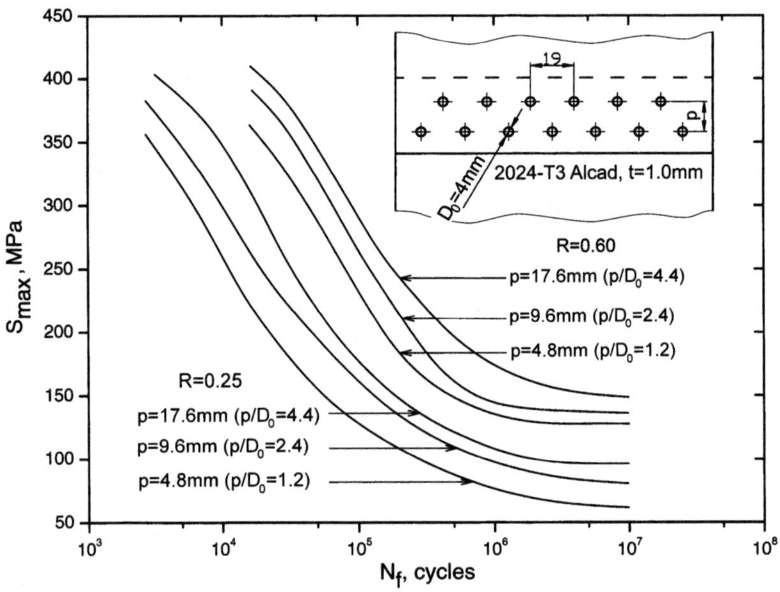

Fig. 4.7 Effect of rivet row spacing on *S-N* curves for double-row lap joints (Schijve 1956)

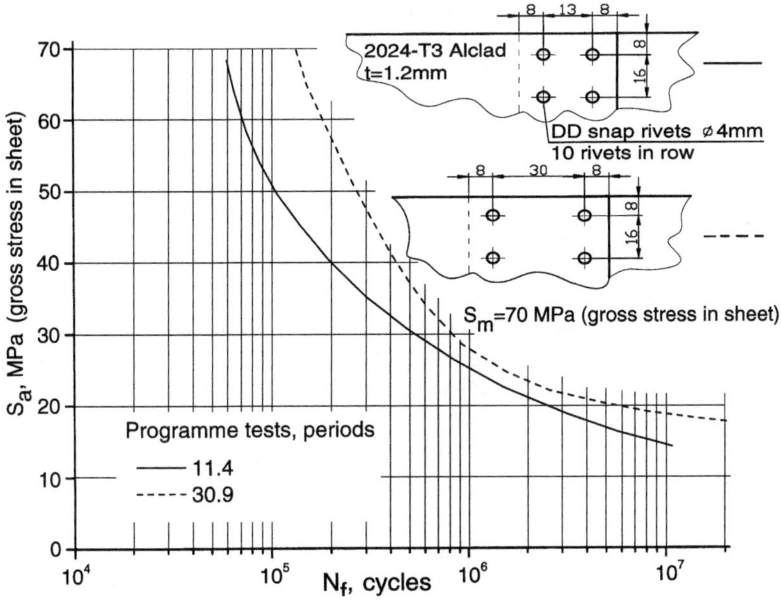

Fig. 4.8 Effect of rivet row spacing on fatigue curves of double-row lap joints (Hartman and Schijve 1969)

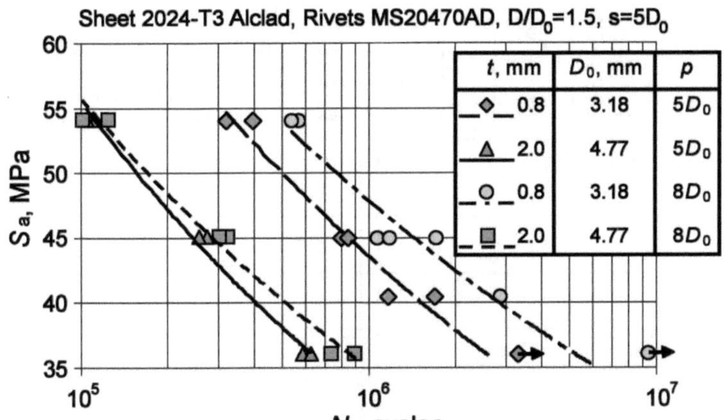

Fig. 4.9 Effect of rivet row spacing and sheet thickness on lap joint fatigue lives at $R = 0.1$ (Skorupa et al. 2011)

the fatigue life of a three-row lap joint (2024-T3 sheet 1.6 mm thick) riveted using NACA rivets (4.8 mm dia). An increase of spacing from 25.4 to 50.8 mm extended fatigue life from 1,010 to 3,734 kcycles.

It is recognized that benefits from applying a larger row spacing in a lap joint are associated with a reduction in secondary bending, as considered in more detail in Chap. 6. It cannot be overlooked that a larger distance between rivet rows has also an unfavourable implication, namely an increase in joint weight due to the longer overlap.

4.3 Rivet Pitch in Row

In general, the ratio between the rivet pitch in row and the rivet diameter (s/D_o) is determined based on static strength calculations, and larger variations in s/D_o do not occur in practice (Schijve 1956). According to photoelastic analysis for a row of holes loaded with closely fitted pins, the lowest stress concentration factor is associated with $s/D_o = 2.5$ (Smith 1960). This optimum spacing value is supported by experimental results according to Bürnheim (ref. XVII/4 in Hertel 1969) for protruding head rivets shown in Fig. 4.10 which indicate that s/D_o ratios in the range from 2.5 to 3.75 are favourable for both fatigue and static loads. Figure 4.10 further demonstrates that an increase in s/D_o yields a decrease in fatigue strength. The latter trend is also backed up by results according to Hartman and Klaassen

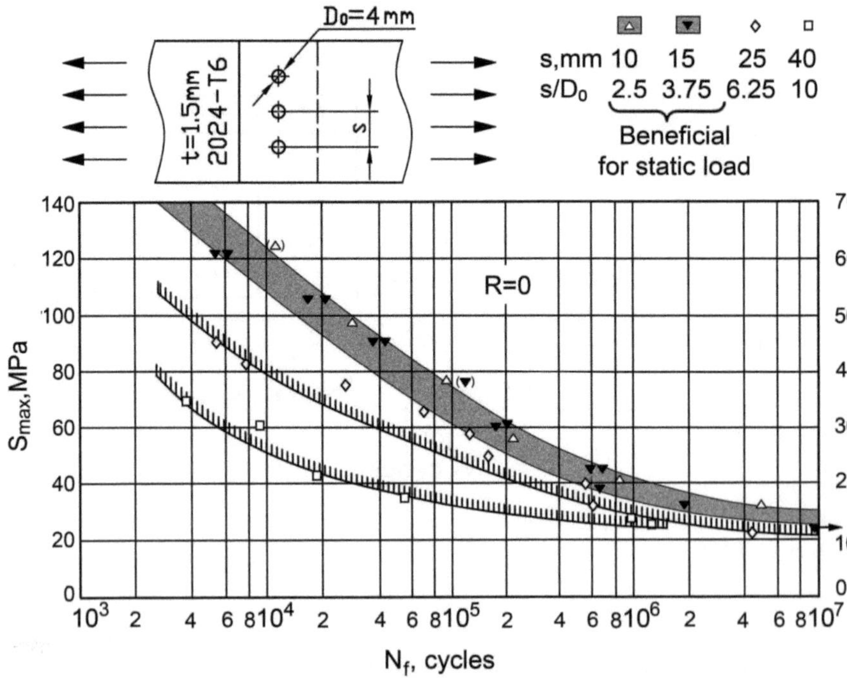

Fig. 4.10 Effect of rivet pitch in a row on *S-N* curves for double-row lap joints (Hertel 1969)

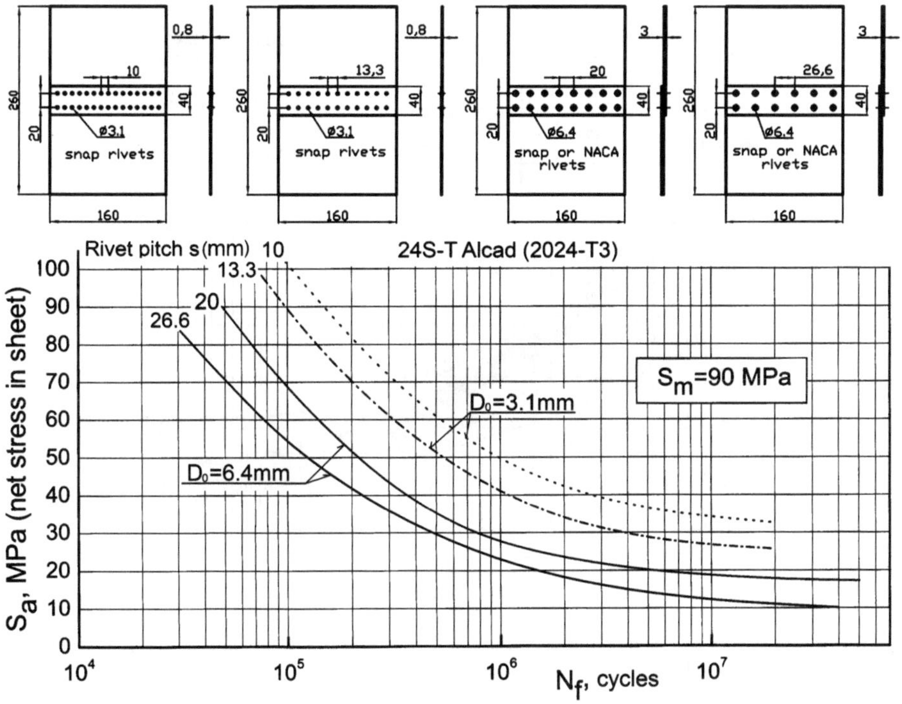

Fig. 4.11 Influence of the ratio of rivet diameter to rivet pitch on fatigue strength of double-row lap joints (Hartman and Klaassen 1956)

(1956) shown in Fig. 4.11 where the stress amplitude is plotted as net section stress. Another piece of similar experimental evidence is presented by Vlieger (1994) who observed specimens with a smaller rivet spacing in a row to show longer lives.

4.4 Distance of the Rivet from the Sheet Edge

The distance of the outer rivet column from the closest sheet edge (s_e) seems to have little effect on fatigue strength (Schijve 1956). Usually, this distance is equal to half the rivet pitch ($0.5s$). According to Niu (1988) and in agreement with Fokker specifications, the outer rivet row centre line should be at least $2D_o$ distant from the sheet edge, which is the static strength requirement.

The minimum permissible sheet edge distance from the centre of the closest rivet hole can be diminished to $1.5D_o$ if the edge is parallel to the load direction. The latter rule has been already postulated by Bürnheim (ref. XVII/4 in Hertel 1969) based on his fatigue test results shown in Fig. 4.12. However, it should be noticed that the reduction of s_e to $1.5D_o$ reduces also the static strength of a joint.

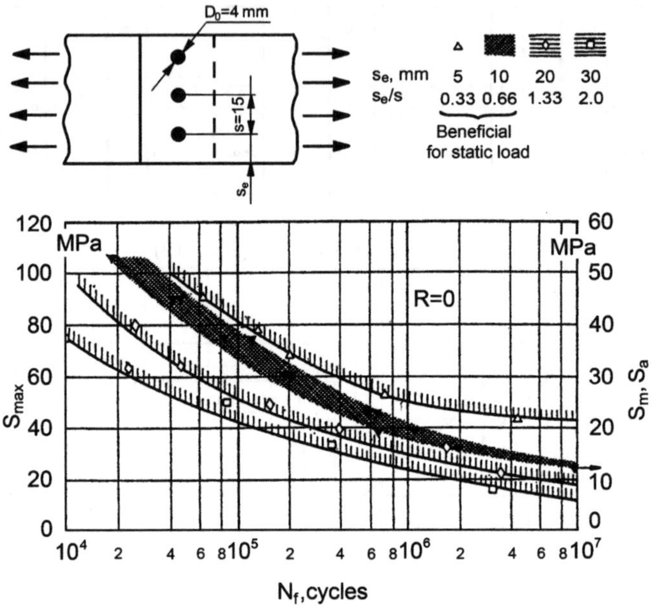

Fig. 4.12 Influence of distance between the outer rivet and the sheet edge on fatigue strength of single-row lap joints (Hertel 1969)

4.5 Rivet Pattern

Two types of rivet pattern are possible, namely a zigzag (staggered) and in-line (chain) riveting. The effect of the rivet pattern was studied by Hartman and Duyn (1952). The test pieces were double-row lap joints of 24 Alclad (2024-T3) sheet 0.8 mm thick with 17S (2017) snap rivets 3.1 mm in diameter. In Fig. 4.13, where their results are presented, no marked difference in fatigue strength can be noticed for the three rivet configurations considered if the rivet row spacing is preserved. The same conclusion can be drawn from another experiment series conducted by the same authors and involving protruding head rivets, dimple hole rivets and countersunk rivets. It can be observed in Fig. 3.11 that the rivet pattern has no significant influence on either static strength or fatigue strength at endurances of 10^6 and 10^8 cycles.

Another investigation of the influence of the rivet pattern was performed by Hartman and de Rijk (ref.14 in Oldersma 1992), who used specimens reinforced in the critical section with adhesive bonded sheets, and by Russel et al. (ref. 28 in Schijve 1956). The final conclusion from both works is that there is no difference in fatigue strength of double-row lap joints whether in-line or zigzag riveted.

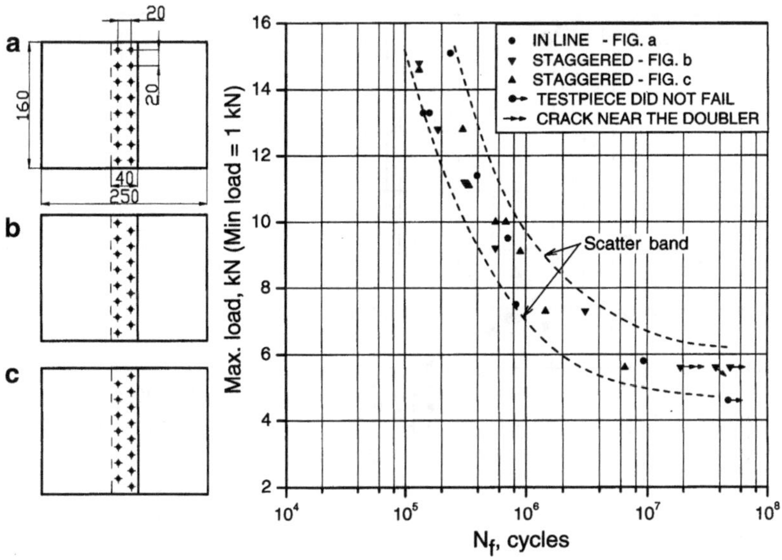

Fig. 4.13 Influence of rivet pattern on fatigue strength of double-row lap joints with protruding head rivets (Hartman and Duyn 1952)

4.6 Sheet Thickness

Essentially, the effect of sheet thickness cannot be considered separately from the effect of other variables, especially of rivet diameter and rivet row spacing. The sheet thickness is increased when the joint has to sustain higher loads. In that case, however, larger rivets are used. The thickness of sheets ranging from 0.8 to 1.6 mm was the only variable in fatigue tests conducted by Hartman (1954) on double-row lap joints with protruding head rivets (sheet material 24 ST Alclad, rivet material 17S, rivet diameter 3.1 mm). The *S-N* data corresponding to these experiments depicted in Fig. 4.14 reveal that a small change in thickness, e.g. from 0.8 to 1 mm, has little effect on the fatigue properties of a lap joint. However, compared to these thinner specimens, those from 1.6 mm sheets exhibit systematically shorter fatigue lives, especially at higher stress levels. The stress range in Fig. 4.14 is based on gross section stress, but the same trend was observed when net section stress was employed. In line with these results are fatigue test data for three-row joints of sheets with four different thicknesses, Fig. 4.15. Unlike in the case of the joints used by Hartman, the rivet diameters and absolute lengths of the overlap are different for different thicknesses. Figure 4.15 demonstrates that the data points for the thinner specimens ($t = 0.6$, 0.8 and 1.2 mm) appear to lie within a single scatter band, while the thickest joints of $t = 2$ mm show somewhat shorter fatigue lives. The latter trend is more evident at lower stress levels, in contrast to Hartman's results in Fig. 4.14.

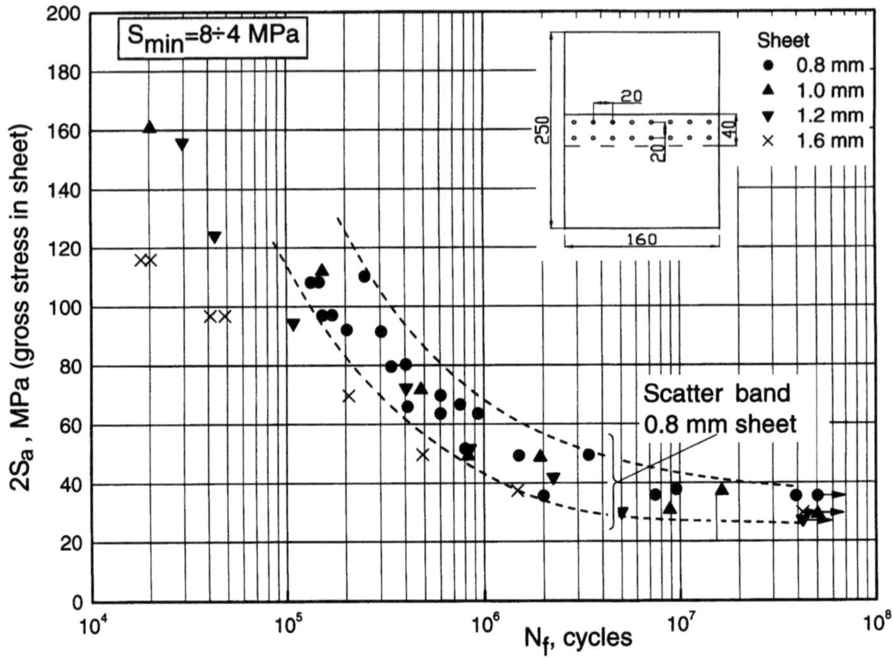

Fig. 4.14 Effect of sheet thickness on fatigue strength of lap joints (Hartman 1954)

Fig. 4.15 Effect of sheet thickness on three-row lap joint fatigue lives at $R = 0.1$ (Skorupa et al. 2011)

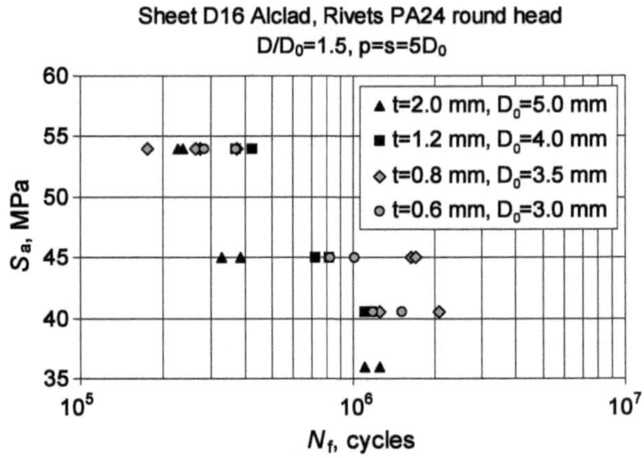

The Engineering Science Data Unit (ESDU 1979) analyzed fatigue test results obtained from published sources on simple lap joints riveted using protruding head and countersunk rivets. It was found that for the same conditions specimens with a larger D_0/t ratio exhibited a longer fatigue life. A detrimental influence of increasing sheet thickness on the fatigue strength of riveted lap joints was confirmed in a numerical study by Fung and Smart (1997) who modelled the riveting process and subsequently applied loading on the joint. The results showed that a local effective stress (at hole edge in the mid plane of the model) was reduced with increasing the D_0/t ratio, which should yield a longer fatigue life. This effect was more pronounced for a protruding head rivet than for a countersunk rivet.

A considerable influence of sheet thickness is also revealed in Fig. 4.9 where again the different thicknesses involve different rivet diameters and different absolute rivet row spacing (p) values. Interestingly, the thickness effect is more significant for the larger p of $8D_0$ than for $p = 5D_0$. The decrease of fatigue life with increasing sheet thickness can be contributed by a more pronounced secondary bending due to the larger joint eccentricity, see Chap. 6.

4.7 Size Effect

The fatigue response of geometrically similar specimens can be influenced by size effects, such as a decrease in the stress gradient associated with the larger component size, the increased possibility of occurrence of a weak spot, and enhanced fretting due to the larger absolute dimensions of the mating component parts. Also, as said earlier, the magnitude of secondary bending increases with increasing sheet thickness. If the size effect occurs in riveted joints, then all these factors should have a negative effect on joint fatigue properties.

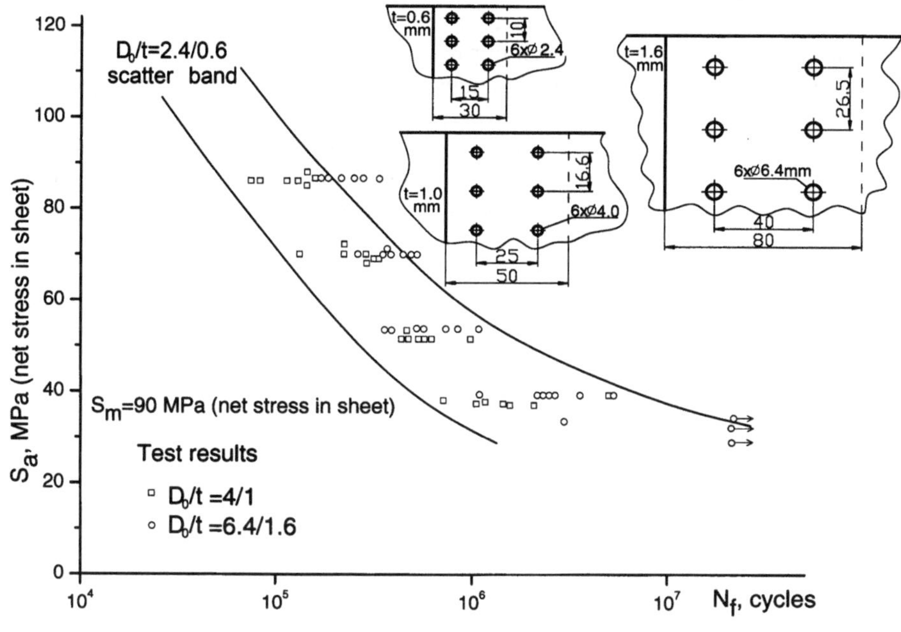

Fig. 4.16 Effect of joint size on fatigue strength of lap joints (Hartman et al. 1962)

Hartman et al. (1962) examined three geometrically similar specimen series (2024-T4 Alclad sheet and 2117-T snap rivets) with ratios between all dimensions of 1:1.66:2.66, Fig. 4.16. All test pieces were manufactured by the same person using the same method (pneumatic riveting) to minimize the manufacturing effects. As can be seen in Fig. 4.16, these tests did not reveal any significant difference in endurances although, in general, fretting was more severe for the larger specimens than for the small specimens. It should be emphasized that because in this experiment all joint dimensions were increased with the same factor, the secondary bending factor (bending stress/nominal applied stress) did not change.

4.8 Summary of This Chapter

Fatigue behaviour of a riveted lap joint depends on design parameters. Increasing the number of rivet rows significantly improves the joint fatigue strength. A typical longitudinal lap joint consists of three rivet rows, though double- and four-row configurations are also applied. A single-row riveted joint should not be accepted due to its very poor fatigue properties.

A larger rivet row spacing yields a higher fatigue strength of a lap joint. This effect is associated with a reduction in secondary bending. The larger distance between rivet rows has also unfavourable implications, namely an increase in the joint weight due to the longer overlap. A decrease in fatigue life of lap joints with increasing sheet thickness can be mainly attributed to more severe secondary bending due to a larger joint eccentricity.

The ratio between the rivet pitch in row and the rivet diameter is based on static strength calculations and larger variations of this ratio do not occur in practice. However, a smaller rivet spacing is favourable to the fatigue strength of a lap joint.

The distance of the outer rivet row from the closest sheet edge seems to have little effect on the fatigue behaviour of a lap joint. Also, no marked difference in the lap joint fatigue strength is shown depending on the rivet pattern, namely zigzag or in-line.

Chapter 5
Load Transfer in Lap Joints with Mechanical Fasteners

The role of a fastener is to transfer the load from one sheet to another sheet in the overlap region. For a configuration with more than one row of fasteners, the applied force P is split at the first row into the bypass load (T_{BP}), which remains in the sheet, and the transfer load (T_{TR}) transmitted to the other sheet, as schematically shown in Fig. 5.1. The T_{TR}-load is comprised of a bearing force (T_{BR}) resulting from the bearing pressure exerted by the rivet shank on the hole surface and a friction force (T_{FR}) induced by friction between the mating sheets. Friction is localized mainly beneath the rivet heads where maximum clamping occurs. Works reviewed in the present chapter reveal that, for well-tightened fasteners, frictional forces can transmit a significant portion of the transfer load.

The distribution of loads acting in the joint must be known to approach the problem of stress conditions at rivet holes. In this chapter the methodology and reported results on analytical and experimental determination of load transmission in riveted joints are considered. Special attention is also paid to investigations of rivet flexibility in view of its importance for load transfer computation.

The consideration of load transfer in this chapter is limited to the case of sound joints. Crack growth at rivet rows significantly alters load transmission through the joint, as addressed in Sect. 9.3.

5.1 Simple Computation of Axial Forces in the Sheets

Axial forces in the sheets can be calculated by considering the displacement compatibility in the overlap region of the joint. Figure 5.2 shows a cut out from the multi-layer joint overlap comprising sheet k and $k+1$ within a segment i located between fastener row i and $i+1$. The following relationship between the deformations of the sheets and fasteners can be observed:

$$l_{k,i} + \Delta_{k,i} + \delta_{k,i+1} = l_{k+1,i} + \Delta_{k+1,i} + \delta_{k,i} \tag{5.1}$$

A. Skorupa and M. Skorupa, *Riveted Lap Joints in Aircraft Fuselage*, Solid Mechanics and Its Applications 189, DOI 10.1007/978-94-007-4282-6_5,

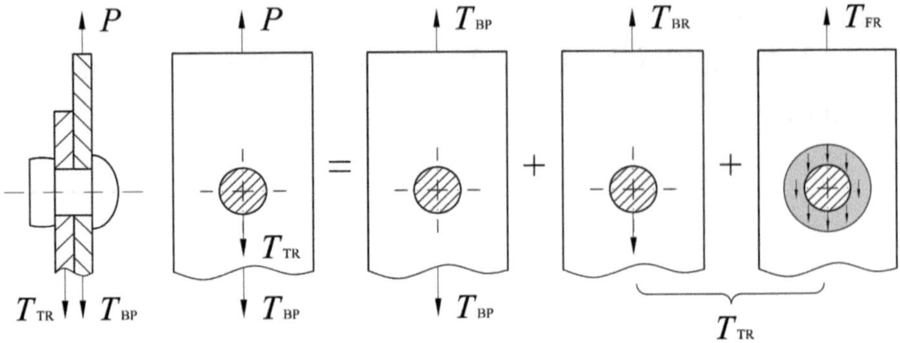

Fig. 5.1 Axial forces in sheets of a two-layer lap joint

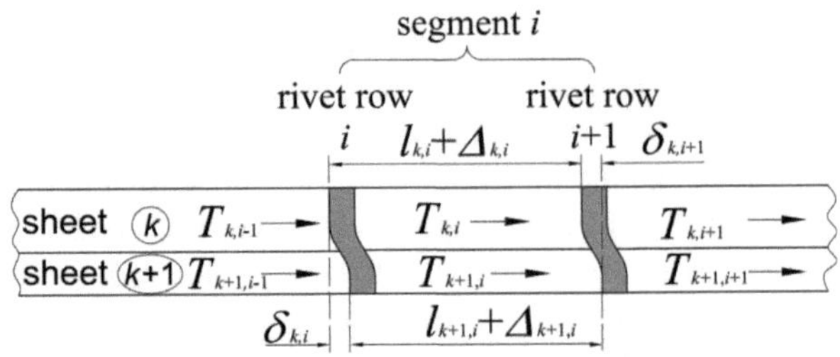

Fig. 5.2 Cut out from a multi-layer lap joint

where l and Δ denote the initial lengths and elongations of the sheets, while δ are the fastener deflections between sheets caused by rivet flexibility.

If the joint deformation is assumed to be elastic, the sheet elongations can be expressed in terms of the internal forces as

$$\Delta_{j,i} = \frac{T_{j,i}\, l_{j,i}}{(EA)_j} \tag{5.2}$$

where $T_{j,i}$ denotes the internal axial force in sheet j of segment i and $(EA)_j$ is the longitudinal rigidity of sheet j.

For a given row i the fastener deflection between sheets k and $k+1$ $(\delta_{k,i})$ is assumed to be linearly related to the load passing through that row from sheet k to $k+1$:

$$\delta_{k,i} = \sum_{j=1}^{k} (T_{j,i-1} - T_{j,i})\, f_i \tag{5.3}$$

where the f_i parameter represents the flexibility of fasteners in row i.

For an m-layer overlap of a joint with r rivet rows, substitution of Eqs. 5.2 and 5.3 into Eq. 5.1 yields a system of $(m-1)(r-1)$ equations containing $m(r-1)$ unknown internal forces. The lacking $r-1$ equations needed to solve for T_{ji} are the equilibrium conditions which must be satisfied for each transverse section of the overlap:

$$\sum_{j=1}^{m} T_{j,i} = P \quad i = 1 \text{ to } r - 1. \tag{5.4}$$

Barrois (1978) postulates that Eq. 5.2 underestimates the elongation of a sheet with a hole. He proposes that Δ should be split into a component depending on the stress in the net section along the rivet row centre line ,

$$\Delta' = \frac{Tk_1 d}{EA(1 - d/s)},$$

and another component depending on the uniform stress far from the holes equal to

$$\Delta'' = \frac{T(l - k_2 d)}{EA}$$

where d is the hole diameter, s is the rivet pitch in a row, and the k_1 and k_2 coefficients depend on the d/s and l/s ratios.

The summation yields

$$\Delta = \Delta' + \Delta'' = \frac{Tl}{EA}\left(1 + \frac{d}{l}\left(\frac{k_1}{1 - d/s} - k_2\right)\right). \tag{5.5}$$

According to Barrois, when interference between the rivet and the hole increases, the term in the brackets on the right-hand side of the above equation should become lower due to lowering the net stress. Based on FE analysis results, he suggests $k_1 = 1.3$ and $k_2 = 1$ for tight fitted rivets within the limits $1 \leq l/s \leq 2$ and $0.125 \leq d/s \leq 0.375$. For example, with $d/l = d/s = 0.2$ Eq. 5.5 yields a Δ-value 12% higher than Eq. 5.2.

Discrepancies between the solution presented above and the actual force distribution in a lap joint may stem from various simplifications involved in the computation procedure. First, bending deflections of the sheets, which can be of significance especially at the outer rivet rows, are not accounted for. Second, the contribution of friction is neglected, which yields an overestimation of rivet deflections, especially if the clamping between the sheets is good. Further, uncertainties exist about rivet flexibility determination, as detailed in Sect. 5.2. Finally, the load distribution becomes affected when cracks start to grow at the rivet holes, which can occur early in joint fatigue life. This can cause load shedding from the cracked to uncracked rows of rivet holes, thereby changing the load transfer.

In spite of its limited accuracy, the solution for the forces in the sheets derived from the procedure described in this section can be useful for studying qualitatively the influence of different variables, especially those associated with joint geometry. As an example, Fig. 5.3 according to Swift (1990) shows the computed dependence

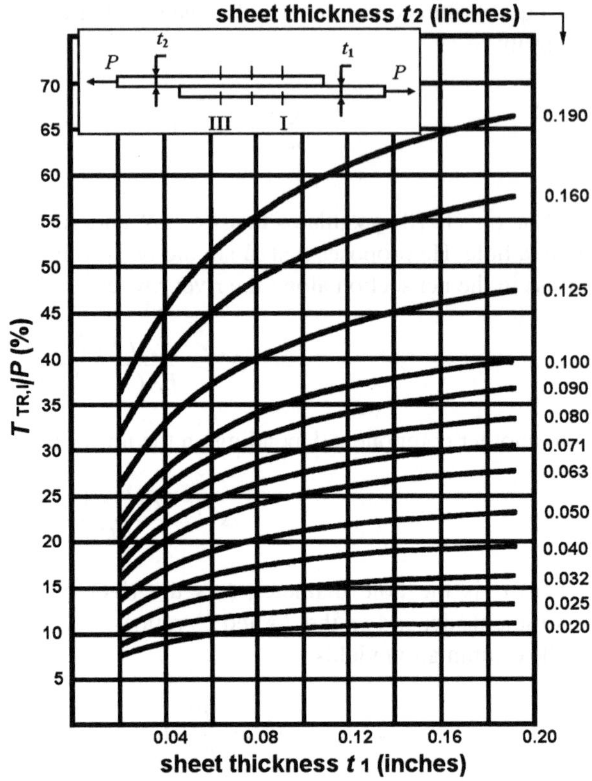

Fig. 5.3 Effect of sheet thicknesses on load transfer by rivet row I of a lap joint (Swift 1990)

between sheet thicknesses and the fraction of the applied load transmitted from sheet 1 to sheet 2 by rivet row I of a lap joint with three rivet rows. An Al alloy material of the sheets and rivets and a 3/16 inch diameter of the rivets are assumed. It is seen that the load transfer increases with increasing thickness of either sheet and that the maximum load transfer occurs in row I if $t_1 < t_2$. For $t_1 = t_2$ both outer rows transfer the same load and when $t_1 > t_2$, the maximum transfer is by row III. For the same type joint with $t_1 = t_2$, Fig. 5.4 demonstrates that the load transfer by the outer rivets becomes lower with increases in the number of rows (n). However, fatigue tests do not reveal benefits from applying more than three rivet rows (cf. Fig. 4.2).

Because stress concentration associated with the bearing load is more severe than that due to the bypass load, attempts at improving load transfer in lap joints typically include diminishing load transmission by the outer rows. As said earlier, these rows are fatigue critical due to a still significant bypass load and the maximum moment associated with secondary bending. An exemplary constructional solution adopted to reduce the bearing stresses in the outer rivet rows of a fuselage longitudinal lap joint is shown in Fig. 5.5. Fingering the doubler effectively reduces the doubler area between the outer rivet row and the neighbouring inner row, thus relieving the outer row. This results in a lower bearing stress in the outer rows. Alternative

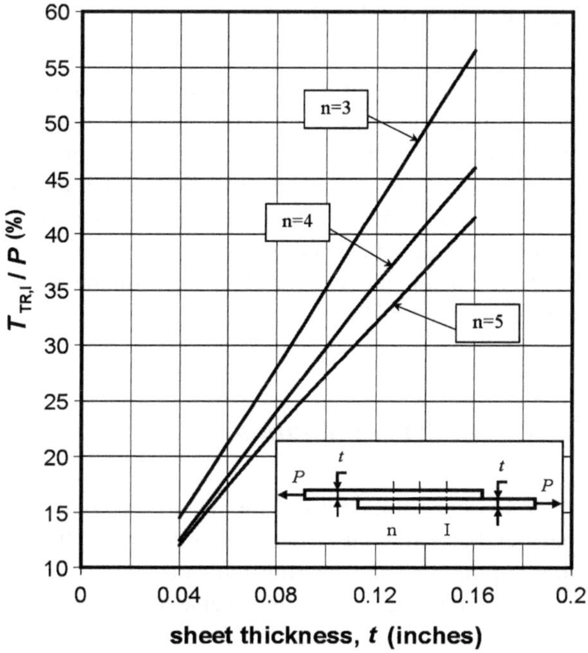

Fig. 5.4 Effect of the number of rivet rows in a lap joint on load transfer by rivet row I (Swift 1990)

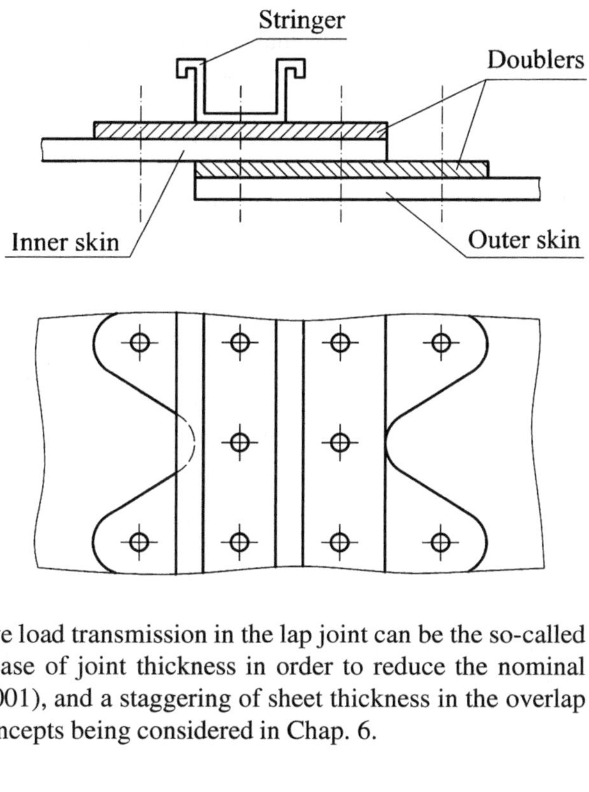

Fig. 5.5 Fuselage lap splice configuration with fingered doublers applied in DC-9/MD-80

design approaches to improve load transmission in the lap joint can be the so-called "padding", i.e. a local increase of joint thickness in order to reduce the nominal tensile stresses (Das et al. 2001), and a staggering of sheet thickness in the overlap area (Schijve 2006), both concepts being considered in Chap. 6.

5.2 Fastener Flexibility

The interest in fastener flexibility has been caused by the desire to calculate the load transfer distribution in joints with multiple fastener rows. The fastener flexibility data is the input necessary for this calculation both when the simple procedure described in the previous section and more elaborated FE computations are applied. The literature evidence cited below indicates that the fastener flexibility depends on the sheet and fastener materials, the joint geometry and the squeeze force. The f-parameter can be determined experimentally or analytically, both approaches being considered below.

5.2.1 Analytical Solution

An analytical solution for f has been proposed by Barrois (1978). The fastener shank embedded in the sheet is modelled as the flexible beam on elastic foundation. This implies that the bearing unit load between the fastener and the hole (q) is proportional to the fastener transverse deflection (y), Fig. 5.6:

$$q = ky. \tag{5.6}$$

From the displacement analysis for a hole in a finite width strip loaded by a tight fit pin, the coefficient k is expressed as

$$k = E(1 - d/s)/0.8 \tag{5.7}$$

where E is the sheet material elastic modulus, d denotes the pin diameter and s is the strip width.

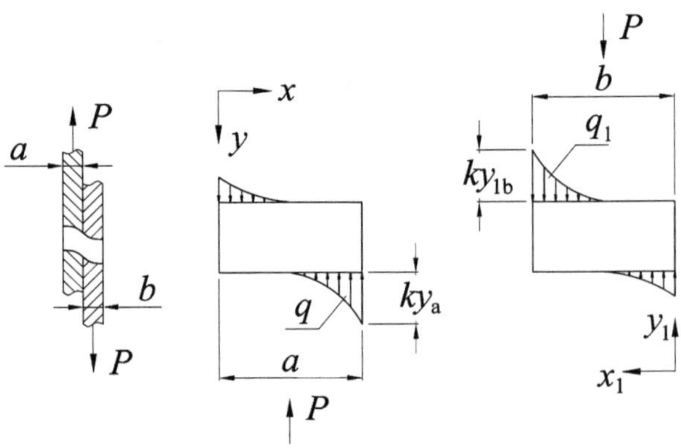

Fig. 5.6 Nomenclature for Barrois' (1978) solution for rivet flexibility

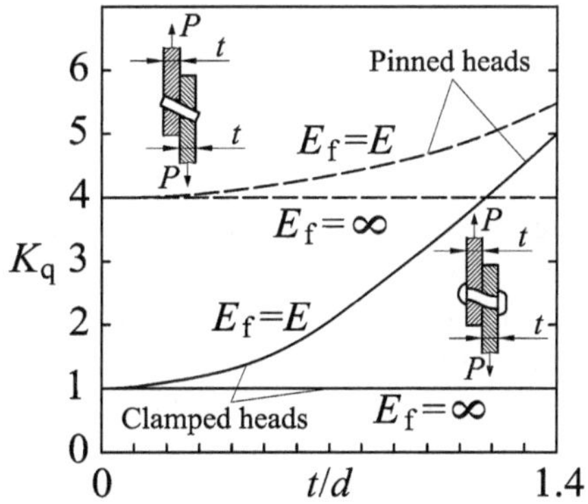

Fig. 5.7 Example of Barrois' (1978) solution, see text

The approach by Barrois accounts for both the bending and shear deformation of the fastener, which results in a 4th-order differential equation for y. The latter is solved analytically considering the following extreme conditions at the contact surface between the rivet head and the sheet surface, see Fig. 5.6:

(1) zero bending moment at $x = 0$ and $x_1 = 0$, which implies a pin without heads, to model a fastener installed without clamping or loosened after some service operation time,
(2) $(dy/dx)_{x=0} = 0$ and $(dy/dx)_{x_1=0} = 0$ to model a rivet ensuring a strong clamping of the sheets by large rigid heads.

For a real lap joint, relative displacements of the mating sheets along the faying surface are possible due to flattening of the fastener shank by the bearing action of the sheets. With Barrois' model, however, the shank is assumed to remain of a circular cross section. Therefore, although the slope continuity at the sheet interface is retained, which implies $(dy/dx)_{x=a} = (dy/dx)_{x_1=b}$, different deflections of the fastener in either sheet occur, i.e. $y_a \neq y_{1b}$ (see Fig. 5.6). Consequently, the relative displacement of the sheets equals

$$\delta = y_a + y_{1b} \tag{5.8}$$

and the rivet flexibility can be computed as

$$f = \frac{\delta}{P} = \frac{y_a + y_{1b}}{P}. \tag{5.9}$$

Figure 5.7 presents the Barrois solution for equal thicknesses of both sheets ($a = b = t$) in the case of the same fastener and sheet material ($E_f = E$), and for the perfectly rigid fastener ($E_f = \infty$). The non-dimensional coordinate K_q of the vertical axis equals

$$K_q = y_t \frac{kt}{P} \tag{5.10}$$

where y_t is the rivet deflection at the interface and k is defined by Eq. 5.7.

Because the mean bearing load per unit length $q_m = P/t$ and, from Eq. 5.6, the maximum bearing unit load at the interface $q_{max} = ky_t$, K_q can be alternatively expressed as

$$K_q = q_{max}/q_m, \tag{5.11}$$

thereby it can be considered as a measure of the bearing pressure non-uniformity caused by the lap joint eccentricity.

The immediate conclusion from Fig. 5.7 is that, with technologically usual *t/d*-values ranging from 0.4 to 0.6, K_q is dramatically increased if a manufacturing defect or a loosening in service annihilates or makes negligible the holding effect of the rivet heads. However, very long rivets (large *t/d* ratio), even if well tightened, tend to show a flexibility close to that for the free end pins.

A merit of Barrois' approach is that it can be applied to more complicated joint configurations. However, its simplifications, such as retaining the circular shape of the rivet cross section and the "neat fit" between the rivet and hole, mirror conditions when a rivet is installed with a low squeeze force. It can be anticipated that in the case of a more severely squeezed rivet, otherwise favourable for the joint fatigue performance, larger discrepancies between the model results and the actual joint flexibility may occur because the interference fit and plastic deformation invalidate the model assumptions.

5.2.2 Experimental Determination

Typically, the fastener flexibility for a lap joint is determined based on the measurement of the specimen elongation (Δl) over a gage length (l) comprising the whole overlap region, as shown in Fig. 5.8a. If the measured Δl is expressed in terms of the contributing extensions of the sheets (Eq. 5.2) and the rivet deflections (Eq. 5.3), an additional equation can be gained for the system considered in Sect. 5.1. Under the assumption that the fastener flexibility is the same for all rows, the resulting system of $m(r-1)+1$ equations enables solving for the $m(r-1)$ unknown tractions in the sheets and, in addition, for the fastener flexibility. For an r-row lap joint of two sheets A and B, the specimen extension measured within the gauge length of $\sum_{i=0}^{r} l_i$ (for notation see Fig. 5.2 and Sect. 5.1) equals

$$\Delta l = \sum_{i=0}^{r-1} \Delta_{A,i} + \delta_r + \Delta_{B,r}, \tag{5.12}$$

which implies that the total measured joint flexibility $f_{measured}$ can be related to the rivet flexibility through the equation

$$f_{measured} = \frac{\Delta l}{P} = \frac{1}{P}\left(\sum_{i=0}^{r-1}\left(\frac{T_{A,i}l_i}{(EA)_A}\right) + T_{A,r-1}f + \frac{Pl_r}{(EA)_B}\right) \tag{5.13}$$

where f represents the average flexibility for all fastener rows, i.e. $f = \frac{1}{r}\sum_{1}^{r} f_i$.

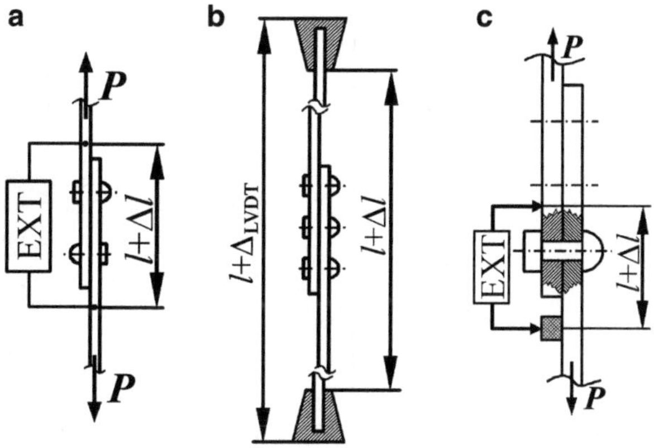

Fig. 5.8 Rivet deflection measurement methods: (**a**) according to MIL-HDBK-5H (2001); (**b**) using LVDT displacement transducer; (**c**) for outer rivet rows

It should be noted that even if fasteners in all rows are identical, the assumption that all fasteners have the same flexibility is a simplification because the proportion of T_{BR} to T_{TR} can be different for different rows, as explained later on.

The middle fastener rows of a lap joint undergo much more uniform bearing pressure than the outer rows where the maximum moment due to secondary bending occurs. Because this effect is not covered in the computation of f from measurement data, the fastener flexibility measurements should ensure, as far as possible, that the measured Δl is not influenced by sheet bending. This requirement is approximately fulfilled if, in agreement with guides set out in MIL-HDBK-5H (2001), the gauge is positioned as in Fig. 5.8a. For longer overlaps, however, such a gauge positioning is not feasible when standard extensometers are used because of their too short gauge length. In that case, the specimen extension between the clamping edges (Δl) and then the corresponding f_{measured}-value can be derived from records of the machine actuator using the LVDT (Linear Variable Differential Transformer) transducer gauge, Fig. 5.8b, provided that the component resulting from the compliance of machine elements is subtracted from the measured displacement value (Δl_{LVDT}).

In order to measure flexibility for the end fastener rows, the extensometer can be positioned as shown in Fig. 5.8c. At that short gauge length, however, the sheet curvature due to secondary bending should be accounted for because the difference between the displacements measured at the outer surface and those occurring at the faying surface can be of the same order of magnitude as the rivet deflection. Figure 5.9a and b show the principle of optical measurements of rivet flexibility which are not affected by sheet bending. The technique, also called an edge method and illustrated in Fig. 5.9a, enables determination of the rivet deflection in the end row from the measured relative displacement between the mating sheets at the overlap edge (Δy) as

$$\delta = \Delta y - \Delta l_1 = (y_{\max} - y_{\min}) - \frac{\Delta P l_1}{EA} \tag{5.14}$$

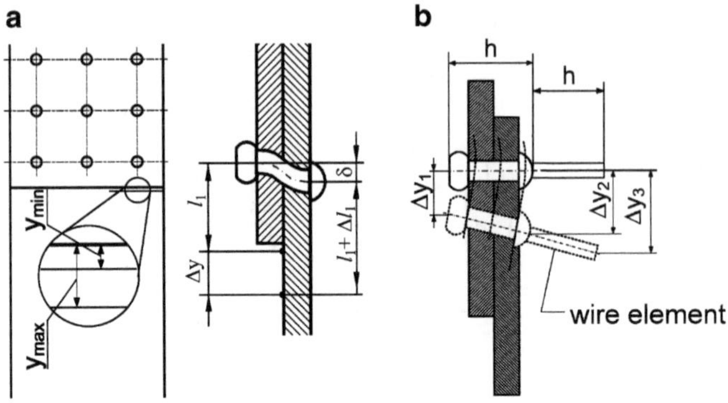

Fig. 5.9 Optical measurements of rivet flexibility: (**a**) edge method; (**b**) protruding wire method (Skorupa et al. 2010a)

where the distances y_{max} and y_{min} correspond to the maximum and minimum load of the fatigue cycle and ΔP is the cyclic load range.

The method presented in Fig. 5.9b is applicable to rivets in both the outer and inner rows. Sheet bending is compensated by using a piece of wire bonded to the sheet surface near the rivet head. Rivet flexibility is determined from measurements of the rivet head tip displacements on both sides of the overlap during the loading cycle (Δy_1 and Δy_2) and the displacement of the bonded wire tip (Δy_3). The difference between Δy_1 and Δy_2 stems from rivet deflection and rivet rotation associated with sheet bending. It is assumed that rotation angles of the rivet and of the wire element are equal. Then, if the wire length protruding above the rivet head is the same as the total length of the installed rivet (h), the rivet deflection can be obtained as

$$\delta = 2\Delta y_2 - \Delta y_1 - \Delta y_3. \qquad (5.15)$$

Figure 5.10 illustrates the determination of $f_{measured}$ (cf. Eq. 5.13) for a riveted joint prescribed by the MIL-HDBK-5H (2001). Here, $f_{measured}$ is derived as the slope of the linear part of the P-Δl hysteresis loop loading branch (J-G). However, the literature data (Jarfall 1986; Huth 1983; Müller 1995; Morris 2004) indicate that the hysteresis loop may not show a linear section or may show more than one linear part. Müller (1995) has observed that the non-linear P-Δl behaviour is more pronounced for rivets installed with a higher squeeze force and attributes it to variations of the rivet/hole interference during the loading cycle. Choosing the P-Δl diagram portion representative for the rivet flexibility involves subjective decisions which can heavily influence the results. A further complication is dependence of the P-Δl diagram shape on the type of loading. The corresponding results are, however, contradictory. Huth (1983, 1987) noted that for a quasi-static loading on a riveted or bolted specimen involving a series of downward-upward excursions stepped to reach a higher force each time, the linear slope of the corresponding P-Δl loops indicated a lower $f_{measured}$-value earlier in the test than later in the test. This implies that a

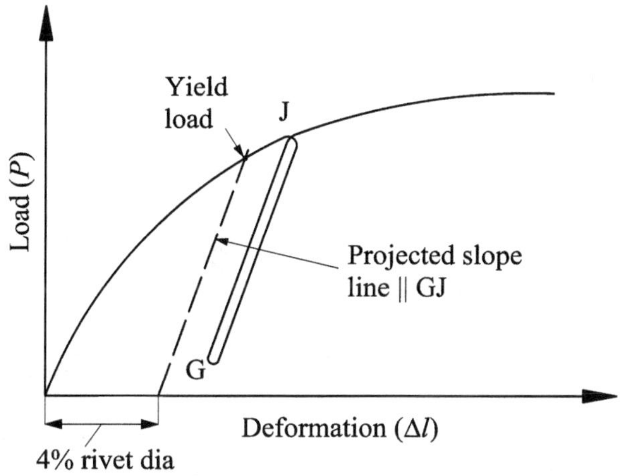

Fig. 5.10 Determination of specimen flexibility from load-deformation diagram (MIL-HDBK-5H 2001)

higher rivet flexibility is measured after a number of quasi-static loops than after a single loop. Going through a number of loops in the tests by Huth can be considered as a particular case of VA fatigue loading. Results by Jarfall (1986) indicated, however, that from the very beginning of a fatigue test, be it under a CA fatigue loading or under a VA programme or random (flight simulation) loading, the joint deformation was systematically decreasing with the number of cycles. Moreover, his measurement data recorded under a flight simulation loading revealed that the deformation of a riveted single strap butt joint represented by variations of the gap between the sheets (Δg) showed a strongly non-linear relation to the applied load and that for the largest load variations of the load history f_{measured} was about 40% higher than for the smallest ones, Fig. 5.11a. Jarfall further observed that f_{measured} associated with a given load variation of this particular load sequence decreased during the life by not more than 15%, which was much less than under CA loading on the same type of specimen when a 37% decrease of the measured flexibility occurred in 10^5 cycles, as depicted in Fig. 5.11b.

At least some of the above trends can be qualitatively understood if it is assumed that the fastener deflection is controlled by the bearing force rather than the total transfer force. Load transfer measurements for Hi-Lok bolts reviewed in Sect. 5.3 demonstrate that while the load transfer ratio (T_{TR}/P) remains essentially invariable during the fatigue loading, the friction contribution can significantly increase with the cycle number, causing the bearing component to decrease. This explains the apparent decrease in joint flexibility, and hence in fastener flexibility, observed under fatigue loading conditions. Because the T_{FR} force does not depend on the applied load level, the T_{BR}/T_{TR}-value, and hence the measured fastener flexibility, become higher with increasing P.

The literature evidence on the effect of the rivet squeeze force on rivet flexibility in lap joints is not fully consistent. For Glare specimens with countersunk 4 mm diameter AD rivets, Müller (1995) observed f to decrease considerably with increasing F_{sq} only up to a certain level corresponding to the D/D_o ratio of about 1.5,

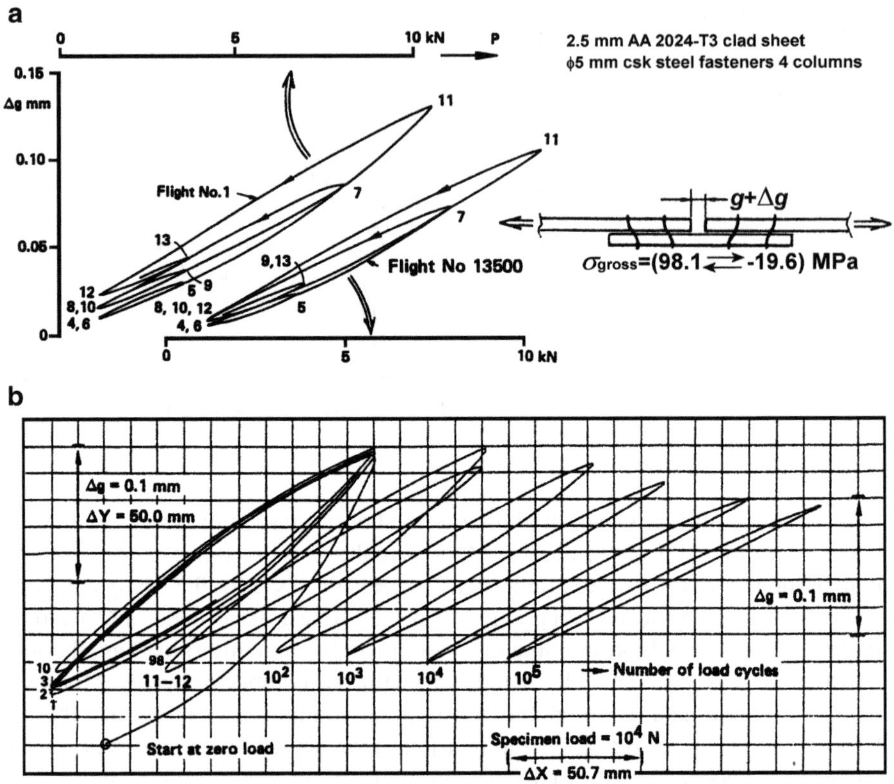

Fig. 5.11 Load-deformation records: (**a**) flight simulation loading; (**b**) CA loading (Jarfall 1986)

above which *f* remained constant. Müller explained these results by a rapid increase in rivet-hole interference up to this limiting F_{sq}-value and a more gradual increase in interference above this value. For the same type of 4.8 mm diameter rivet and Al 2024-T3 sheet material Wang (1998) noted, however, a very moderate decrease in *f* with increasing F_{sq} only up to $D/D_o = 1.32$, while a still higher squeeze force caused an increase in *f* attributed by this author to a possible damage of the interface.

Evidently, due to the lack of a standard procedure for fastener flexibility measurements, experimental results for a given riveted specimen obtained in various laboratories may be quite different depending on the measurement method, the loading type and the recorded data processing. For example, Morris (2004) has observed that while the total flexibility of a multi-row riveted joint becomes lower with more rows of rivets, the average flexibility of the rivet increases. Huth's (1987) measurements for a lap joint with four fastener rows indicate, however, the same fastener flexibility as for a double-row joint. It is, therefore, not surprising that several semi-empirical formulae, quoted elsewhere (see e.g. Jarfall 1986; Müller 1995; Morris 2004), which have been developed to calculate fastener flexibility, show a large spread in absolute values.

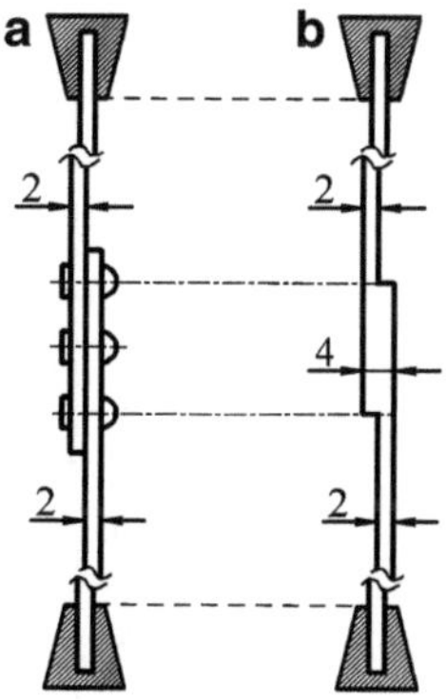

Fig. 5.12 Riveted (**a**) and dummy (**b**) specimen used by Skorupa et al. (2010a)

Jarfall (1986) demonstrates a large overestimation of his experimental results on f by several formulae available in the literature. Morris (2004) reports, however, an opposite behaviour, namely that almost all formulae by other authors, including those considered in Jarfall's work, nearly halve the f-values determined from his own measurements performed according to Fig. 5.8b. The reasons for the overestimation of rivet flexibility by the Morris formula are considered later on in this section.

The present authors with co-workers (Skorupa et al. 2010a) compared rivet flexibility results derived using the conventional techniques sketched in Fig. 5.8b and c and the optical methods from Fig. 5.9a and b. The measurement data were acquired during CA fatigue loading ($S_{min} = 12$ MPa, $S_{max} = 120$ MPa) on simple lap joint specimens of 2 mm thick D16 Alclad sheets connected with three rows of 5 mm diameter, round head PA24 (AD) rivets, Fig. 5.12a. For either of two specimens tested, the rivets were squeezed with a different force to obtain the D/D_o ratio of 1.3 and 1.5. The axial forces in the sheets needed to compute the sheet elongations and the rivet flexibility using Eq. 5.13 were determined from strain gauge measurements, as detailed in the next section.

Only with the method employing the bonded wire element (Fig. 5.9b) could the rivet deflection be directly derived. When the other techniques were used, two different approaches were applied to extract the rivet deflection from the measured data. One of these, referred to as analytical compensation, involved subtracting from the measured extension appropriate sheet elongations computed from Hooke's law (cf. Eq. 5.2). In the case of the LVDT transducer records, this analytical compensation can only eliminate the effect of sheet extensions between the specimen clamping edges (Δl), but it is not capable of eliminating the deformation of fatigue machine components included in the Δ_{LVDT} data. With the other approach referred to as experimental compensation, the rivet deflection was obtained as the difference between the elongations acquired for the riveted specimen and those measured for the monolithic dummy specimen shown in Fig. 5.12b. It was believed that the experimental compensation should enable elimination of the machine parts' deformation from the LVDT transducer records.

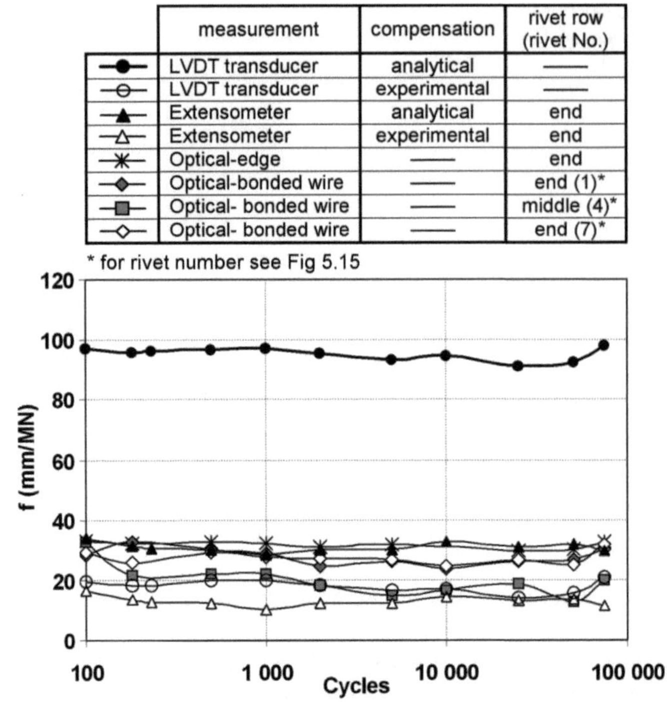

Fig. 5.13 Results on rivet flexibility from several measurement techniques (Skorupa et al. 2010a)

Comparisons between some results on the rivet flexibility parameter derived for the $D/D_o = 1.3$ specimen are presented in Fig. 5.13. Compared to the f-values shown in Fig. 5.13, those measured for the $D/D_o = 1.5$ specimen (higher squeeze force) were consistently lower. The measurements indicated a good repeatability of the results obtained for various rivets in a given row and symmetry of the flexibility values for both end rows. The latter observation is illustrated in Fig. 5.13 by a close agreement between the results for rivets 1 and 7 located in different end rows. Both optical methods and the extensometer measurements coupled with the analytical compensation yield similar results for rivets in the end rows. The measurements indicate the f-values consistently lower for rivets in the middle row than for the outer rows, as evidenced by the results for rivet 4 (middle row) and rivets 1 and 7 (end rows). The latter observation can be explained by the load transfer measurement results (see next section) showing a lower load transfer force in the middle row compared to the outer rows. This implies a lower bearing force percentage in the total load transfer for the middle row as the friction force should be approximately the same for all rows.

The effect of fatigue loading on rivet flexibility was found to depend on the squeeze force level. In the case of the $D/D_o = 1.3$ specimen an insignificant decrease of flexibility during fatigue loading was observed only for rivets in the middle row, Fig. 5.13. For the $D/D_o = 1.5$ specimen a moderate decrease in f occurred also for

Table 5.1 Rivet flexibility values according to various sources and corresponding computed load transfer ratios for a lap joint with three rivet rows

		f_{end} (mm/MN)	f_{middle} (mm/MN)	$T_{TR,outer}/P$	$T_{TR,middle}/P$
Skorupa et al. (2010a)[a]	$D/D_o = 1.3$	26.7	16.4	0.293	0.414
	$D/D_o = 1.5$	18.1	12.8	0.314	0.372
	$D/D_o = 1.3$	23.3		0.344	0.311
	$D/D_o = 1.5$	15.5		0.349	0.301
Huth (1987)	–	33.4		0.341	0.318
Swift (1971)	–	26.1		0.343	0.313
Morris (2004)	$D/D_o = 1.3$	79.4		0.337	0.327
	$D/D_o = 1.5$	75.6		0.337	0.326
Barrois (1978)	Pinned heads	57		0.338	0.324
	Clamped heads	17.6		0.348	0.305

[a]Averaged within initial 75 kcycles

the outer rows. After 100 kcycles the rivet flexibility for the latter specimen became reduced by 20%, which was less than observed in the experiments by Jarfall (1986) cited above (cf. Fig. 5.11b).

While the extensometer measurements and both optical methods considered in Fig. 5.13 provide flexibility for a specific rivet row, the results derived from the LVDT transducer represent the average rivet flexibility of the joint. Figure 5.13 demonstrates that compared to the other measurement techniques, the LVDT transducer measurements, if coupled with the analytical compensation which does not cover deformation of the machine parts, yield a more than threefold overestimate of rivet flexibility. These results offer an explanation for overly high f-values according to the Morris (2004) formula based on the same measurement technique. On the other hand, the data in Fig. 5.13 imply that experimental compensation leads to an underestimate of flexibility because the average results on f obtained from the LVDT transducer records and, especially, the extensometer measurement results for the outer rows are even below the flexibility measured for the middle row. Evidently the deformation of the dummy specimen cannot properly represent the deformation of the sheets in the riveted specimen. One possible reason can be insignificant differences in mechanical properties of the riveted specimen and the dummy specimen for which the sheets cannot come from the same material batch because of their different thicknesses. In Table 5.1, the rivet flexibility averaged within the initial 75 kcycles of the fatigue loading measured by Skorupa et al. (2010a) for the outer and middle row rivets using the optical method from Fig. 5.9b as well as the average f-values for the joint resulting from these measurements are compared with the f-values according to three semi-empirical equations reported in the literature and with the Barrois (1978) solution. It is seen that the results according to the formulae by Swift and Huth and the averaged f-values for the lower squeeze force ($D/D_o = 1.3$) according to Skorupa et al. (2010a) are encompassed by two extreme boundary conditions considered in the Barrois solution. Presented in the last two columns of Table 5.1, load transfer ratios computed for the outer and middle rivet row, as described in Sect. 5.1, demonstrate a surprisingly small dependence on the f-value. For example, an over threefold increase in f causes the load transferred by the outer rivet rows to decrease by less than 3% and that transmitted by the middle row to increase by about 6%. Computation results not included in Table 5.1 indicate that applying Eq. 5.5 rather than Eq. 5.2 for the sheet elongation has a negligible effect on the resulting solution for $T_{j,i}$. The derivation of the semi-empirical formulae considered in Table 5.1 implies that they represent the rivet flexibility averaged over the joint. It can be seen in Table 5.1 that if the same flexibility is assumed for all rivet rows, the computed load transmission by the middle row is always lower than by the outer row. The reverse is true, namely the middle row transmits more load than the outer row, if rather than the averaged f-values, the actual flexibilities for either rivet row measured by Skorupa et al. (2010a) are accounted for in the computation procedure. The latter results on load transfer cannot be, however, considered reliable because the phenomena responsible for the differences in the flexibility of the middle and outer rows are not covered in the load transfer computations.

As the semi-empirical equations for f are obtained from experiments on simple geometries, they are not capable of accounting for cases when the overlap consists of several sheets. An example of a pronounced effect of the joint configuration can be results produced by Jarfall (1986) showing that the fastener deflections for the three overlap configurations from Fig. 5.14 are different, though the load transfer by the fastener is the same for all cases because the third layer added in Fig. 5.14b and c to a standard two-layer configuration from Fig. 5.14a is non-load-carrying.

5.3 Measurement Results on Load Transmission

The internal axial force in a sheet of a lap joint can be determined by integrating the stress field that is calculated from measurements of the longitudinal strains in the cross section of the sheet. Most commonly, strain gauges are used with this purpose. For joints with eccentricities, like lap joints, the measured axial strains indicate a combined effect of axial loads and secondary bending. If the gauge installations on the outer surface of the sheet are duplicated on the faying surface of this sheet, the strains associated with the axial load can be obtained as the average of the strains measured at the same location on either side of the sheet. An example of overlap instrumentation with strain gauges in order to determine the axial loads in two sections of a sheet adjacent to the rivet manufactured heads is presented in Fig. 5.15a. Bonding the gauges at the faying surface requires that a small recess for the gauge should be machined in the mating sheet, as shown in Fig. 5.15a for gauge B2. Also a groove must be created in this sheet in order to lead out the gauge contact wires. The need for machining of the opposite plate implies that this plate must be metallic because cutting the carbon fibres of a composite material involves a risk of introducing delaminations. Strain measurements on the outer sheet surface that, except for areas hidden under the rivet heads, is accessible for direct observations can also be

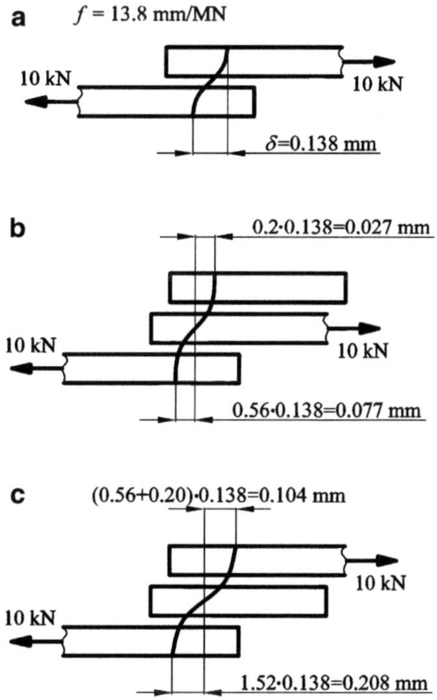

Fig. 5.14 Fastener deflection for a two-layer (**a**) and three-layer (**b**, **c**) overlap (Jarfall 1986)

carried out using more advanced methods, for example digital imaging correlation or a thermo-elastic camera, which enable a continuous mapping of the strains.

Measurement results reported for lap joint specimens with two rows of Hi-Lok bolts reveal a highly non-uniform axial strain distribution in the cross section midway between the bolt rows with the minima (close to zero) at the bolt columns (Huth 1987; Palmberg 2002; Starikov 2002). The minima are due to local compression caused by the bearing pressure on the fasteners. Starikov (2002) found the strain distribution to be hardly affected by the amount of clamping which, for this type of fastener, can be controlled by the initial pretension in the bolt attained by applying an appropriate tightening torque. As observed by Terada (1995) using a thermo-elastic analyzer, riveted joints show a different response, since increasing the squeeze force level smoothes away the stress distribution. Terada presumes that the presence of deep minima observed for a joint shown in Fig. 5.16 when rivets in the critical top row are tightened with a 67% of the nominal squeeze stress (case B and C in Fig. 5.16) proves that a large portion of the load is transferred by the bearing pressure. On the contrary, the smoothed stress distribution in that row corresponding to a 100% squeeze stress, case A in Fig. 5.16, is evidence that for well-tightened rivets a large portion of the load is transmitted by friction. Terada further noted that whilst for the 100% squeeze stress the T_{TR} load transmitted by the top rivet row equalled about 56% of the applied load, it increased to 70% of P if the squeeze stress was reduced to the level of 67%.

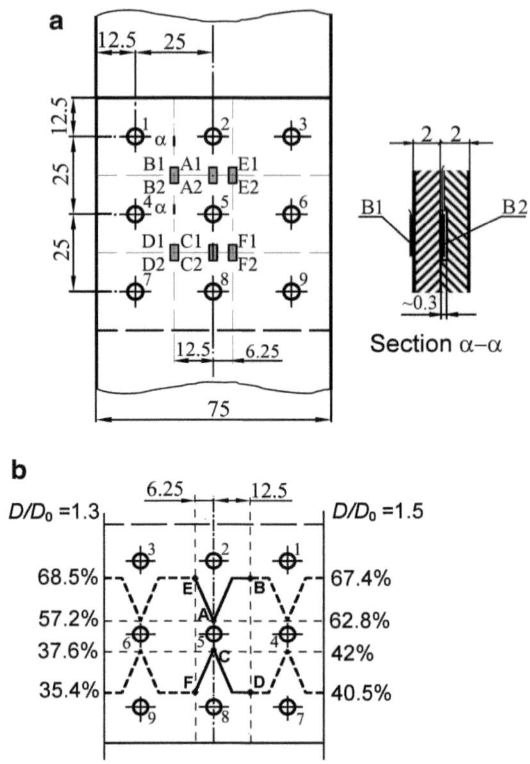

Fig. 5.15 Measurements of stresses contributed by axial forces in the sheets: (**a**) overlap instrumentation; (**b**) schematization of stress distributions (Skorupa et al. 2010a)

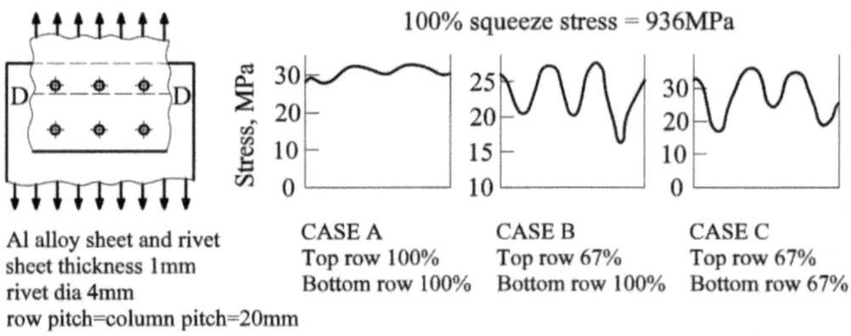

Fig. 5.16 Stress distribution in cross section D-D of a riveted lap joint for various squeezing conditions (Terada 1995)

Though the increased friction contribution to the load transfer at a higher squeeze stress is incontrovertible, it seems that the trend exhibited in Fig. 5.16 is mainly due to a more severe interference fit between the rivet shank and the hole at the higher squeeze stress. For bolted joints, a higher clamping force is paired with a more severe prestraining of the bolt, which leads to a reduction of the initial interference fit or to an increase in the clearance fit between the bolt and hole. Hence, though for the bolted joints a higher clamping force leads to higher load transmission by the

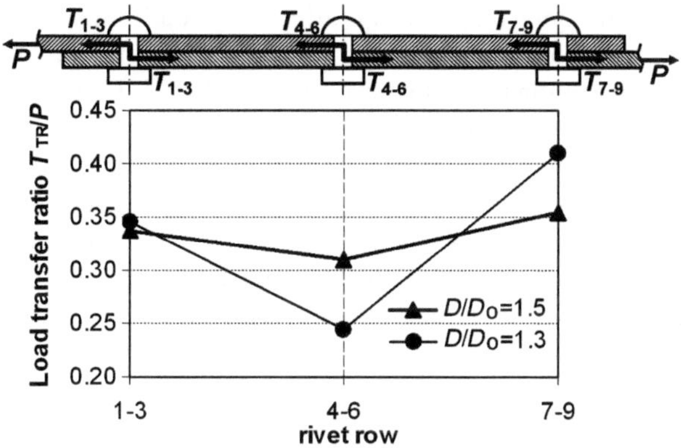

Fig. 5.17 Measured load transfer ratios for a lap joint (Skorupa et al. 2010a)

friction forces (see Sect. 5.4), the stress distribution remains unaffected, as reported by Starikov (2002).

Skorupa et al. (2010a) observed that in either cross section considered in their measurements (cf. Fig. 5.15a) the gauge located midway between the rivet columns indicated nearly the same stress as the gauge located at a distance of 6.25 mm from the column, i.e. the stress measured at location B equalled that measured at location E and the stress at location D was the same as at location F. Consequently, the stress distributions associated with the axial forces could be schematized as shown in Fig. 5.15b where the measured stresses are specified as a percentage of the applied tensile stress. In agreement with Terada's results (cf. Fig. 5.16), Fig. 5.15b indicates that, compared to the D/D_o ratio of 1.3 (lower squeeze force), the stress distribution corresponding to $D/D_o = 1.5$ (higher squeeze force) is more levelled off. This is evidenced by, in the case of $D/D_o = 1.5$, lower differences between the stress at the rivet column and the stress midway between the columns observed in either cross section. Integrating the measured stress field from Fig. 5.15b yielded the distribution of the load transfer ratio through the joint shown in Fig. 5.17. Here, the most striking feature is non-symmetrical load transmission by the end rivet rows. For both the $D/D_o = 1.3$ specimen and the $D/D_o = 1.5$ specimen, load $T_{1\text{-}3}$ transferred from the sheet adjacent to the rivet manufactured head to the sheet adjacent to the driven head is lower than load $T_{7\text{-}9}$ transmitted at the other end row, i.e. from the sheet next to the driven head to the sheet under the manufactured head. The difference in load transfer distribution by the outer rows is more pronounced for the $D/D_o = 1.3$ specimen than for the $D/D_o = 1.5$ specimen for which load transmission through all three rows is more homogenous. The non-symmetrical load transfer is not reflected by the rivet flexibility measurements which for either specimen yield identical results on f for both outer rivet rows, see Sect. 5.2.2. Presumably, the differences in load transmission by the outer rows seen in Fig. 5.17 are associated with large differences in hole expansion within the sheet under the rivet driven head and the sheet next to the manufactured head revealed in experimental and numerical studies (cf. Sect. 3.3.5.3). Note in Table 5.1 that the computed load transfer ratios show an opposite behaviour from the observed results, namely a higher load transfer by the middle row than by the outer rows. It can be concluded that because the rivet flexibility is not the only factor which affects the load transmitted by the rivet and because of simplifications involved in the displacement compatibility equations (Eq. 5.1), the computation procedure presented in Sect. 5.1 may provide a quantitatively and even qualitatively incorrect solution for the internal forces in the sheets.

In spite of the latter observation, it is believed that reliable measurement results on the rivet deflections as well as any other experimental data on joint deformation behaviour, such as the Δy_1, Δy_2 and Δy_3 quantities acquired when the technique from Fig. 5.9b is applied, can still be most useful for verification of numerical stress analyses of lap joints.

5.4 Frictional Forces

As said earlier, it has been recognized that friction at the faying surface accounts for some of the load transfer through the joint; hence, the actual load transmitted by the fasteners (T_{BR}) is lower than the total transfer (T_{TR}), cf. Fig. 5.1. For riveted joints, strain gauge measurements offer no means of extracting the T_{BR} force actually transmitted by the rivet as only T_{TR} can be determined. The determination of T_{BR} is, however, feasible when the bolts are used. In an attempt to measure T_{BR}, several researchers replaced ordinary Hi-Lok fasteners by instrumented bolts equipped with an external or internal rosette strain gauge to record shear strains in the plane of the fatigue surface (e.g. Jarfall 1986; Palmberg 2002; Starikov 2002). The part of the load transmitted by friction (T_{FR}) was obtained as the difference $T_{TR} - T_{BR}$. As could be expected, the percentage of the applied load transferred by friction increased with the amount of clamping. For simple lap joints in an Al alloy with two fastener rows under a quasi-static loading involving several CA load cycles, Starikov (2002) observed the stabilized T_{FR} to attain 16–40% of P, depending on the bolt initial pretension. Under VA fatigue loading (standard flight simulation FALSTAFF sequence) on specimens with a high degree of bolt pretension, T_{FR} increased with the cycle number and became dominant already after the first block of flights. Initially, T_{FR} at P_{max} was higher than T_{FR} at P_{min}, in accordance with observations by Jarfall (1986) for a double shear joint with three rows of instrumented fasteners subjected to CA loading. After a certain number of blocks, however, Starikov (2002) noted that the same portion of load was transferred by the friction forces at both the maximum and minimum load peaks. At this stage, up to 80–90% of P was carried by friction.

Experiments to measure frictional forces in a riveted joint have been carried out by Hartman (1961). Specimens assembled from Al 2021-T3 plates using a single row of Al 2117 snap rivets, Fig. 5.18a, were tested up to various percentages of the fatigue life (N_f) under fluctuating tension with a load amplitude corresponding to N_f of 3×10^6 cycles. Afterwards, the specimens were sawn into strips (Fig. 5.18a) and subjected to static three-point bending using the setup shown in Fig. 5.18b. The permanent rotation of the sheets, revealed by a sudden change of slope of the load-displacement diagram recorded during the test, indicated that the interfacial friction forces have been overcome. Hartman observed only the middle sheet to

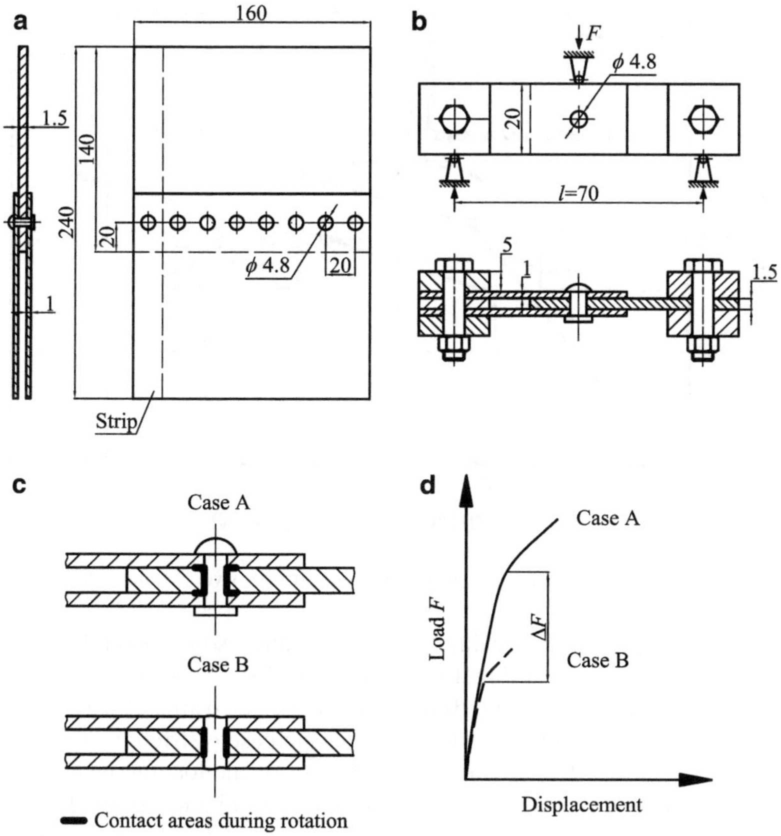

Fig. 5.18 Experimental determination of friction force by Hartman (1961): (**a**) specimen geometry; (**b**) setup for the rotation test; (**c**) contact areas during sheet rotation; (**d**) force-displacement diagram

rotate, whereas the rivet did not rotate. Consequently, only the areas marked with a thick line in Fig. 5.18c, case A, contributed to friction. The three-point bending test was repeated after milling both heads of the rivet to eliminate the clamping force and, thus, to extract the contribution of friction between the rivet shank and the hole according to case B in Fig. 5.18c. The difference ΔF between the F levels corresponding to the knees of the diagrams for case A and B was assumed to be a measure of friction between the mating surfaces of the sheets, Fig. 5.18d. Assuming a uniform shear stress distribution in the annular contact region restricted to a zone under the rivet head, the possible load transmission by friction per rivet has been determined as

$$T_{FR,rivet} = \frac{3l(D^2 - D_o^2)}{4(D^3 - D_o^3)} \Delta F \tag{5.16}$$

where l is the span between the supports (Fig. 5.18b).

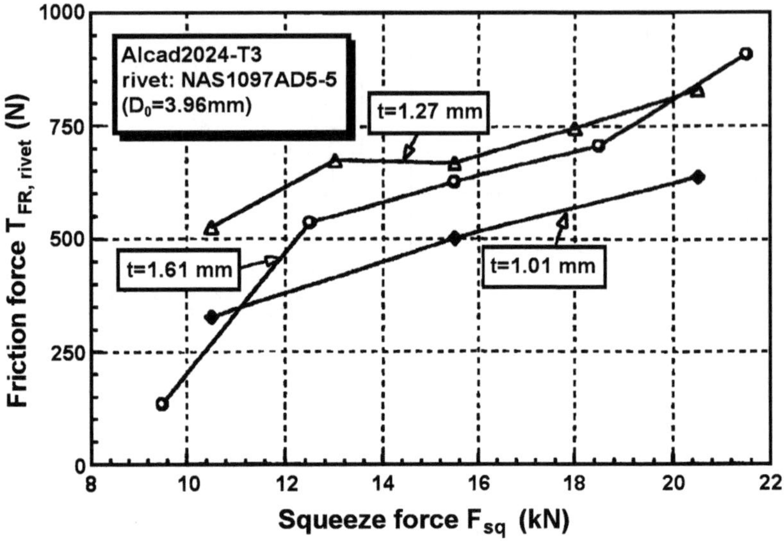

Fig. 5.19 Effect of squeeze force on measured friction force per rivet (Müller 1995)

Though the $T_{\text{FR,rivet}}$-values exhibited a large scatter, some general trends could be clearly observed. Over a threefold increase in $T_{\text{FR, rivet}}$ compared to the value determined for the as-manufactured condition was detected after 25% of N_{f}, whereas the subsequent cycling did not cause a further increase of friction. The stabilized $T_{\text{FR,rivet}}$ was equal to 2.1 kN, which implied that, for the loading near to the joint fatigue limit, 80% of the maximum applied load could be transmitted by friction. It is worthwhile digressing at this point to note that a very high squeeze force must have been applied during the riveting in order to obtain an extra large D/D_0 ratio of about 1.88 reported in Hartman's work.

Investigation into the influence of the squeeze force on friction between the mating sheets for lap joints with flush rivets was also carried out by Müller (1995) who adopted the same methodology and assumptions as Hartman did to compute the friction force. The measurements were made only on specimens in the as-manufactured condition. As could be expected, a higher squeeze force led to an increased friction force, Fig. 5.19. The effect of the sheet thickness on T_{FR} revealed in Fig. 5.19 is not fully systematic and difficult to be readily explained. Müller also carried out a simple test to determine the coefficient of friction. Three partly overlapping sheets were clamped together with an instrumented clamp which could be adjusted to set the clamping stress σ_{n}, as shown in the top of Fig. 5.20. The middle strip was pulled out of the clamping by the outer two sheets. Assuming that the shear stress at the mating surface (τ_{FR}) is homogenously distributed, the coefficient of friction can then be determined as

$$\mu = \frac{\tau_{FR}}{\sigma_n} = \frac{P}{2F_{cl}} \tag{5.17}$$

where F_{cl} is the clamping force.

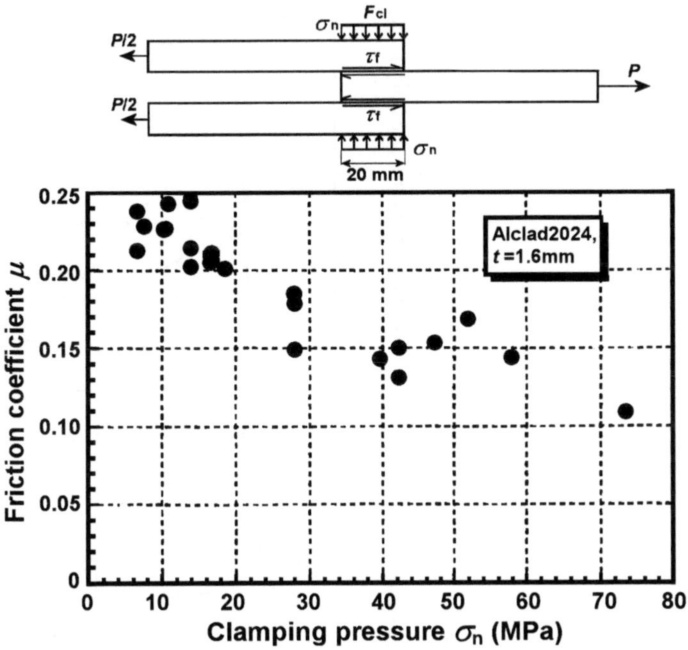

Fig. 5.20 Measurements of the coefficient of friction by Müller (1995)

Müller's measurements have indicated that, against the Coulomb dry friction theory, the friction coefficient for the Al 2024 material decreases with an increase in the clamping pressure, as shown in Fig. 5.20. The μ-values at the σ_n stress of approximately 20% of S_y were 0.12 and 0.15 for the clad and bare mating sheet surfaces respectively. The latter value is the same as reported by Szolwinski (1998), cf. Sect. 7.2.1.

Figure 5.19 implies the T_{FR}/F_{sq} ratio between 0.013 and 0.05, the values corresponding to $t = 1.01$ mm being close to 0.03. The ratio of the clamping force to the squeeze force F_{cl}/F_{sq} can then be estimated utilizing Müller's results on the coefficient of friction as:

$$\frac{F_{cl}}{F_{sq}} = \frac{1}{\mu}\frac{T_{FR}}{F_{sq}}. \tag{5.18}$$

Taking from Fig. 5.20 μ equal to 0.23 and 0.12 for the lowest and highest clamping pressure respectively, Eq. 5.18 and Fig. 5.19 yield F_{cl}/F_{sq}-values ranging from 0.13 to 0.25 for the 1.01 mm thick sheet and from 0.19 to 0.33 for the 1.6 mm thick sheet.

Because the residual pressure between mating sheets after releasing the riveting force cannot be measured, FE analyses were pursued to determine F_{cl}. The available literature evidence indicates that the pressure at the mating surface under the maximum squeeze force considerably exceeds the residual clamping pressure

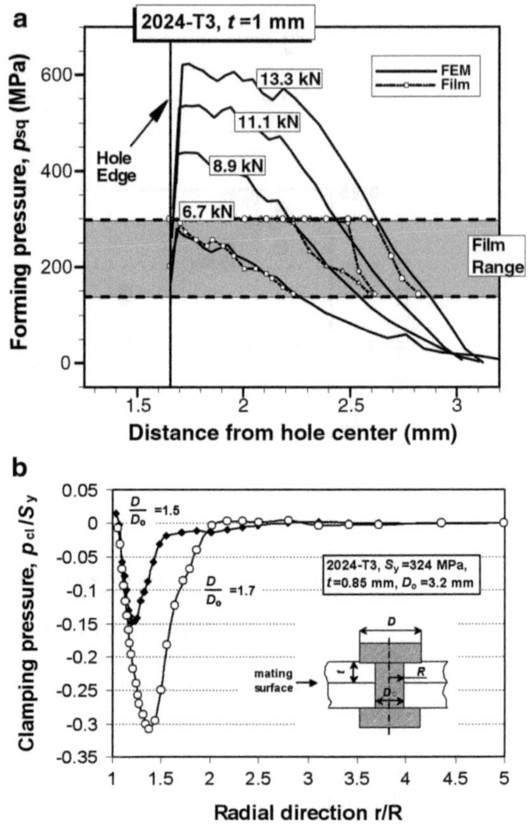

Fig. 5.21 Contact pressure at the mating surface according to FE analyses: (**a**) during rivet forming at several squeeze force values (Brown and Straznicky 2009); (**b**) after releasing the squeeze force (Müller 1995)

occurring after releasing the squeeze force. This is evident from comparisons between distributions of the maximum forming pressure reported by Brown and Straznicky (2009) at several squeeze force values, Fig. 5.21a, and FE results on the residual clamping pressure by Müller (1995), Fig. 5.21b. Also shown in Fig. 5.21a are measurement results obtained using a pressure sensitive film inserted between the sheets prior to their assembling. In Müller's simulations of the riveting process the sliding of surfaces in contact occurred without friction, whereas the friction coefficient of 0.178 was used in the analyses by Brown and Straznicky. It follows from Rans's (2007) work, referred to by Brown and Straznicky (2009), that the F_{sq}-values of 8.9 and 13.3 kN in Fig. 5.21a correspond to $D/D_o = 1.5$ and 1.7 respectively. Hence, almost direct comparisons can be made between these results and the clamping pressure distributions presented in Fig. 5.21b, even though somewhat different sheet thicknesses have been assumed in either study. The maximum clamping pressure values in Fig. 5.21b are several times lower than the forming pressure values in Fig. 5.21a. It is also seen in Fig. 5.21a that the distance from the hole centre at which the forming pressure drops to zero significantly exceeds the driven head radius ($D/2$), while Fig. 5.21b indicates that the clamping pressure occurs within the area beneath the rivet head.

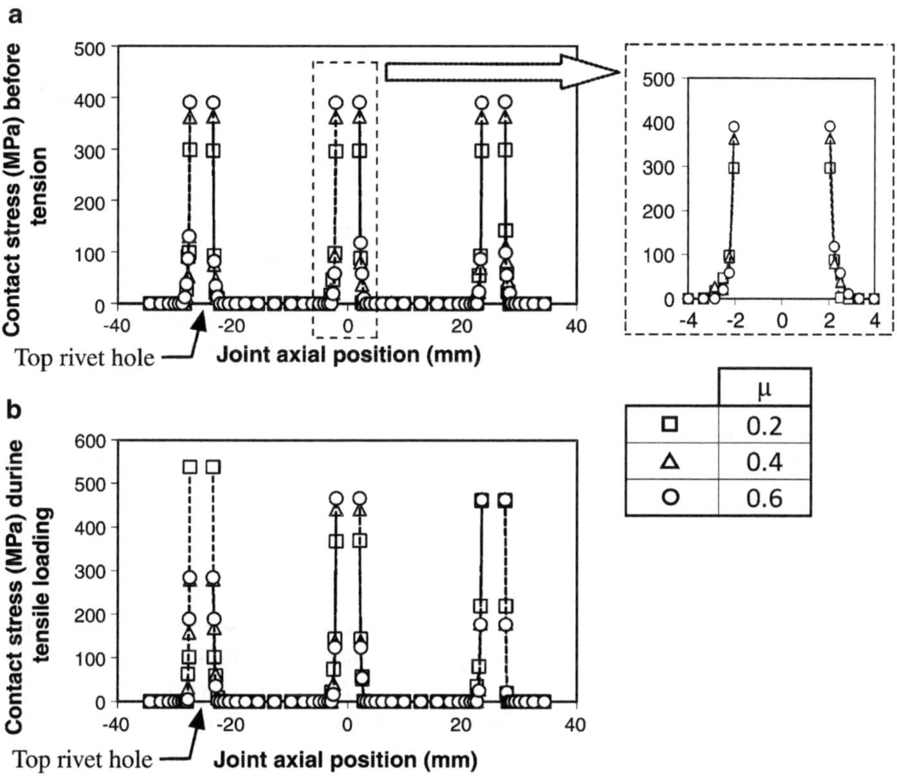

Fig. 5.22 Effect of friction coefficient on contact stress distribution on the faying surface for the joint from Fig. 3.73a: (**a**) after releasing squeeze force; (**b**) under 98.6 MPa applied tensile stress (Li et al. 2011).

More elaborated FE simulations of the riveting process carried out by Li et al. (2011) and addressed earlier in Sect. 3.3.5.4 provide a clamping pressure distribution very different from that in Fig. 5.21b. The joint from Fig. 3.73a was considered for which the squeeze force of 18 kN applied to model the rivet installation yielded the D/D_0 ratio of 1.66. Figure 5.22a indicates that after releasing the squeeze force the maximum contact stresses on the faying surface occur at the hole edge and decay very rapidly. It is further seen that the larger the friction coefficient, the higher maximum contact stresses are generated. Figure 5.22b demonstrates that when the joint is subjected to a remote tensile stress of 98.6 MPa, the maximum contact stresses become different for the individual rivet rows. Whether the applied load causes an increase or a decrease in the peak contact stress compared to the unloading condition depends on the rivet row position and the friction coefficient value. For example, Fig. 5.22b shows that for the lowest friction coefficient of 0.2 the largest contact stress occurs at the top rivet hole, while for the highest μ of 0.6 that rivet hole is a location of the lowest contact stress. Altogether, the results in Fig. 5.22b

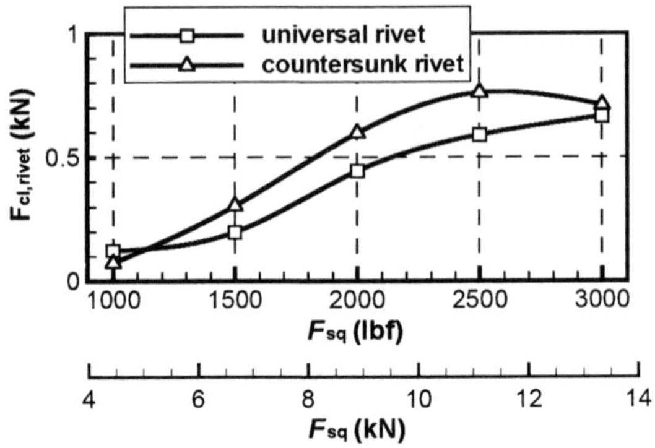

Fig. 5.23 Rivet clamping force as a function of squeeze force (Rans 2007)

evidence an extremely complex local stress response of the lap joint to the applied loading due to a multitude of contributions involved. These include residual stress state, load transmission, secondary bending, stress concentration, friction between the contact pairs, and material nonlinearities.

Figure 5.23 gives the values of the clamping force per rivet F_{cl} reported by Rans (2007) from FE simulations of joining 1 mm thick Al 2024-T3 sheets using 3.2 mm diameter universal and countersunk Al 2117-T4 rivets. It is seen in Fig. 5.23 that, in agreement with Müller's results in Fig. 5.21b, the rivet clamping action becomes enhanced by increasing the F_{sq}-value. For a given F_{sq} level, a larger F_{cl} is observed for the countersunk rivet than for the universal rivet. From Fig. 5.23, the F_{cl}/F_{sq}-values for the countersunk rivet fall between 0.02 and 0.07, which implies that at intermediate and higher squeeze force levels they are several times lower than the values deduced earlier in this section from Müller's (1995) experiments.

Among published FE analyses of the riveting process, the one according to Deng and Hutchinson (1998) brings particularly detailed information on the mechanics of the clamping force. A perhaps most remarkable finding in this work is that the average compressive stress in the rivet shank has to drop below the yield stress of the rivet material ($S_{y,r}$) prior to releasing the squeeze force if a residual clamping stress is to exist. The maximum compressive stress that develops in the shank during the riveting process reaches $-2S_{y,r}$. However, in the final stage of the riveting these stresses become substantially lowered as the highly compressed material in the shank flows out of the rivet hole into the rivet head of which the material can flow radially outward without any constraint. In agreement with other numerical studies, the residual clamping stress in the rivet shank ($\sigma_{cl,r}$) was found to be small compared to the maximum stress occurring during the rivet squeezing. It can be seen in Fig. 5.24 showing the computed effect of the rivet material yield stress ($S_{y,r}$) on $\sigma_{cl,r}/S_y$, where S_y is the sheet material yield stress, that the clamping stress in the rivet will never exceed $0.2S_y$. Moreover, Fig. 5.24 reveals that reducing $S_{y,r}$ substantially reduces the clamping stress. (We digress at this point to say that the $\sigma_{cl,r}/S_y$-value of 0.2 is the same as that according to Fig. 5.23 for $F_{sq} = 8.9$ kN which in the analyses by Rans corresponds to $D/D_o = 1.5$.) The general relevance

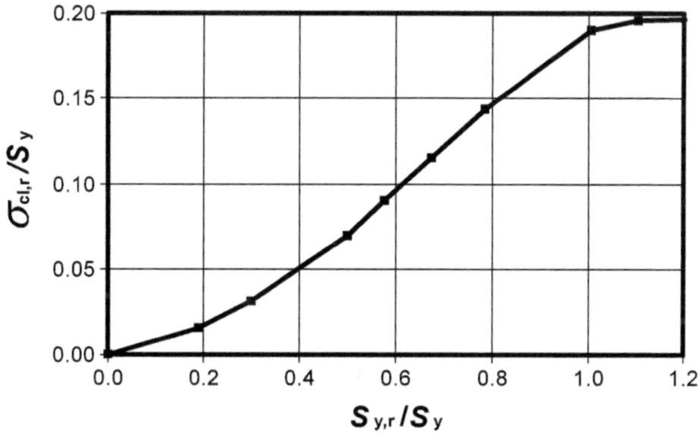

Fig. 5.24 Influence of the rivet yield stress on the clamping stress in the rivet shank according to FE analyses (Deng and Hutchinson 1998)

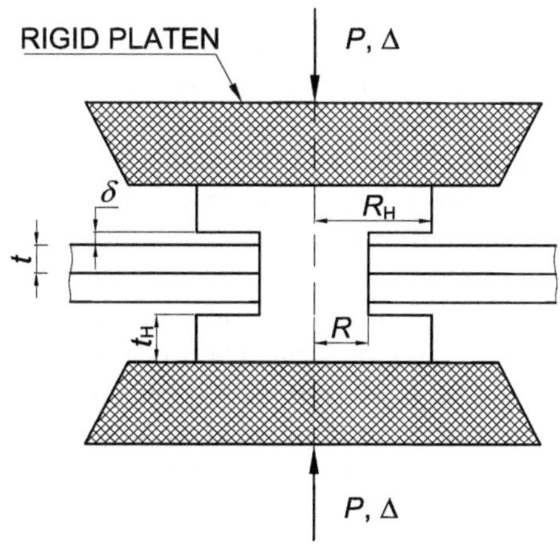

Fig. 5.25 Geometry of Deng and Hutchinson's model (Deng and Hutchinson 1998)

of Deng and Hutchinson's results is, however, limited in the quantitative sense due to simplifications involved in their modelling and illustrated in Fig. 5.25 where the geometry at the start of the riveting process simulation is shown. Unlike under real riveting conditions, the model implies an immediate contact between the already formed rivet head and the sheet once the imposed displacement Δ reaches the gap δ. The manufactured head is not modelled and any friction effects are neglected. The analyses have been performed assuming elastic-perfectly plastic rivet and sheet materials, but the authors emphasize that allowing for the material hardening does appear quantitatively important, though it does not change the mechanism underlying the development of the clamping stresses.

The results shown in Fig. 5.21b imply that the outer radius of the faying surface contact area nearly equals the rivet driven head radius, which is consistent with fractographic observations by Hartman (1961) addressed in Sect. 7.2.1. However, the contact stress distribution in Fig. 5.22 as well as FE results by Brown and Straznicky (2009) indicate that under fatigue loading conditions the contact area for the lap joint is much smaller. Their analyses were performed for a single

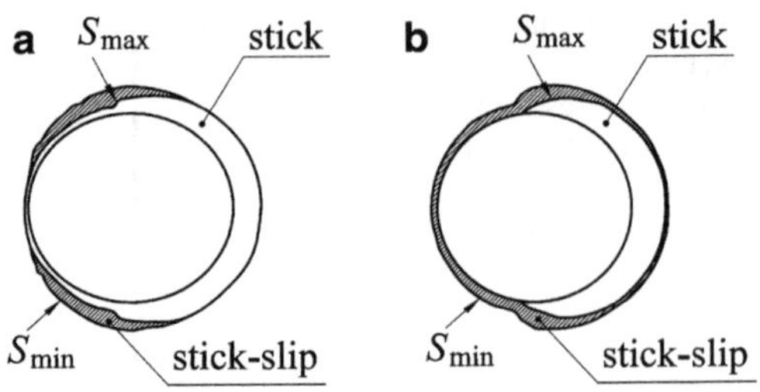

Fig. 5.26 Predicted contact areas on the faying surface of a lap joint under fatigue loading for universal (**a**) and countersunk (**b**) rivet (Brown and Straznicky 2009)

shear lap joint of Al 2024-T3, 1 mm thick sheets consisting of two rows of 3.2 mm diameter AD rivets. Several values of the squeeze force were considered (compare Fig. 5.21a). In the simulations of fatigue loading on the joint ($R = 0.1$, $S_{max} = 100$ MPa) the friction coefficient of 0.65 at the faying surface of the sheets was assumed. Predicted contact regions at the S_{max} and S_{min} stresses for the universal and countersunk rivet are plotted in Fig. 5.26a and b respectively. It can be noted that the furthermost range of the so-called stick region, which is the area common for the loading and unloading condition, is much smaller than the radius of the rivet driven head equal to 1.5 the rivet shank radius for the F_{sq}-value of 8.9 kN considered in Fig. 5.26. The same is true for other values of F_{sq}. Moreover, the predicted contact area does not generally increase with F_{sq}, and hence with the driven head diameter, though the clamping pressure does. In Fig. 5.26, the so-called stick-slip region located between the contours of the contact areas corresponding to S_{max} and S_{min} is the location where the most severe fretting damage should occur and where, therefore, microcracks should form. The analyses further revealed that on unloading the predicted contact area did not always return to the symmetrical annular shape created after the rivet forming step.

According to Szolwinski et al. (1997), the manner in which the cyclic loading is transferred through the joint in the presence of friction can be illustrated, as shown in Fig. 5.27. In the unloading state a contact area exists between the mating surfaces around the rivet hole due to the clamping action of the rivet heads. Initially, when the applied loading starts increasing, only the friction force transmits the load. No relative motion between the contact surfaces occurs at that stage, which defines the stick condition of the contact area (stage A, Fig. 5.27). The rivet does not experience any loading until the friction forces are overcome by the external loading and the relative motion of the sheets begins. Only when this global slip at the faying surface occurs (in the stick-slip region, cf. Fig. 5.26) does the rivet begin to take up the additional load (stage B, Fig. 5.27). The decrease of the load transmitted by the rivet does not start immediately when the applied load is released (stage C, Fig. 5.27), but only when the slip in the direction opposite to the slip due to the increase of the applied loading occurs in the plate interface (stage D, Fig. 5.27). Finally, during subsequent reloading the slip again has to reverse in the stick-slip region before the rivet takes up any more load. Therefore, the fraction of the contact area in slip varies cyclically similar to the schematic shown in Fig. 5.27.

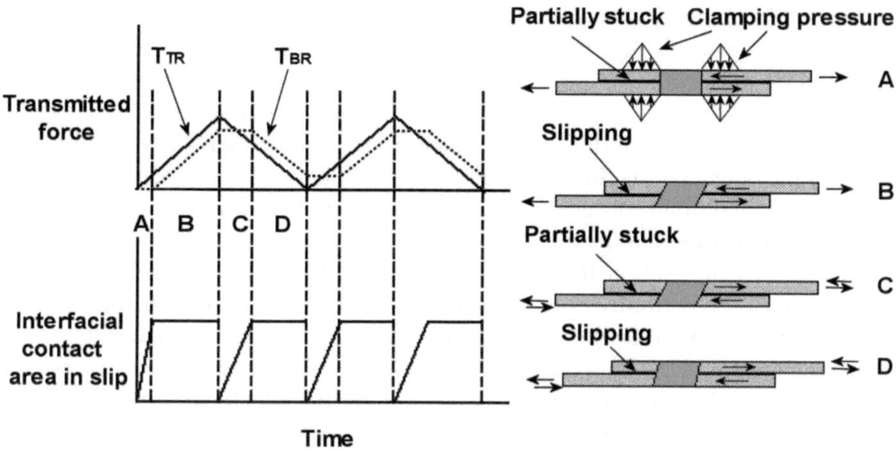

Fig. 5.27 Schematic of load transfer mechanism under fatigue loading (Szolwinski et al. 1997)

Load transfer by friction has a dual effect on joint fatigue properties. It works favourably by a decrease of the bearing force, but it works unfavourably by an increase of the probability of fretting corrosion between the sheets. Vlieger and Ottens (1997) noted a beneficial effect of friction at the interface between the mating sheets on the fatigue life of riveted lap joints. The riveted 2024-T3 Alclad specimens with degraded bonding (fully bonded, then torn apart and riveted) showed a 25–30% longer fatigue life compared to the specimens riveted without previous bonding. The remains of bonding on the faying face caused extra friction which contributed to an increase in fatigue life. Implications of the load transfer by friction for the failure mode of riveted lap joints are considered in Chap. 7.

5.5 Summary of This Chapter

Axial forces in the sheets of a lap joint with mechanical fasteners can be determined analytically by considering the displacement compatibility in the overlap region or numerically from FE analyses. For either methodology, fastener flexibility data is the necessary input.

Rivet flexibility can be derived either utilizing an analytical solution by Barrois suitable only for a "neat fit" between the fastener and hole or experimentally. Due to the lack of a standard procedure for fastener flexibility measurements, experimental results for a given riveted joint specimen obtained in various laboratories may be quite different depending on the measurement method, the loading type and the recorded data processing. Consequently, a large spread occurs in fastener flexibility values according to different semi-empirical formulae proposed in the literature.

If rivet flexibility is determined based on specimen compliance measurements, it should be ensured that the effect of secondary bending is avoided and the contribution of sheet/machine parts' elongations is properly eliminated from the measurement data. If the above requirements are met, experimental results reveal a dependency of rivet flexibility on the squeeze force and indicate that flexibility of rivets in the outer rows is different than in the middle row for a three-row lap joint. All experimental trends observed can be explained qualitatively by considering the contribution of friction to load transmission through the joint.

The analytical solution for axial forces in the overlap area obtained from the displacement compatibility conditions shows a surprisingly small dependence on rivet flexibility and does not correlate with strain gauge measurement results. These measurements indicate non-symmetrical load transmission in a three-row riveted lap joint, while rivet flexibility measurements show identical flexibility for the outer rivet rows. Such results indicate that rivet flexibility is not the only factor which affects load transfer and suggest that differences in hole expansion in the sheet adjacent to the rivet driven head and the sheet next to the manufactured head may play a role.

Friction at the faying surface accounts for some of the load transfer through the lap joint. Both direct measurements of frictional forces for lap joints with instrumented bolts and experiments to indirectly estimate the friction contribution indicate that during fatigue loading load transmission by friction initially increases with the cycle number until, after a certain portion of fatigue life, a stabilized level is attained. At this stage, up to 80–90% of the transfer force is transmitted by friction.

The percentage of the applied load transferred by friction increases with the amount of clamping, and therefore with the squeeze force. Clamping force values evaluated by means of experiments are several times higher than those derived from FE analyses. Also, compared to fractographic examinations, numerical analyses indicate a smaller region of contact between the mating sheets on the faying surface beneath the rivet head. According to FE analyses, under tensile loading applied on a three-row riveted lap joint the peak contact stresses on the faying surface are different at the individual rows.

Chapter 6
Secondary Bending for Mechanically Fastened Joints with Eccentricities

6.1 The Phenomenon of Secondary Bending

In the case of joints eccentric with respect to the load path, like lap joints or single strap butt joints, out-of-plane deformations of the sheets referred to as secondary bending (SB) occur under nominally tensile loading. Maximum moments due to SB are induced at load path eccentricities, namely at the rivet rows. As said earlier, for a lap joint with more than two rivet rows, the maximum bending moment is always induced at the outer rows. The out-of-plane deformation of a lap joint with three rivet rows associated with SB is shown in Fig. 6.1a. The largest tensile stresses due to SB are produced along the faying surface at location A of sheet 1 and location B of sheet 2. Because outside the outer rivet rows either sheet carries the full applied tensile load, A and B are also locations of the maximum combined tensile stresses and, therefore, the most commonly observed fatigue crack initiation sites.

The severity of SB at a given location is often quantified by the bending factor which can be defined in several ways:

$$K_b = \frac{S_b}{S}, \tag{6.1a}$$

$$K_b = \frac{\sigma_b}{S}, \tag{6.1b}$$

or

$$K_b = \frac{\sigma_b}{\sigma} \tag{6.1c}$$

where S_b is the nominal bending stress, i.e. the stress determined disregarding stress concentration near the rivet holes, σ_b is the local bending stress, S is the remotely applied tensile stress and σ is the local membrane stress in the direction of loading. Note that only K_b according to Eq. 6.1c evaluates the amount of SB at any location.

A. Skorupa and M. Skorupa, *Riveted Lap Joints in Aircraft Fuselage*, Solid Mechanics and Its Applications 189, DOI 10.1007/978-94-007-4282-6_6,

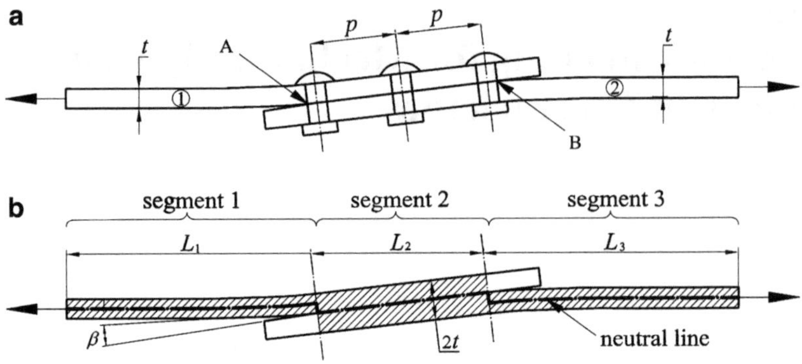

Fig. 6.1 Secondary bending in a riveted lap joint: (**a**) joint deformation and critical locations; (**b**) model of a joint according to Schijve (1972) and Das et al. (2001)

An experimental investigation by Schütz and Lowak (1974) into the presence of SB in real aircraft structures indicated that out of 150 aircraft structural details considered, 86% exhibited SB. The literature evidence on the effect of various factors on the severity of SB in mechanically fastened joints and available results on the significance of this phenomenon for fatigue properties of joints are reviewed in the present chapter.

6.2 Analytical Investigations

6.2.1 Models

Analytical models enable us to compute nominal bending stresses in a joint, which implies that the magnitude of SB is determined in terms of the bending factor according to Eq. 6.1a. Two very different analytical approaches to estimate SB have been proposed in the literature. One of these is the solution by Barrois (1978), considered in more detail in Sect. 5.2.1, that enables us to compute the through-the-thickness distribution of the bearing load $q(x)$ between the flexible rivet and the rivet hole (cf. Fig. 5.6). If SB occurs, $q(x)$ is assumed to consist of two components:

$$q(x) = q_{\mathrm{m}} + q_{\mathrm{b}}(x) \tag{6.2}$$

where

$$q_{\mathrm{m}} = T_{\mathrm{TR}}/t \tag{6.3}$$

is the bearing pressure, which is uniformly distributed over the thickness and is associated with transmission by the rivet of the T_{TR} force, and the pressure $q_{\mathrm{b}}(x)$ is a result of tilting of the rivet shank by SB.

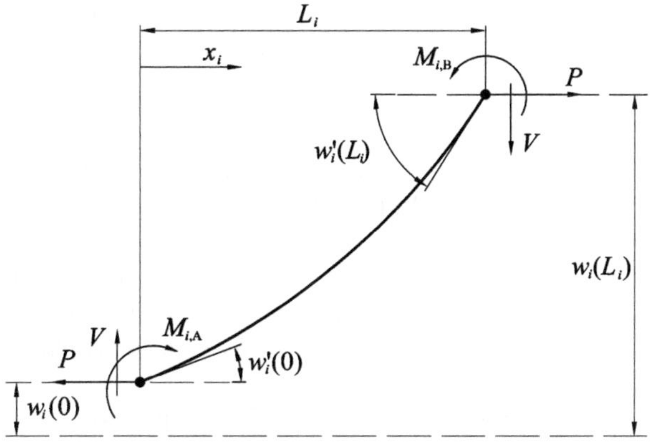

Fig. 6.2 Segment i of a riveted joint

From Eqs. 5.11 and 6.2, the maximum unit bearing pressure at the sheet interface associated with SB can be estimated as

$$q_{b,\max} = (K_q - 1)\, q_{\mathrm{m}} \tag{6.4}$$

Figure 5.7 implies that $q_{\mathrm{b,max}}$ depends on the t/d ratio. For a two-layer overlap, the influence of the joint geometry (sheet thicknesses and the spacing between the rivet rows) can only be accounted for by considering its effect on the T_{TR}-value. Application to a multiple layer overlap is possible if axial forces in the sheets required to determine the $q(x)$ distribution are known. Note in Fig. 5.7 that Barrois' model predicts a very large effect of the amount of clamping on the SB magnitude, which is against the available experimental and FE results for mechanically fastened joints, as reviewed further on in this chapter.

While with Barrois' model SB is estimated from deflections of the flexible fastener, the presence of fasteners is disregarded in models according to Schijve (1972) (see also Schijve et al. 2009) and Das et al. (2001). According to both concepts, further referred to as beam models, the nominal bending stresses are computed from the out-of-plain deformations of the sheets determined based on the theory of elastic beams or flat shells under bending. The overlap region between the axes of the end fastener rows is considered as an integral beam of which the flexural rigidity corresponds to the combined thickness of the sheets, Fig. 6.1b. For computational purposes, the joint is decomposed into segments of a constant flexural rigidity connected at the ends, Fig. 6.1b. The bending moment for segment i sketched out in Fig. 6.2 can be computed from the differential equation for the deflection $w_i(x_i)$ of the mid-plane surface, i.e. the neutral line of the beam, at any point x_i along the segment

$$M_i(x_i) = G_i w_i''(x_i) \tag{6.5}$$

with

$$M_i(x_i) = M_{i,A} + V x_i + P\,[w_i(x_i) - w_i(0)]\,, \quad i = 1 \text{ to } n \tag{6.6}$$

where P is the applied load, V is the fixing reaction, G_i is the bending stiffness (plane strain condition) of segment i, $w_i''(x_i) = \frac{d^2 w_i(x_i)}{dx_i^2}$, moment $M_{i,\mathrm{A}}$ represents the action of segment $i-1$, and n is the number of segments.

In the case of hinged clamping of the sheet ends, the fixing moments $M_{1,\mathrm{A}}$ and $M_{n,\mathrm{B}}$ (cf. Fig. 6.2) and the reaction V add up to zero. In agreement with FE computations by Müller (1995), results of both models reveal that the influence of clamping conditions on the bending factor vanishes when the distance between the clamping edge and the outer rivet row reaches a certain multiple of the sheet thickness. For example, for sheet thicknesses below 2 mm, the distance of $50t$ suffices to make the change in the S_b-value due to the change of the type of clamping less than 1%.

The general solution of Eq. 6.5 is in the form

$$w_i(x_i) = A_i \cosh(\alpha_i x_i) + B_i \sinh(\alpha_i x_i) + C_i x_i + D_i \tag{6.7}$$

where

$$\alpha_i = (P/G_i)^{1/2}, \quad C_i = -V/P \quad \text{and} \quad D_i = -M_{i,A}/P. \tag{6.8}$$

The unknowns, namely the constants A_i, B_i, C_i and D_i, the reaction V, and the moment $M_{i,\mathrm{A}}$, can be solved by considering the equilibrium conditions of the individual segments and by setting the boundary conditions at the segment intersections and at the joint clamped ends, i.e. for $x_1 = 0$ and $x_n = L_n$. Matching the slopes at the intersection of segment i and $i+1$ is governed by the equation

$$w_i'(L_i) = w_{i+1}'(0) \tag{6.9}$$

where $w_i'(x_i) = \frac{dw_i(x_i)}{dx_i}$.

According to Schijve's model, the deflected neutral axis is discontinuous at the eccentricities, which in the case of eccentricity e_i between segment i and $i+1$ leads to

$$w_{i+1}(0) = w_i(L_i) \pm e_i. \tag{6.10}$$

On the contrary, Das et al. consider $w(x)$ to be a continuous line, which implies

$$w_{i+1}(0) = w_i(L_i), \tag{6.11}$$

while the occurrence of eccentricity is covered by introducing an additional moment

$$M_{i+1,A} = M_{i,B} \pm P e_i. \tag{6.12}$$

In either model different assumptions about the deflections of the joint overlap are adopted. With Schijve's model, Eq. 6.5 holds also for segments within the

Table 6.1 Bending moments on the rivets for two-layer lap joint with three rivet rows

Rivet row, j	M_j	$M_{\mathrm{TR},j}$	$\Delta M_j = M_{\mathrm{TR},j} - M_j$
I	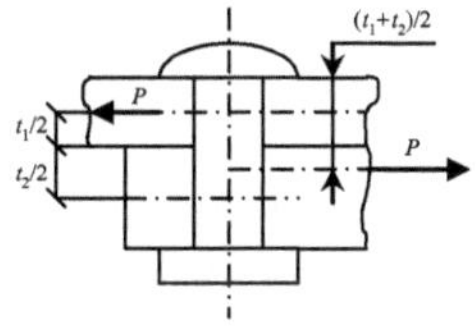 $Pt_2/2$	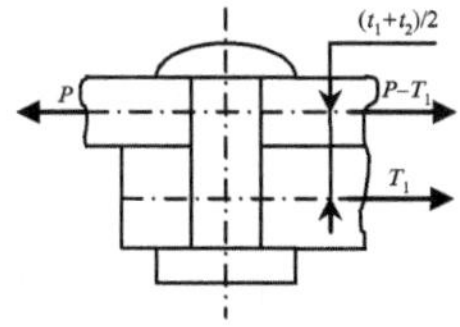 $T_1(t_1+t_2)/2$	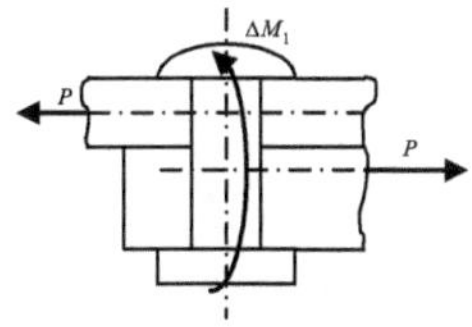 $T_1(t_1+t_2)/2 - Pt_2/2$
II	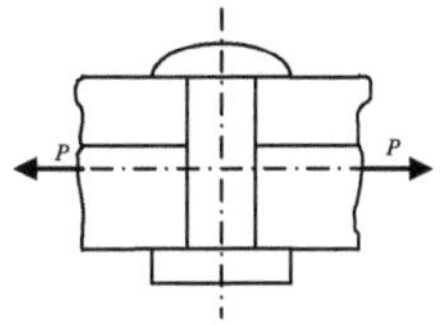 0	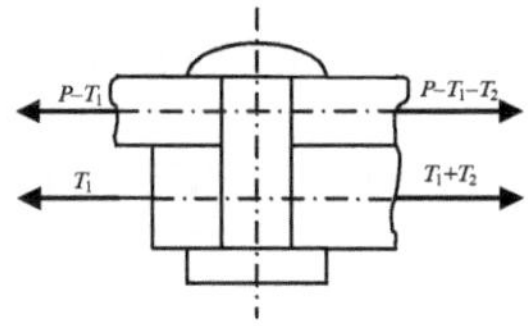 $T_2(t_1+t_2)/2$	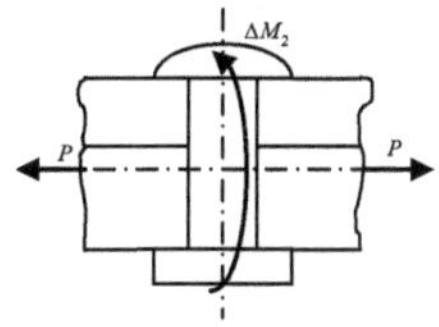 $T_2(t_1+t_2)/2$
III	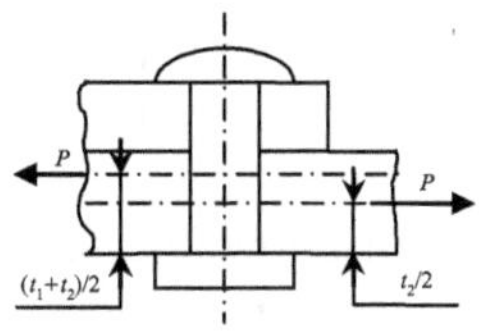 $Pt_1/2$	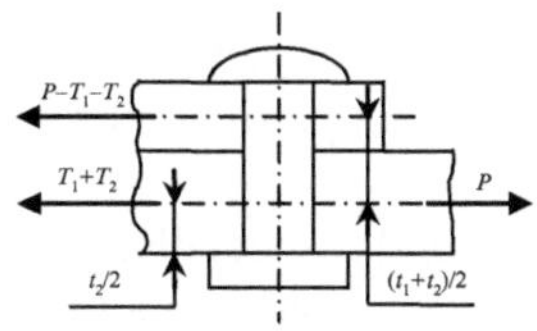 $(P-T_1-T_2)(t_1+t_2)/2$	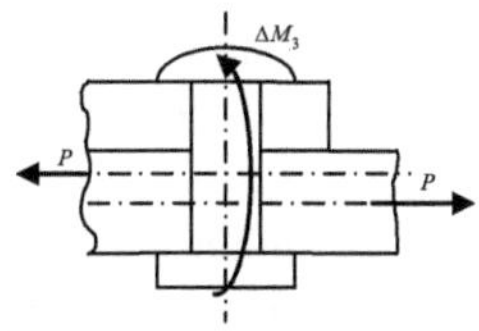 $(P-T_1-T_2)(t_1+t_2)/2-$ $Pt_1/2$

overlap, whilst Das et al. consider the overlap to be perfectly stiff. If segment i within the overlap region is connected to segments i–1 and $i+1$, the latter assumption yields

$$w_i(L_i) = w_i(0) + L_i w_i'(0) \tag{6.13a}$$

and

$$w_i'(0) = w_{i-1}'(L_{i-1}) = w_i'(L_i) = w_{i+1}'(0). \tag{6.13b}$$

Neither of the models can cover the influence of load transmission through the joint because the resultant internal axial force in any transverse section of the overlap (equal to the applied load P) is assumed to act at the section gravity centre, i.e. along the mid-plane of the joint. The associated bending moments on the rivet rows for a simple lap joint of two sheets with three fastener rows are given in Table 6.1 (notation M_j). The moment in the middle fastener row II equals zero, which implies that for a lap joint with more than two rows only the outer rows take part in load transmission. It is obvious that if load transfer through the inner rivet rows occurs due to rivet flexibility, the resultant force acts at a distance from the neutral axis. The bending moments on the rivets computed for the latter case

($M_{\mathrm{TR},j}$) are also presented in Table 6.1. In order to account for load transmission by the inner rivet rows, de Rijck (2005) proposed applying on each rivet row j an additional bending moment ΔM_j satisfying the condition

$$M_j + \Delta M_j = M_{TR,j}. \tag{6.14}$$

The ΔM_j-values are given in the last column of Table 6.1. The axial forces in the sheets necessary to determine $M_{\mathrm{TR},j}$ can be computed as described in Sect. 5.1, provided the rivet flexibilities are known.

Incorporating the concept by de Rijck into the models according to Schijve or Das et al. requires that Eq. 6.6 should be accordingly modified:

$$M_i(x_i) = M_{i,A} + Vx_i + \sum \Delta M_j + P\left[w_i(x_i) - w_i(0)\right] \tag{6.15}$$

where the summation of the ΔM_j moments is over the rivet rows located between $x_1 = 0$ and x_i (see Fig. 6.2).

Generally, the modified model yields slightly lower bending stresses than the original model, the differences getting more pronounced with decreasing sheet thickness.

At higher applied loads and, especially, in the case of a low interference fit or a clearance between the fastener and the hole, the fasteners can considerably tilt in the holes. This leads to enhanced contact stresses between the fastener and the hole at critical locations A and B (cf. Fig. 6.1a), which may cause a small permanent rotation of the sheets. An indication of the effect of this phenomenon on SB can be obtained from the beam model by assuming a certain small angle of β between the overlap and the sheet, as shown in Fig. 6.1b. The boundary condition for matching the slopes between segments 1 and 2 is then formulated as (cf. Eq. 6.9 and Fig. 6.2)

$$w_1'(L_1) = w_2'(0) + \beta. \tag{6.16}$$

A considerable sensitivity of the model to the value of β can be demonstrated by exemplary results given by Schijve (1972). For a simple lap joint $\beta = 1°$ reduces SB by a factor of 0.52 and 0.74 for t of 1 and 2 mm respectively compared to the case with $\beta = 0$.

For circumferential fuselage joints the beam model considered above is no longer appropriate because, due to the sheet curvature, both the hoop and longitudinal stress components as well as the internal pressure contribute to the out-of-plane deformation. An analytical 3-D model for estimating SB in circumferential joints of the fuselage has been developed by Wit (1993). The joint is modelled by cylindrical segments placed in rows, Fig. 6.3. As in the beam model, the overlap of the sheets is represented by an integral element of a combined thickness of the sheets. Considering the joint segment as a cylindrical shell of a constant thickness loaded by an internal pressure (Δp) and accounting for both membrane stress components (σ_l and σ_h), a fourth-order non-homogenous differential equation on the out-of-plain

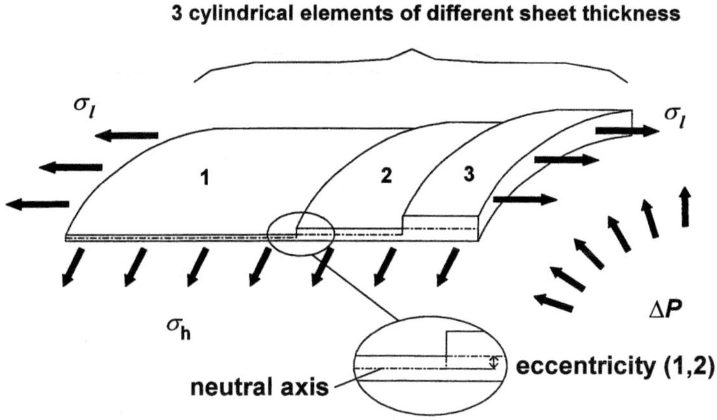

Fig. 6.3 Cylindrical elements modelling the transverse fuselage joint (Wit 1993)

deformation of the joint neutral axis is obtained. The general solution of this equation contains four unknown constants which can be solved from appropriate boundary conditions at the segment intersections. Subsequently, bending stresses at the outer fibres of the sheets can be calculated. Exemplary computations produced by Wit for a circumferential single strap joint, assuming a fuselage diameter of 4 m, sheet and strap thicknesses of 1 mm, and a longitudinal stress of 50 MPa, indicate that the 3-D model gives approximately 25% higher bending stresses at the critical skin location (outer fastener rows) than the beam model. The critical locations in the strap (inner fastener rows), however, are bent by 33% less severely compared to the beam model.

6.2.2 *Exemplary Applications to Lap Joints*

6.2.2.1 Standard Geometry

Both the model by Schijve and the one by Das et al. correctly predict that the largest deflections and, hence, the peak bending moments and bending stresses occur at the end fastener rows. Plotted in Fig. 6.4 are the neutral line deflections derived from Schijve's model for a simple lap joint sketched out in the top of the figure. Also shown in Fig. 6.4 are variations of the nominal bending stresses along the lower edge of the joint (thickened line) computed for two specimen thicknesses using the computer implementation of the models by Schijve and Das et al. (notation [S] and [D] respectively). It is seen that for $t = 2$ mm both models give very close results on S_b at the outer rows, while for $t = 0.8$ mm the peak S_b-values differ quite significantly ($S_b = 99.1$ MPa from Schijve's model and 66.5 MPa from

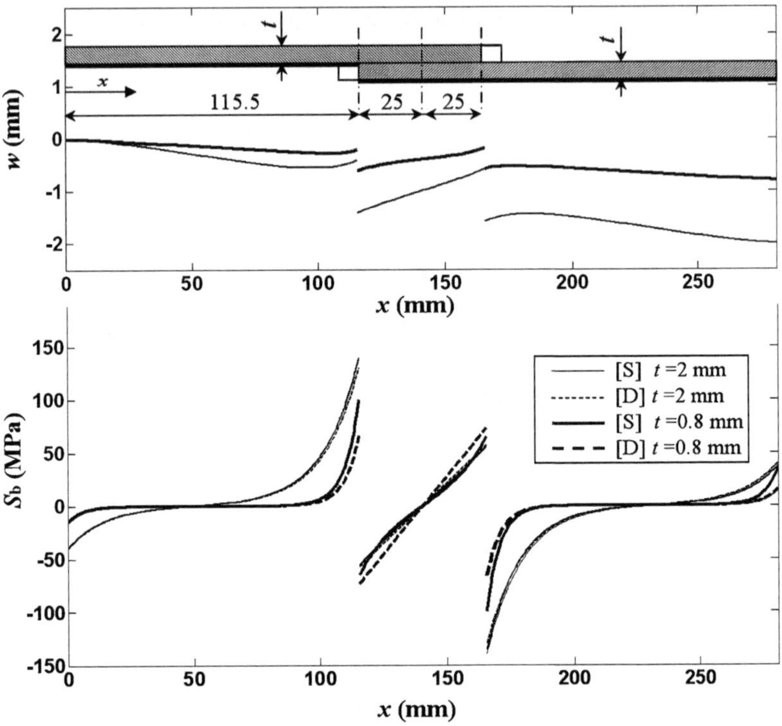

Fig. 6.4 Computed neutral line deflections and variations of peak bending stresses along a lap joint at applied stress $S = 120$ MPa according to models by Schijve [S] and Das et al. [D] (Skorupa and Korbel 2008)

the model by Das et al.). The more pronounced discrepancies in the S_b stresses observed for thinner sheets stem from distinct assumptions about the overlap deformation adopted in either model. Obviously, the lower the sheet thickness, the more meaningful differences between the deflections computed for the flexible (Schijve's model) and perfectly rigid (the model by Das et al.) overlap. Including de Rijck's modification in Schijve's model yields the peak S_b-values lower by 5% and 12% for $t = 2$ and 0.8 mm respectively in comparison to the original approach.

According to both models, the magnitude of SB becomes considerably reduced with a decrease in the sheet thickness, as already seen in Fig. 6.4, and with an increase in the spacing p between the fastener rows. This could well be anticipated since thinner sheets imply smaller eccentricities and because for a longer p the joint out-of-plane deflections are smaller. The computed bending factor is a non-linear function of the applied stress and becomes higher at lower S-values. Discrepancies between both models' predictions on the SB dependence on several variables for the lap joint shown in Fig. 6.4 are quantified in Figs. 6.5 and 6.6. Generally, the Das et al. model yields lower estimates on SB compared to Schijve's model, the differences considerably increasing for thinner sheets, as already revealed in Fig. 6.4 and further

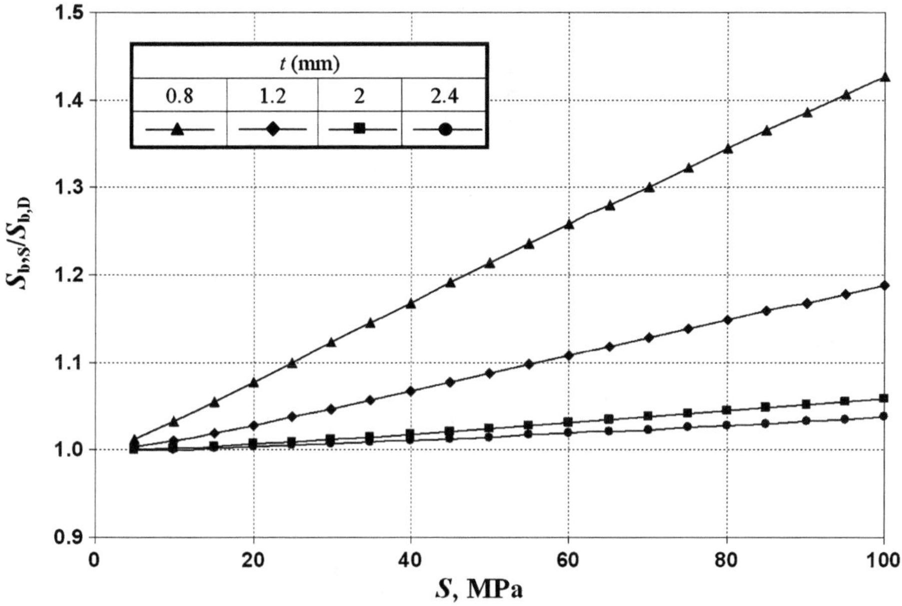

Fig. 6.5 Comparisons between peak bending stresses at end rivet rows according to models by Schijve ($S_{b,S}$) and Das et al. ($S_{b,D}$) for the lap joint from Fig. 6.1 with $p = 25$ mm (Skorupa and Korbel 2008)

backed up in Fig. 6.5 by plots of the $S_{b,S}/S_{b,D}$ ratios, where the subscripts S and D refer to the model according to Schijve and Das et al. respectively. In addition, Fig. 6.5 demonstrates that $S_{b,S}/S_{b,D}$ increases with an increase in the applied stress level.

In Fig. 6.6, both models' estimates on the reduction in SB for the lap joint from Fig. 6.1 caused by increasing the fastener row pitch are compared for a range of t-values. Here, $S_b(40)$ and $S_b(25)$ denote the bending stress at the critical location for p of 40 and 25 mm respectively. Schijve's model predicts fewer benefits from increasing p, as visualized by the plots corresponding to his model lying above those according to the Das et al. model except at very low S levels. Interestingly, the results from either model show an opposite behaviour with respect to S and t. While according to the model by Das et al. increasing the row pitch becomes the more beneficial the thinner the sheet and the higher the stress level, the reverse is true in the case of Schijve's model. Consequently, the largest discrepancies in the $S_b(40)/S_b(25)$ ratio from both models are exhibited for the 0.8 mm thick sheet at the highest S of 100 MPa. In that case the Das et al. model predicts the bending factor for $p = 40$ mm twice lower than Schijve's model.

It is often thought that SB for a joint in an actual structure should be lower than for simple specimens, because the structural joint benefits from lateral support by frames, e.g. Terada (2009, private communication). Hartman and Schijve (1969) considered this issue by comparing results on K_b, computed using Schijve's model, for two single strap butt joints without and with a rigid support in the centre of the

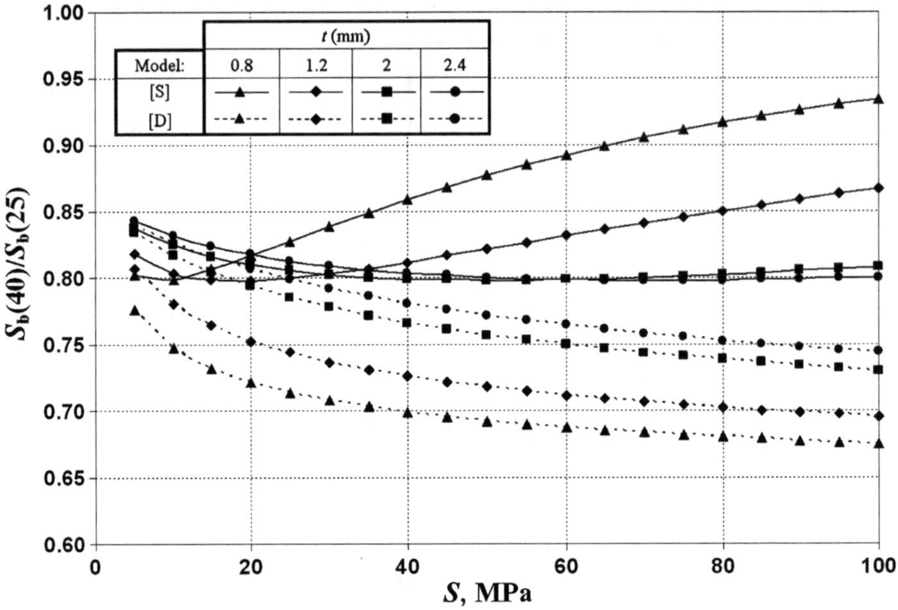

Fig. 6.6 Reduction in peak bending stresses for the lap joint from Fig. 6.1 due to increasing rivet row spacing from 25 to 40 mm according to models by Schijve [S] and Das et al. [D] (Skorupa and Korbel 2008)

specimen. The joints' configurations, namely type C and E exhibit a significant and small amount of SB respectively (for dimensions see Fig. 6.25) and the computation results are shown in Fig. 6.7. The K_b factor is presented for the maximum bending stresses in the joint which occur in the strap at location D for configuration C and in the sheet at location A for configuration E. The location of the support along the vertical centre line of the strap adjacent to the support is defined by "a". The results in Fig. 6.7 indicate that a misorientation of the support, i.e. $a \neq 0$, can lead to either an increase or a reduction in bending, depending on whether the strap deflection is supported or diminished. The effect mainly occurs at low applied stresses and decays rapidly for high S-values. The support located at $a = 0$ yields approximately the same bending as in the case of the unsupported joint. The explanation of the latter observation is that the support does not eliminate the eccentricities and does not change the bending stiffness of the joint. The results in Fig. 6.7 correspond to a distance of 200 mm between the specimen clamping edge and the outer row of fasteners, which Hartman and Schijve consider to be realistic for actual structures. For a smaller distance the effect of the support would be larger. A fairly small influence of the frame on bending stresses at the critical location of a circumferential single strap joint has also been reported by Wit (1993) from computations using the 3-D model (cf. Sect. 6.2.1).

On the basis of their FE analysis of composite lap joints, Ekh and Schön (2005) note that suppressing the out-of-plane deformation of the plates by external support,

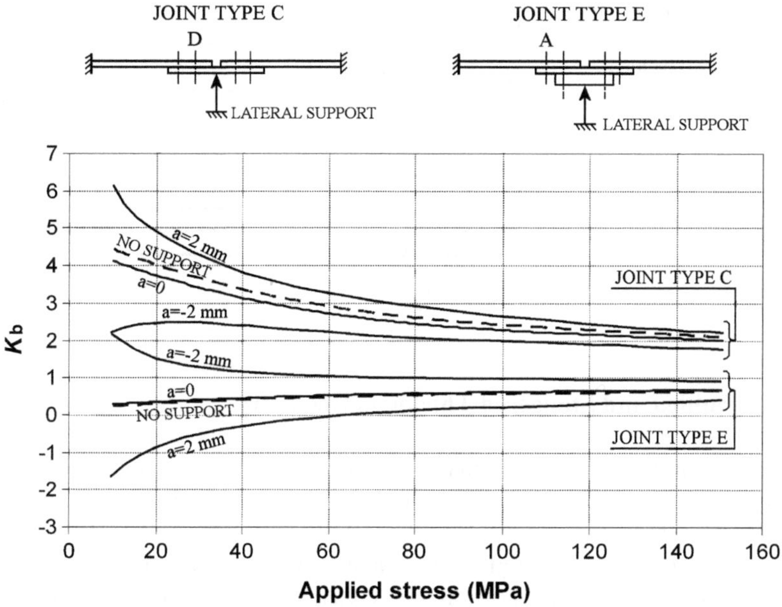

Fig. 6.7 Computed effect of lateral support on secondary bending in two strap joints (Hartman and Schijve 1969)

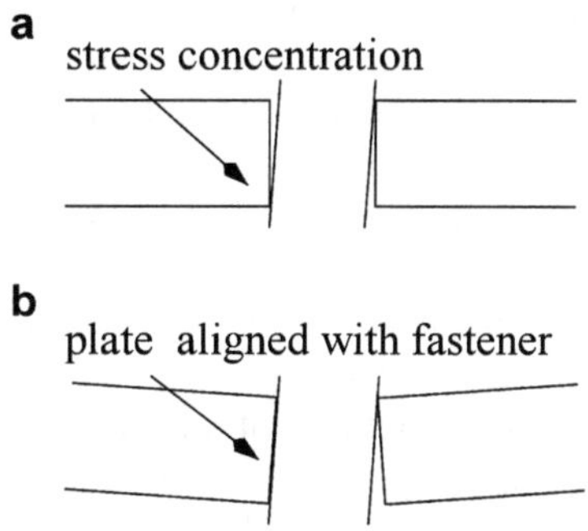

Fig. 6.8 Contact area between fastener and hole without (**a**) and with (**b**) the out-of-plane deformation of the plate (Ekh and Schön 2005)

if associated with maintaining the load eccentricity, may have a negative impact on the joint fatigue performance. As considered in detail in Chap. 5, the fasteners tilt in their holes due to the load transfer and SB. Figure 6.8 demonstrates that if the plate is bent, the fasteners are closer aligned with the hole than in the case when the out-of-plane deformations are restrained. A better alignment yields an increased fastener/hole contact area and, consequently, lower bearing stresses on the hole. Suppressing the plate deformation leads to a reduced contact area and, hence, to more severe bearing stresses. These authors claim that in design it is more important to minimize eccentricity of the load paths than to minimize SB by means of external forces.

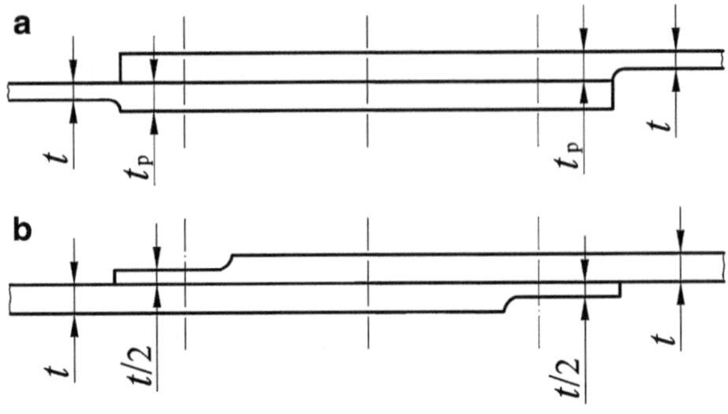

Fig. 6.9 Two examples of a modified lap joint configuration: (**a**) padded joint; (**b**) staggered thickness joint

6.2.2.2 Padded and Staggered Thickness Geometry

Despite the simplifying assumptions implicit in the beam model, it is still instructive to apply it in order to learn how different modifications in the joint design, proposed to improve its fatigue behaviour, affect SB and load transmission. Two examples of the modified lap joint configuration, namely by the so-called "padding" and by staggering the thickness, are shown in Fig. 6.9a and b respectively. In the case of the padded lap splice design used at Boeing (Das et al. 2001), the thickness of the sheets at the joint region is locally increased in order to reduce the nominal tensile stress at the critical fastener rows. Padding can be obtained either by bonding additional sheets in the joint area or by milling the sheets down away from the joint. The bonded doublers are superior because the adhesive layer acts as a crack growth retarding element. Quantitative estimates on the bending stresses are required in order to learn if the beneficial effect of pads is not negated by the bending moments introduced due to the increased eccentricities at the critical, end fastener rows.

More recently, Schijve (2006) proposed to reduce high stresses at the outer fastener holes by reducing SB at that location. This can be achieved by a staggering of the sheet thickness in the overlap region, as shown in Fig. 6.9b. Halving the original thickness at the end fastener row regions yields an eccentricity twice lower compared to the conventional solution, i.e. without the thickness staggering. In addition, because of the lower stiffness of the thinned part of the sheet, the thickness staggering is expected to lower the load transmitted by the outer rows in a similar way as fingering the end of the sheet (cf. Fig. 5.5).

Figure 6.10 presents variations of the bending stresses along the lower surface (thickened line) of a padded lap joint sketched in the top of the figure. Two different pad thicknesses, $t_p = 1.5\,t$ and $2\,t$, are considered. For comparison, the stress distribution for the standard geometry ($t_p/t = 1$) is also plotted. The S_b-values have been derived using the computer implementation of the model by Das et al. (2001) assuming $t = 1.2$ mm. It is seen that padding yields a considerable release of S_b in fastener row I. Because at that location the tensile stress associated with the axial force (S_t) becomes also lower due to the increased thickness, row I is no longer critical for this specific joint configuration. At the same time, padding causes a significant increase of S_b in the other end row (row III, location B) and,

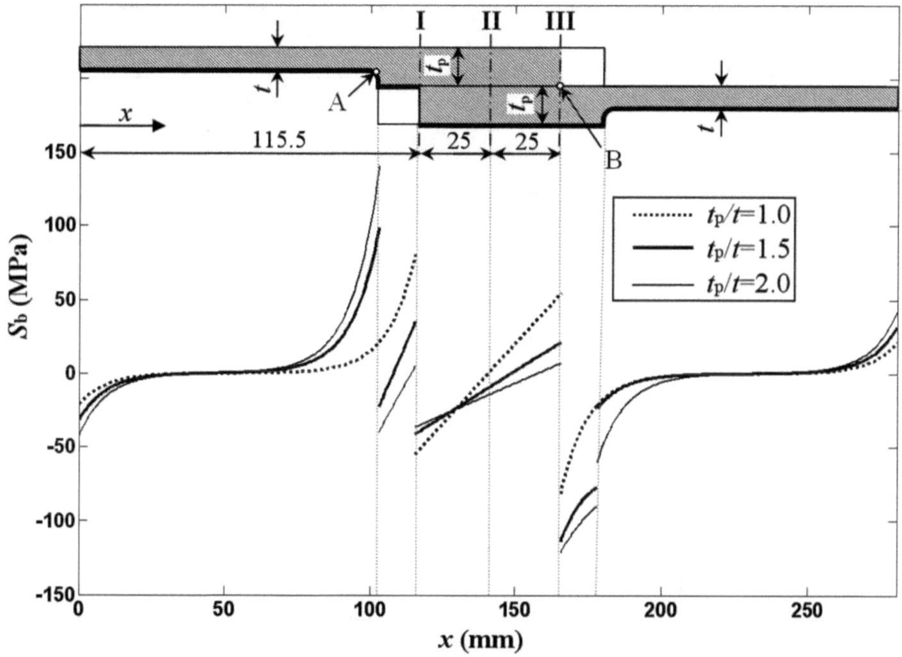

Fig. 6.10 Variations of peak bending stresses along a padded and unpadded lap joint at applied stress $S = 100$ MPa

Table 6.2 Effect of the pad thickness on stresses at critical locations of the lap joint

		t_p/t			
Location	Stress (MPa)	1	1.5	1.8	2.0
A	S_t	100	100	100	100
	S_b	81.4[a]	99.3	126.7	141
	$S_t + S_b$	181.4[a]	199.3	226.7	241
B	S_t	100	66.7	55.6	50
	S_b	81.4	113.6	120.5	121.2
	$S_t + S_b$	181.4	180.3	176.1	171.2

[a]Location A according to Fig. 6.1a

in addition, it induces a pronounced bending at location A due to the eccentricity caused by the sheet thickness transition. The dependence of the axial, bending, and combined stress (S_t, S_b, $S_t + S_b$) at both critical locations on the pad thickness for the joint considered in Fig. 6.10 is quantified in Table 6.2. It can be seen that padding results in only an insignificant decrease in the combined stress in row III (location B) compared to the unpadded geometry.

Figure 6.11 presents the effect of staggering the sheet thickness for a lap joint sketched in the top of the figure. The peak bending stress variations along the joint surface plotted by a thick line have been computed for two fastener row spacings

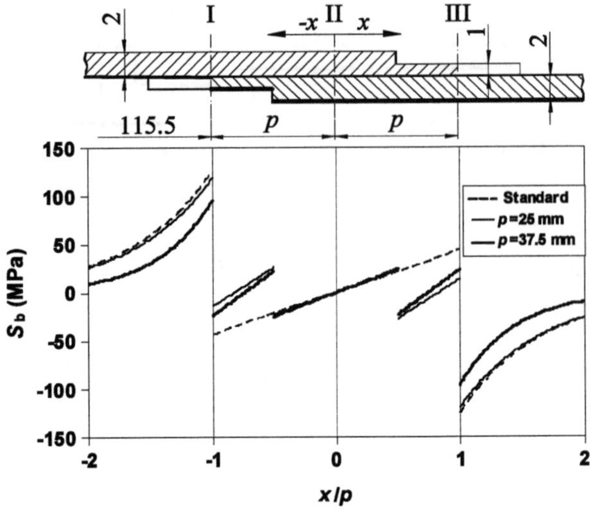

Fig. 6.11 Variations of peak bending stresses along a lap joint with staggered thickness and standard configuration at applied stress $S = 100$ MPa

using Schijve's model. The results for the non-staggered geometry with $p = 25$ mm are also shown for comparison. The staggered thickness joint is lighter compared to the non-staggered specimen at the same p of 25 mm because the staggering of the thickness implies a small weight reduction. If the row pitch for the staggered thickness configuration is increased by a factor of 1.5 to obtain $p = 37.5$ mm, the weight of the staggered lap joint becomes the same as that of the standard joint. It is seen in Fig. 6.11 that for the larger row spacing the peak bending stresses become more effectively reduced than when the same spacing as for the standard geometry is retained. For $p = 25$ mm, the peak S_b at the end fastener rows in the staggered geometry is only 6% lower than in the standard joint, but for $p = 37.5$ mm a 24% reduction in S_b is observed. If de Rijck's modification is applied, the absolute S_b-values get several per cent lower than according to the original model, but the proportion between the bending stresses for the standard and modified geometry is retained. Application of the procedure presented in Sect. 5.1 to compute the axial forces in the sheets yields an about 13% lower load transfer by the outer rivet rows for the staggered geometry in comparison with the standard joint.

6.3 Finite Element Modelling

SB in lap joints was determined in relatively few FE analyses, and in most cases bolted joints in metallic (Änggård 2000), composite-to-aluminium (Ekh et al. 2005) or composite (Ireman 1998) plates have been considered.

A FE modelling of riveted aluminium lap joints covering the effect of SB has been carried out by Müller (1995), Newman et al. (1997) and Rans (2007). Because of a simplified representation of the joint in Müller's elastic analysis, a stress singularity in the sheet occurs near the rivet. Consequently, stress concentration near the rivet cannot be properly determined and only the results on bending midway between the rivets can be considered realistic. Müller's results on the bending moment distribution in a simple three-rivet row joint indicate that SB is localized mainly at the rivets, while between the rivets it is significantly reduced. Compared with the values resulting from Schijve's model, the K_b factors (Eq. 6.1b) computed by Müller midway between the rivets are in the range of 44–74% depending on the joint configuration, the higher values being associated with a larger spacing between the rivets in a row.

FE analyses of SB by Newman et al. (1997) for a two-row lap joint (neat-fit rivet, no clamp-up) indicated that a discrepancy between the elastic and elastic–plastic solution for local peak bending stresses was noted already at an applied stress of 50 MPa. With a further increase in S, σ_b became gradually reduced and dropped to zero at S of about 100 MPa. The authors concluded that in fatigue analyses bending effects should be accounted for at a much reduced level compared to the SB magnitude according to elastic or beam model analyses.

A more elaborated FE modelling by Rans (2007) aimed at investigating the dependence of SB in lap joints on the type of rivet (countersunk or universal) and the squeeze force level. A two-row lap joint configuration was chosen, consisting of two 1 mm thick Al 2024-T3 sheets joined by 3.2 mm diameter Al 2117-T4 rivets. The FE analysis was performed in two phases. First, the rivet installation was simulated in order to induce residual stresses due to the riveting process. Subsequently, a tensile load on the joint model was applied. The analyses accounted for friction on contact interfaces between the rivets and sheets, between the sheets along the faying surface, and between the riveting tools and the rivets.

The local bending stresses at the faying surface of each sheet were determined from the faying and free surface stress components (σ_{fay} and σ_{free} respectively) in the loading direction as

$$\sigma_b = \frac{\sigma_{fay} - \sigma_{free}}{2}. \tag{6.17}$$

The through-the-thickness stress distribution from the FE solution by Rans is a combination of the membrane stress due to the applied loading, the residual stress and the bending stress. Because the residual stress has been found to vary nonlinearly through the thickness of the sheet, theσ_b stress given by Eq. 6.17 and, hence, the local bending factor K_b (according to Eq. 6.1b) includes the effect of the residual stresses.

Rans's results suggest that the dependency of SB on both the type of rivet and the squeeze force is mainly related to the geometry of rivet heads. The driven head geometries are nearly identical for the universal and countersunk rivets of the same diameter and length if installed with the same F_{sq}-value. It follows, therefore, that the results for both types of rivet closely agree for the sheet adjacent to the driven head (inner sheet) and are in a good agreement with the results for the sheet next to the universal rivet manufactured head (outer sheet) because this head

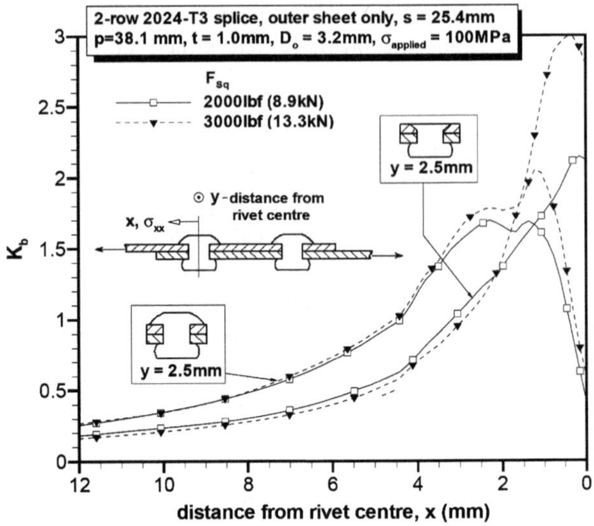

Fig. 6.12 Comparison of FE results on secondary bending in the outer sheet for a universal and countersunk rivet (Rans 2007)

geometry resembles that of the driven head. Consequently, differences between the SB behaviour for the countersunk and universal rivet only occur in the sheet next to the manufactured head, as illustrated in Fig. 6.12. It is seen that at a given F_{sq} level the peak K_b-values in the outer sheet are considerably higher for the countersunk rivet than for the universal rivet. Rans attributes this behaviour to a greater constraint to the rivet rotation provided by the large manufactured head of the universal rivet. The increased flexural rigidity beneath both heads of the universal rivet and beneath the driven head of the countersunk rivet results in a shift of the location of maximum K_b from the rivet row centre line to above the rivet hole, as exemplified in the inset in Fig. 6.13a for the sheet adjacent to the manufactured head of the universal rivet. Such a migration is not revealed, however, in the outer sheet for the countersunk rivet, as shown in the inset in Fig. 6.13b. In addition, Fig. 6.13a and b indicate that for both types of rivet the shift of the location of maximum K_b from the net sections is sustained away from the rivet as well. The latter region is, however, of minor interest as, due to stress concentration, crack nucleation and growth occur in close proximity to the rivet hole.

The FE results by Rans imply that the primary dependency of SB on F_{sq} only occurs in the peak K_b location, and only in the inner sheet. Figure 6.14 demonstrates that increasing the squeeze force from 8.9 to 13.3 kN causes the location of peak K_b for this sheet to move further away from the hole by a distance equal to the difference in the corresponding rivet driven head radii ($\Delta D/2$). Comparisons between the results of Rans and those from the analytical models considered in Sect. 6.2 are not appropriate because the beam model does not allow for residual stresses and stress concentration which contribute to Rans's solution on the SB stresses in the rivet proximity.

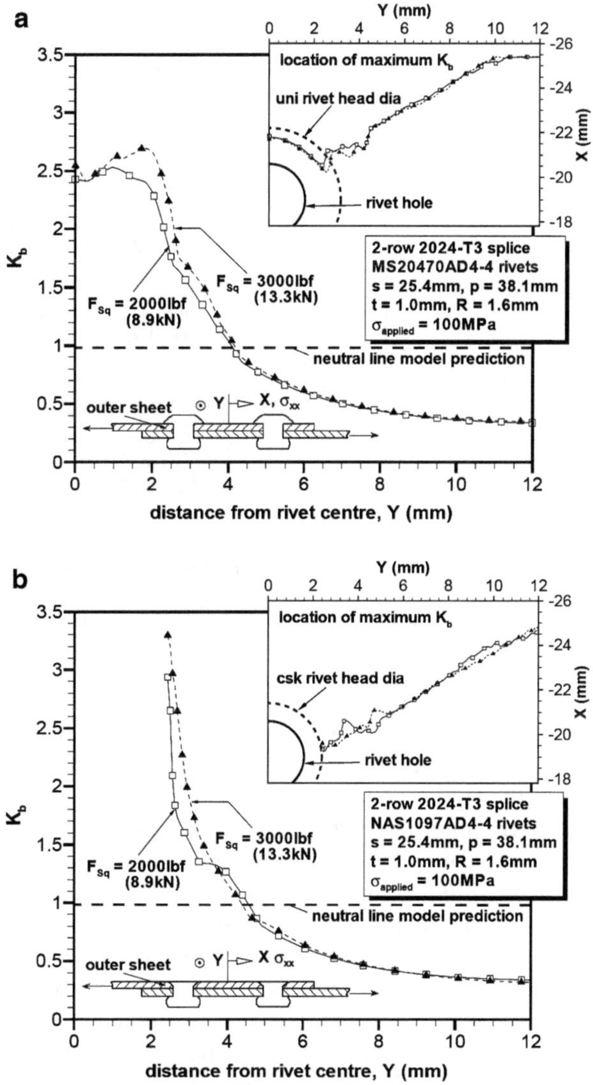

Fig. 6.13 FE results on the distribution of maximum secondary bending along the joint width for the outer sheet: (**a**) universal rivet; (**b**) countersunk rivet (Rans 2007)

As in the case of riveted joints, FE analyses for bolted joints indicate that SB varies in the transverse direction, becoming more pronounced close to the fastener. For example, for a composite-to-aluminium lap joint with four rows and a single column of 6 mm diameter bolts, numerical results by Ekh et al. (2005) indicate that for the outer fastener rows the bending strain at the so-called AGARD point,[1]

[1]The AGARD point is located above the fastener on the tangent to the fastener hole in the direction transverse to the loading at a distance of the hole diameter from the fastener column axis.

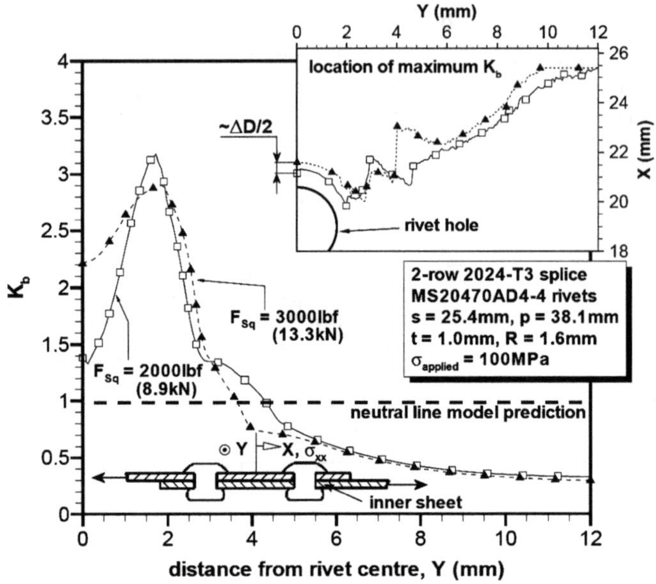

Fig. 6.14 Effect of the squeeze force on the maximum bending factor for the inner sheet (Rans 2007)

which in their study implies a distance of 6 mm from the fastener column axis in the specimen width direction, is a factor of over 1.3 higher than at a distance of 14 mm from the hole. The latter work further reveals that though the bending strain at the AGARD point increases with an increase in the applied load, the bending factor (according to Eq. 6.1a) decreases, thus showing the trend consistent with predictions from the simple beam model (cf. Sect. 6.2.2). At the critical fastener row the location of the peak bending stresses is shifted above the net section, in agreement with the FE analyses by Rans (cf. Figs. 6.13 and 6.14). Regarding the effect of the bolt type (countersunk or protruding head), the SB response of the bolted joint is qualitatively different than for the riveted joint. In agreement with Ireman (1998) experiments, the results by Ekh et al. (2005) demonstrate a slight reduction of SB in the lap joint if the protruding head bolt is replaced by the countersunk bolt, whereas the reverse is true for the riveted joint (cf. Figs. 6.12 and 6.13).

6.4 Measurements of Secondary Bending

6.4.1 Methodology

Stresses due to SB can be determined experimentally in a direct way utilizing strain gauges or indirectly, by measuring a quantity related to the bending stress. When

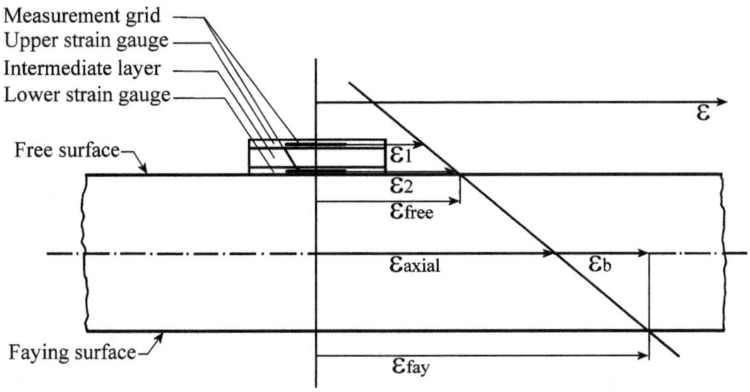

Fig. 6.15 Double-grid strain gauge for secondary bending measurements

strain gauge measurements are applied, usually a pair of gauges bonded at the same locations on the free and on the faying surface of the sheet, in the same way as for the load transfer measurements (see gauges B1 and B2, Fig. 5.16a), is utilized. For reasons explained in Sect. 5.3, this measurement method is mainly applicable to metallic plates. The local bending stresses can then be determined from Eq. 6.17 assuming that σ_{free} and σ_{fay} denote the axial stress measured by the strain gauge on the free and faying surface of the sheet respectively. If the gauges are zeroed after the joint assembling, the measured stresses do not include the residual stress contribution. The accurate positioning of the gauges at the same location on either surface of the sheet is of crucial importance to the measurement accuracy.

Alternatively, a special strain gauge shown in Fig. 6.15, which consists of two measurement grids lying one over the other and joined together by a layer of synthetic resin, is sometimes applied (Schütz and Lowak 1974; Schra et al. 1995). The gauge is bonded on the accessible side of the component. The total strains on the free surface and on the faying surface are determined by linear extrapolation of the measured strains ε_1 and ε_2, as illustrated in Fig. 6.15. Then the bending and axial stress components can be derived in the same way as in the case of gauges bonded at both sides of the sheet. According to Schütz and Lowak (1974), who widely applied this type of gauge in SB measurements for both simple specimens and real aircraft structures, the measurement error in the case of Al alloy sheets of 1–6 mm in thickness was up to -5% and -10% for the lower and upper measuring grid respectively. Because of the extrapolation required, these errors have all the more effect on the measurement results on axial and bending stresses, the thicker the component. Due to its higher accuracy, the determination of SB with the help of two gauges at both sides of the sheet is preferred to measurements with the double-grid strain gauge. The latter, however, can be applied to obtain an indication of SB if only one side of the component is accessible, e.g. in a real aircraft structure. This method can be especially useful for comparative measurements because the error is approximately the same for all results.

An alternative experimental approach to study SB is based on measurements of the joint out-of-plane displacements along lines parallel to the loading direction (x). If the deflection function $w(x)$ is known, the bending stress at the outer fibres of the sheet can be derived, based on Eq. 6.5, as

$$\sigma_b(x) = \frac{M(x)}{Q} = w''(x)E\frac{t(x)}{2} \tag{6.18}$$

where Q denotes the section modulus of bending and $t(x)$ is the sheet thickness at location x.

The displacements $w(x)$ can only be tracked on the outer surface of the joint. It is assumed that the error introduced by using in Eq. 6.18 deflections of the outer surface rather than deflections of the neutral axis can be ignored because the ratio of the sheet thickness to the radius of curvature is very small. Müller (1995) scanned the out-of-plane displacements with the help of a linear voltage displacement transducer in contact with the outer surface of riveted lap joint specimens. The out-of-plane deformations contributed by the applied loading were obtained by subtracting the deflections recorded at zero load from those recorded in the loading condition. In order to obtain the second derivative $w''(x)$ he approximated the measured $w(x)$ data using a mathematical function. The type of equations that best described the deflection curve and did not involve instabilities in $w''(x)$ were found to depend on rivet flexibility and sheet flexural rigidity. Even in the case when a good curve fit was obtained for different functions, the corresponding σ_b-values determined from Eq. 6.18 showed differences in the order of 20%. Ekh et al. (2005) did not report, however, similar accuracy problems in the determination of SB for composite-to-aluminium simple lap joint specimens fastened with a single column of several titanium bolts. They tracked deflections on the outer surface of the composite plate in a contact-free manner using a commercial system based on digital speckle photography. The data were collected within two narrow stripes on either side of the fastener column. Three experiments were conducted for each of several joint configurations considered. The numerical post-processing of the measurement data involved several steps. Special measures were taken to assign all experiments equal weight when forming mean values because the spatial resolution of the measurements was dependent on the applied speckle pattern and, therefore, it could be expected to vary between different experiments. The final step of the data processing involved representing the mean values by cubic splines and differentiating twice with respect to the longitudinal coordinate to arrive at the surface curvature $w''(x)$.

Whatever measurement method is applied, only the area beyond the rivet head is accessible, whereas the peak bending stresses in a joint typically occur closer to the rivet hole, as revealed by the FE analyses addressed in Sect. 6.3. When strain gauges are used, the maximum strain values in the region of a higher stress gradient near the rivet head can be lost as the measured strains are averaged over the relatively large strain gauge grid area. The deflection measurement based method can give adequate results at a sufficient distance from the rivet columns because Eq. 6.18

is valid only when the strain varies linearly through the thickness of the plate, which is not the case close to the holes where the strains are affected by the contact between the hole and the fastener. It can be concluded that neither direct nor indirect experimental techniques offer a means to reliably measure the peak bending stresses in mechanically fastened joints.

6.4.2 Comparisons Between Measured and Computed Results

Figure 6.16a shows the configuration of riveted Al alloy specimens applied in strain gauge measurements of SB by Schütz and Lowak (1974) and Brenner and Hübsch (1969), the latter work cited by Hartman and Schijve (1969). In the experiments by Schütz and Lowak the sheets were assembled using countersunk rivets, and the measurements were made with the help of double-grid gauges bonded on the strap outer surface (cf. Fig. 6.15), while protruding head rivets and two standard strain gauges bonded at both sides of the strap were applied by Brenner and Hübsch. The gauge locations relative to the rivets were the same in both studies and are shown in Fig. 6.16a and b. The measured results on K_b (according to Eq. 6.1b) produced in Fig. 6.16c indicate that the values from Schütz and Lowak are a factor of 1.3–1.8 lower than those from Brenner and Hübsch. In both works measurements of SB were also carried out for the sheet (the outer rivet row) revealing somewhat smaller discrepancies between the K_b-values. It is seen in Fig. 6.16c that results computed from Schijve's model (cf. Sect. 6.2) almost double the K_b-values according to Schütz and Lowak. These authors observed the same order overestimation of the measured results by the model predictions for several configurations of a lap joint assembled with Hi-Lok bolts. Schütz and Lowak attribute this disagreement to permanent deformation of the sheets evidenced in the lap joints used in their experiments by angle β up to 2° (cf. Fig. 6.1b). The corresponding reduction of SB, though possible to be covered in Schijve's model, has not been, however, accounted for in their predictions. At the same time, a fairly good agreement is shown in Fig. 6.16c between the experimental data according to Brenner and Hübsch and the model predictions, especially at lower stress levels. Another result from the work by Brenner and Hübsch reported by Hartman and Schijve (1969) is a good agreement between the measured and calculated value of the maximum out-of-plane deflection of the joint (1.25 versus 1.14 mm respectively). It is likely that errors incurred in measurements with the double strain gauge could additionally contribute to the discrepancies observed in Fig. 6.16c between the experimental results by Schütz and Lowak and the predictions.

The present authors with co-workers (Skorupa et al. 2011) carried out strain gauge measurements of bending stresses in a riveted lap joint specimen at several locations. One of the goals was to compare the experimental results close to the critical, end rivet row with the predictions from Schijve's model. D16 Alclad, 2 mm thick sheets were connected using PA24 round head rivets, 5 mm in diameter. The overlap configuration and positioning of the strain gauges mounted in pairs

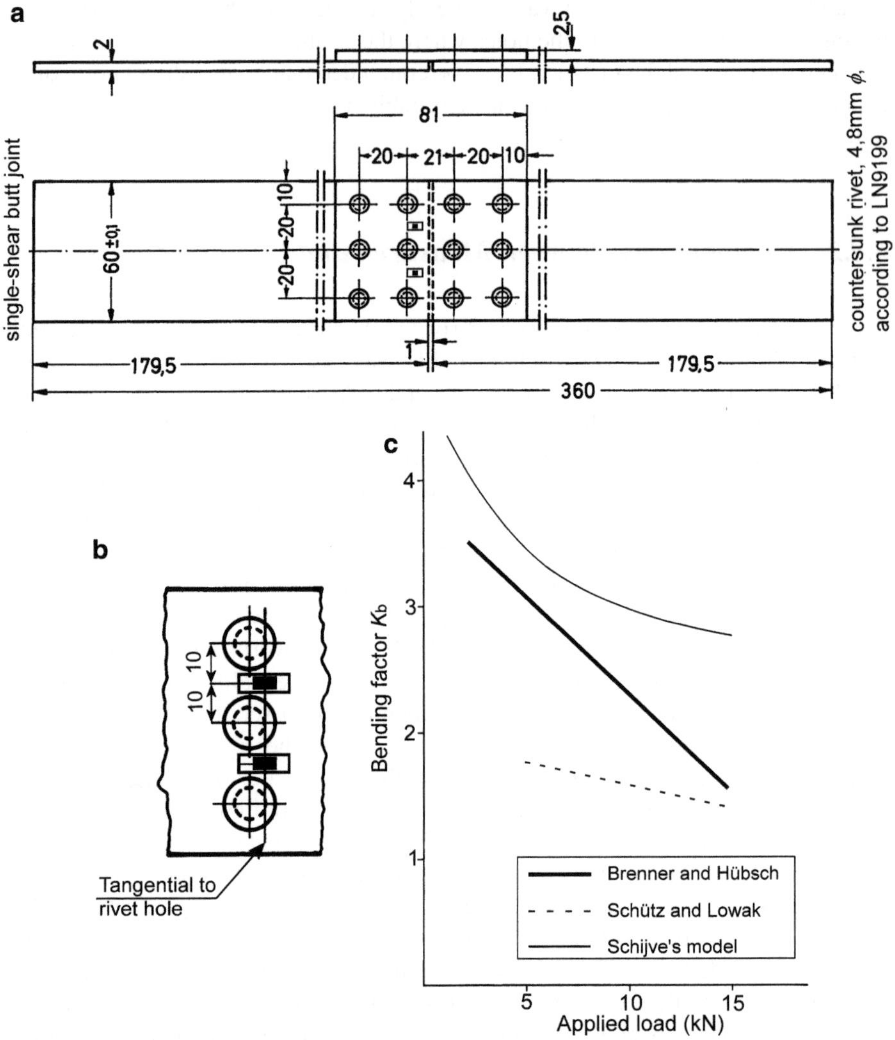

Fig. 6.16 Comparison between the bending factors according to strain gauge measurements by Schütz and Lowak (1974) and Brenner and Hübsch (1969) and predictions from Schijve's model: (**a**) specimen geometry; (**b**) gauge locations; (**c**) K_b versus applied load plots

at the same locations on the free and faying surface of the sheet are shown in Fig. 6.17a. Additionally, a monolithic dummy specimen shown in Fig. 6.17b was also considered. Its geometry conforms to that assumed by Schijve to model SB in a lap joint (cf. hatched area in Fig. 6.1b). The location of the gauges bonded on the dummy specimen was the same as in the case of gauges G1/G2 for the riveted joint. The gauge grid length and width were 2 and 0.9 mm respectively. The bending stresses measured at several applied stress levels for the riveted joint at

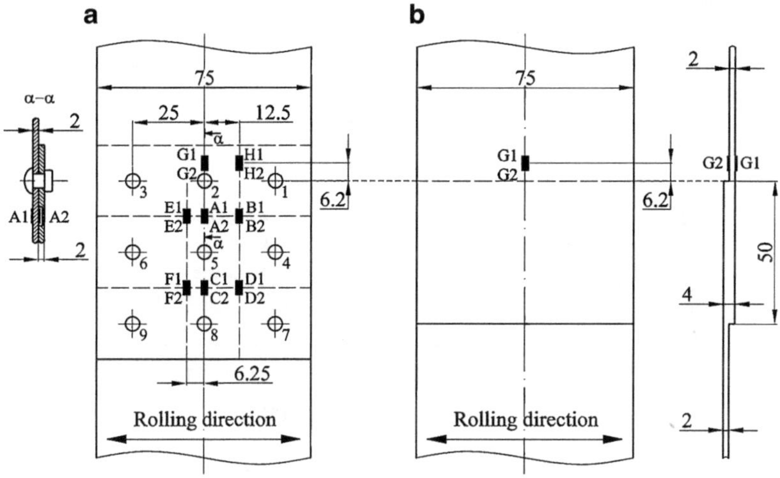

Fig. 6.17 Instrumentation for secondary bending measurements: (**a**) riveted specimen; (**b**) dummy specimen (Skorupa et al. 2011)

the end rivet row and the results for the dummy specimen are plotted in Fig. 6.18a. Though the distance of gauges G1/G2 from the rivet hole centre equals nearly 2.5 rivet shank radius, the concentration of bending at the rivet is still detected because the measured σ_b-values are consistently higher at location G than at location H midway between the rivet columns. Obviously, the bending factor defined according to Eq. 6.1b exhibits a similar behaviour, namely it adopts lower values at location G than at location H, as seen in Fig. 6.18b. At the same time, Fig. 6.18b reveals that the bending factor determined according to Eq. 6.1c, the local σ stress due to the axial force being derived from the strain gauge measurements as $(\sigma_{\text{fay}} + \sigma_{\text{free}})/2$, obtains the same value at G and H at a given applied stress level. A similar behaviour of the local bending factor (Eq. 6.1c) has also been noted in either of the two cross sections between the rivet rows because the K_b-values measured at various locations in a given section turn out to be very alike. Specifically, the local K_b at locations F, C and D ranges between 0.18 and 0.19, and at locations E, A and B it varies between 0.13 and 0.16. A tentative conclusion from the above observations is that at a given location and stress level the multiplication factor due to the stress concentration at the rivet holes is the same for the bending stress and for the stress due to the axial force in the sheet. It is seen in Fig. 6.18a that, as it could be expected, the stresses measured for the dummy specimen are between those at locations G and H for the riveted specimen. Figure 6.18 further reveals that in terms of stresses, the predictions fall between the values measured at location G and H of the riveted specimen. Obviously, the same holds true for the K_b factors defined according to Eq. 6.1b. The largest underestimation by the model of the measured stresses and K_b factors (Eq. 6.1b) at location G is below 7% and occurs at an intermediate applied stress level. Because the local K_b (Eq. 6.1c) is lower than K_b according to

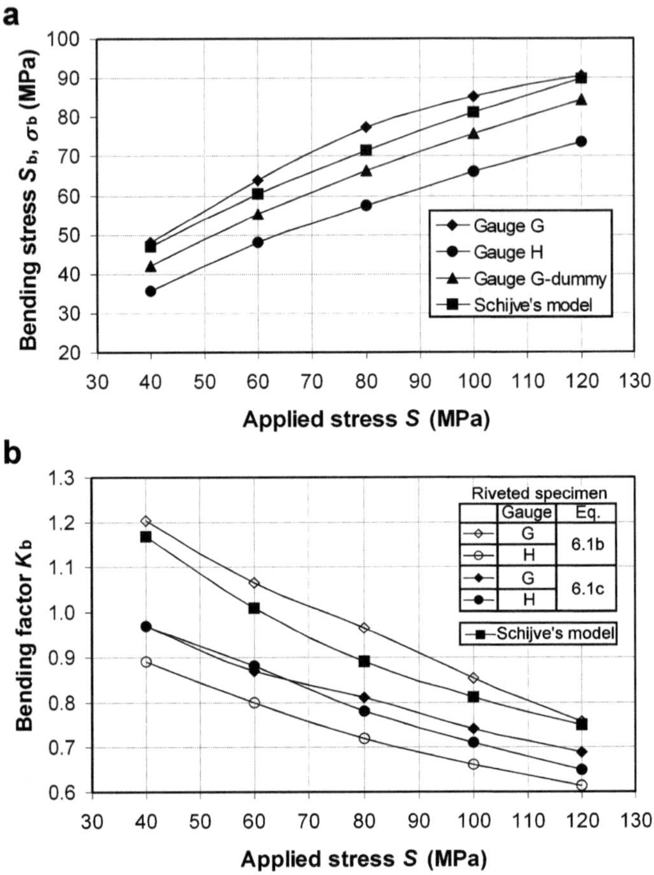

Fig. 6.18 Comparisons between measured and computed secondary bending in terms of: (**a**) bending stresses; (**b**) bending factors (Skorupa et al. 2011)

Eq. 6.1b, the predictions slightly overestimate this local K_b. A drastic overprediction by the model of the bending factor measured midway between the rivets in the critical row, as observed by Schütz and Lowak (cf. Fig. 6.16c), is not confirmed in Fig. 6.18b because the differences between the predictions and the K_b-values (Eq. 6.1b) measured at location H do not exceed 22%.

Somewhat against expectation, Fig. 6.18a indicates that the model predictions exceed the stresses measured for the dummy specimen. An about 5% overestimation by Schijve's model of the bending stresses measured in a bonded single strap joint of Al 2024-T3 2 mm thick sheets was also reported by Fawaz and de Rijck (1999). As in the case of the dummy specimen, the configuration of a bonded joint conforms precisely to that considered in the beam model. Fawaz and de Rijck attributed the discrepancy between the predictions and experimental results to a lower stiffness of the actual joint, due to the presence of the adhesive layer, than had been assumed in the model. The above interpretation is not, however, applicable in the case of the monolithic dummy specimen for which similar discrepancies are exhibited in Fig. 6.18a. A possible reason for overestimation of the bending stresses in the dummy specimen by the model can be simplifications involved in the elementary theory of beams.

In accordance with other theoretical and experimental studies referred to in the present work, strain gauge measurements by Jarfall (1986) for a single strap joint with two fastener rows per sheet and a single fastener column showed a pronounced 3-D distribution: high secondary bending close to the fastener and a rapid decay when moving away from the fastener. The strain gauges were bonded at the AGARD points (cf. footnote in Sect. 6.3) on both sides of the sheet. Even at locations close to the fasteners, Jarfall reported local bending factors (Eq. 6.1c) only half the K_b magnitude calculated from Schijve's model. The fact that the local K_b close to the rivet is always lower than K_b determined in terms of the nominal applied stress could only partly contribute to this discrepancy. Also K_b's (Eq. 6.1b) determined from strain gauge measurements in an effort by Fawaz (2003) for a riveted specimen shown in Fig. 9.11 never exceeded 0.36, which was again much below estimates from the beam model.[2]

All in all, the available literature evidence reviewed above is not conclusive regarding the capability of the beam model to quantitatively estimate the magnitude of SB. Both results indicating a significant overestimation by the model of the bending stresses measured at locations close to the fasteners in the critical row (Jarfall 1986) and results showing a slight underprediction of these stresses (Skorupa et al. 2011) can be found. While reasons for the above inconsistency are not fully clear, it is obvious from the earlier considerations in this chapter that neither the model nor experiments are capable of providing the maximum values of bending stresses in the joint required for predictions of fatigue behaviour. At the same time, both the measurements and the beam model come in useful for obtaining an impression of the effect of various factors on the magnitude of SB. While the beam model is predominantly fit for studying the SB dependency on some geometric variables and the applied load, as amply exemplified in Sect. 6.2.2, the measurements offer, in addition, a means to evaluate the effect of the fastener system.

6.4.3 *Parametric Studies*

Experimental results according to Schütz and Lowak (1974) and Ekh et al. (2005) indicate a pronounced influence of the overlap length on the magnitude of SB for single shear lap joints. Measurements by Schütz and Lowak were performed on Al

[2]According to Fawaz (2012, private communication), possible reasons for the low K_b could be: gauge location at a distance from the rivet, gauge placement not exactly back to back and rivet tilting.

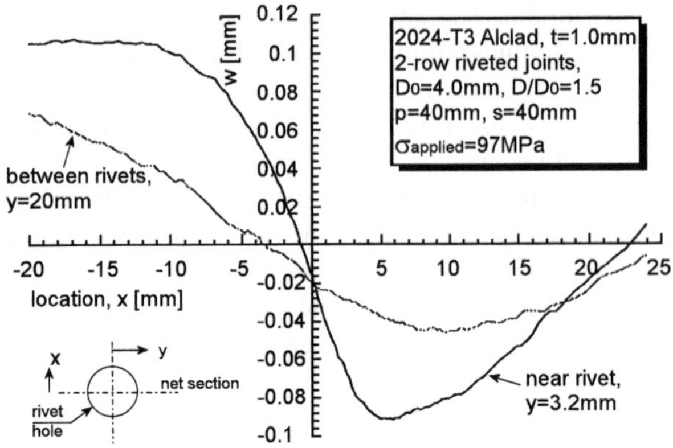

Fig. 6.19 Deflection measurement results at the bottom rivet row close to the rivet and midway between rivets in the row (Müller 1995)

alloy specimens of 3 mm thick sheets connected with two rows and three columns of interference fit 4.8 mm diameter steel bolts. For the distances between the fastener rows of 20, 35 and 60 mm, the double-grid strain gauges bonded between the rivets in a row, as shown in Fig. 6.16b, indicated bending factors defined according to Eq. 6.1b of 1.10, 0.87 and 0.71 respectively. Ekh et al. (2005) presented their results of SB measurements using digital speckle photography (cf. Sect. 6.2.1) for a single shear lap joint specimen 48 mm in width consisting of a 4.16 mm thick composite plate and an Al alloy plate assembled using a single column of titanium bolt fasteners. The reference configuration involved an Al alloy plate thickness of 8 mm and four protruding head fastener rows with row spacing of 32 mm. If the row spacing was reduced to 20 mm, the bending strain $\left(= w''(x)\frac{t(x)}{2}\right)$ increased by about 11% compared to the reference configuration. If the thickness of the Al alloy plate was reduced to 4 mm, the bending strains became 20% lower. Because the bending deflections $w(x)$ were scanned along the line about 20 mm distant from the fastener column, no indication of the SB response close to the fasteners, which is the location of more interest, has been obtained.

In agreement with other experimental and numerical works referred to earlier in this chapter, comparisons between deflection measurements along the line close to a rivet in the outer row and midway between the rivets in that row provided by Müller (1995) indicate a much larger curvature and, hence, a more severe bending near the rivet than in the centre of the rivet pitch, as exemplified in Fig. 6.19. For a lap joint of 1 mm thick 2024-T3 Alclad sheets connected with three rows of 4 mm diameter countersunk rivets, the ratio of σ_b at the rivet to σ_b between the rivets equals about 2.7. For a joint with the same overlap length but only two rivet rows, this ratio amounts to as much as 6.8, which implies still more inhomogeneous bending stress distribution in the width direction. It is worthwhile to note that the same order of inhomogeneity is exhibited by Rans's FE results (cf. Fig. 6.13b). Near the rivets, the K_b factors (according to Eq. 6.1b) from Müller's measurements for a three-row lap joint are slightly above unity, but for a two-row joint they exceed the value of 1.5. Müller's results further reveal that the maximum bending stress at the rivet driven head occurs slightly above the rivet row centre line, i.e. away from the net section of

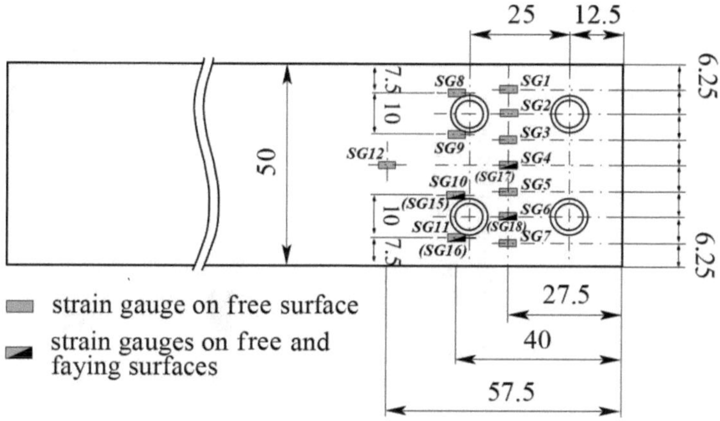

Fig. 6.20 Specimen instrumentation for secondary bending measurements by Starikov (2002)

the sheet. A still larger shift of the location of the peak bending stress from the rivet row centre line occurs midway between the rivets. The above observation on the location of maximum bending is again qualitatively consistent with the FE results by Rans, see the insets in Figs. 6.13 and 6.14.

The aforementioned measurements by Ekh et al. (2005) reveal a negligible influence of the bolt diameter and the amount of clamping as well as a minor effect of the head geometry (countersunk or protruding head) on SB measured remotely from the fasteners.

An investigation by Starikov (2002) aimed at obtaining evidence of the influence of the clamping force and of the anti-buckling fixture on the magnitude of SB in lap joint specimens made of 3 mm thick 7475-T761 sheets connected using titanium alloy countersunk Hi-Lok fasteners with a shank diameter of 6 mm. The specimen instrumentation is shown in Fig. 6.20 where the gauge numbers given in brackets correspond to the faying surface. To reduce friction, Teflon sheets were interleaved between the specimen and the anti-buckling support. The accuracy of the gauge measurements has been confirmed by a good repeatability of the records from three identical specimens. Exemplary results on K_b (Eq. 6.1b) obtained at an applied stress of 150 MPa during a quasi static loading, which involved two full cycles between 0 and 150 MPa, are shown in Fig. 6.21 where the notation finger torque, 3.5 kN (initial pre-tension due to the applied torque) and high torque corresponds to increasingly higher clamping force values. In accordance with the results by Ekh et al. (2005) addressed earlier, no clear dependence between the degree of SB and the amount of clamping can be observed close to the fasteners, as indicated by the results from gauges 10/15 and 11/16. SB is much lower in the overlap area between the bolt rows, where gauges 4 and 17 are located, than measured by gauges 10/15 and 11/16 positioned close to the fasteners, as also observed by Änggård (2000).

Measurements and FE results for riveted and bolted joints reviewed in this chapter suggest certain similarities in the SB behaviour of both types of joint,

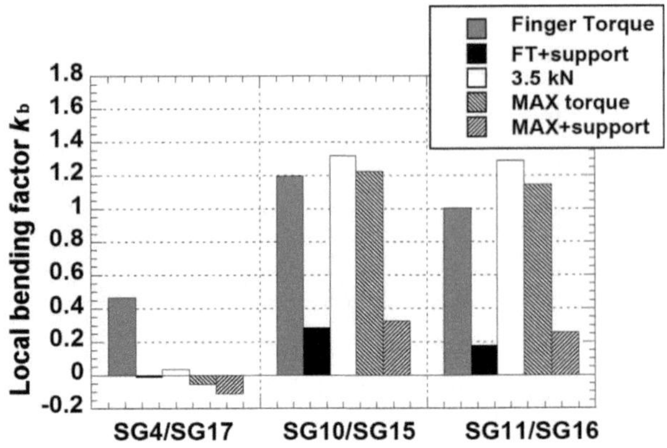

Fig. 6.21 Effect of the clamping force and the anti-buckling support on the bending factor (Starikov 2002)

namely similar variations of the bending stresses in the longitudinal and transverse direction and a minor influence of the clamping force. The available experimental and numerical evidence implies that regardless of the fastener type, the geometry parameters, such as the overlap length between the end fastener rows and the plate thickness, are the most important factors affecting the magnitude of SB in a joint.

An expected trend in Starikov's (2002) experiments revealed in Fig. 6.21 is the radical reduction of SB by the anti-buckling support due to preventing specimen deflections over the whole surface of the joint. The result reported by Änggård (2000) for identical specimens and the same location of strain gauges as in Fig. 6.20 are between the values obtained by Starikov with and without the anti-buckling support. Most probably, this difference stems from different gaps between the specimen and the anti-buckling plates in either study. Measurements by Jarfall (1986), utilizing two strain gauges on both sides of the strap in a single strap joint with two fastener rows per sheet (the fastener type not specified), indicate that a localized lateral reinforcement can cause a redistribution rather than a suppression of SB. In the case of a reinforcement that is unsymmetrical relative to the fastener hole, SB was considerably less severe on the side of the hole close to the reinforcement than on the opposite side of the hole. As seen in Fig. 6.7, simple analytical models indicate an insignificant effect of a symmetrically located reinforcement.

6.4.4 In Situ Measurement Results

Measurements of SB in the actual fuselage structure under service loading conditions are scarce in the literature. Among the few examples is a test program carried out by Schütz and Lowak (1974) which involved bending stress measurements using double-grid gauges (cf. Fig. 6.15) bonded on typical aircraft structural details, including fuselage skin joints. Exemplary results for a fuselage lap joint with a bonded and riveted doubler and a stringer, shown in Fig. 6.22a together with the

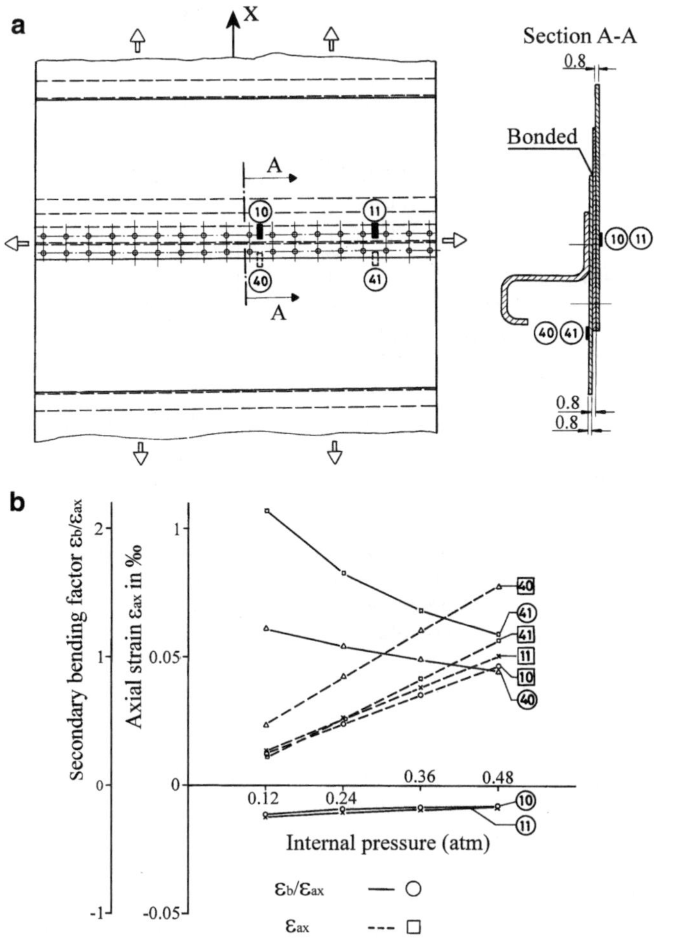

Fig. 6.22 In situ measurements of secondary bending in a fuselage lap joint: (**a**) instrumentation; (**b**) axial strains and bending-to-axial strain ratios (Schütz and Lowak 1974)

gauge locations, are produced in Fig. 6.22b. A low bending factor measured by gauges 10 and 11 indicates that the stringer effectively reduces SB at this location. It is obvious that a pronounced SB effect at the location of gauges 40 and 41, evidenced by the high bending factor values, is caused by the large eccentricity due to the thickness transition. The difference between results from gauge 40 and 41 may stem from a disbond between the sheet and the doubler near gauge 40. Consequently, the inner sheet at that location carries more load than in the case of a complete bonding, as suggested by higher axial strains measured by gauge 40 compared to gauge 41. Because in the disbond region the doubler and the outer sheet do not pick up the load, the location of the load path eccentricity moves away from overlap edge towards the fastener row.

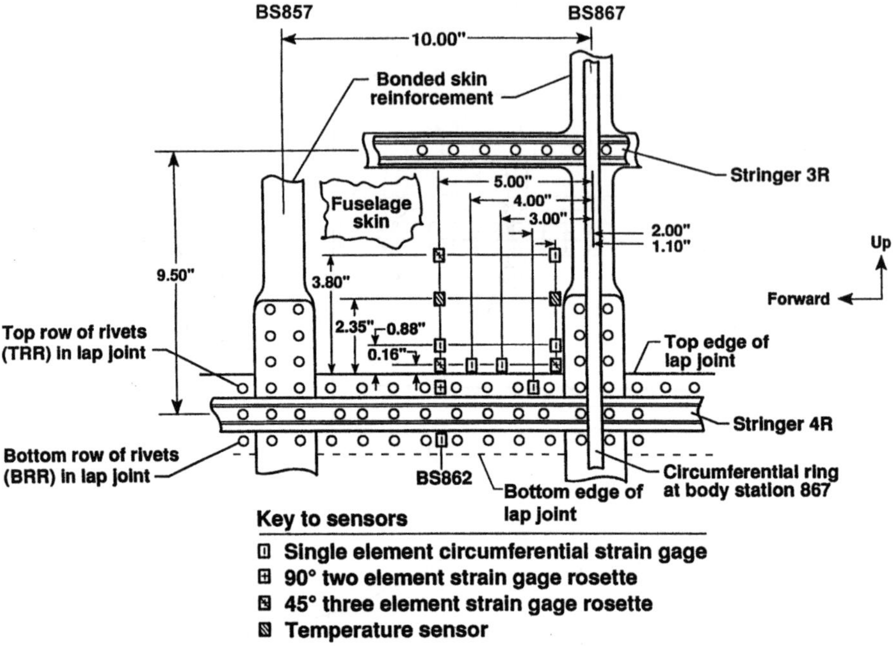

Fig. 6.23 Location of strain gauges on the inside of the fuselage skin (Phillips and Britt 1991)

A closer insight into the results presented by Schütz and Lowak indicates that due to the presence of lateral reinforcements, SB can be locally considerably reduced compared to observations on simple specimens without lateral supports, as exemplified in Fig. 6.22b by the gauge 10 and 11 results. Out of 129 measurement results indicating the presence of SB, in 34 cases the local bending factor was above 0.8, including 14 cases when it was higher than 1.4. Because of, on one hand, locating the gauges midway between the rivets (according to Fig. 6.16b) and, on the other hand, a relatively low accuracy of the double-grid gauge, these results, though instructive and useful for comparisons between various design solutions, cannot be considered to provide quantitative information on the magnitude of SB in a specific structural detail.

Phillips and Britt (1991) installed strain gauges in a small area near a longitudinal riveted and bonded lap joint in the fuselage skin of a B737 aircraft during the pressurization cycle while the aircraft was on the ground. The skin material was a 0.036 in. thick 2024-T3 Alclad sheet; the rivet type and diameter have not been reported. It was assumed that due to complete bonding of the joint, the overlap region responded structurally as a monolithic sheet. Therefore, evaluation of bending strains for the overlap region was based on readings from pairs of the gauges placed at the same locations on the outside and inside surface of the fuselage, as shown in Fig. 6.23. For this configuration, bending strains due to the joint eccentricity and those due to pressure pillowing are of the same sign at the top side of the joint and of the opposite sign on the bottom side. Hence, an unsymmetrical

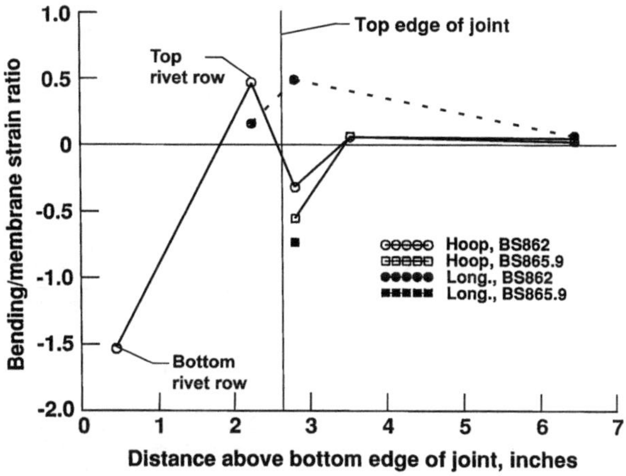

Fig. 6.24 Measurement results on the local bending factors in the hoop and longitudinal direction at body stations BS862 and BS865.9 (1.1 in. from BS867 forward) at a differential pressure of 6.2 psi as a function of vertical location relative to the bottom edge of the joint, cf. Fig. 6.23 (Phillips and Britt 1991)

distribution of the bending about the stringer at the middle of the joint is expected due to enhancement of bending at the top of the overlap and reduction of bending at the bottom. An opposite trend is, however, revealed by measurement results on the bending strain to membrane strain ratio for the hoop direction presented in Fig. 6.24 where the bending strain is defined as (outside strain – inside strain)/2. One of the possible reasons can be areas of disbond detected by ultrasonic inspections on the strain gauged portion of the joint. This is partly substantiated by the membrane strain measurements which show that the hoop strains measured at the bottom row of rivets and the top row of rivets at BS862 differ by a factor of 1.25, a result that is not reasonable for a well-bonded joint. Because the membrane strain measurements produced by Phillips and Britt indicate the hoop to longitudinal strain ratio of about 1.3 at the top row rivet and of about 1.85 for the sheet, the data in Fig. 6.24 suggest significant absolute values of the bending strains in the hoop direction and a much weaker bending effect in the longitudinal direction.

Development of cracks in a riveted lap joint causes a reduction in SB, as follows from strain gauge measurements by Chen et al. (1999) on a fuselage panel addressed in Sect. 10.4. Consequently, relatively low bending factors are assumed in fatigue crack growth analyses, see Chap. 9.

6.5 Fatigue Behaviour of Joints Exhibiting Secondary Bending

6.5.1 Effect of Secondary Bending on Fatigue Life

As said in Chap. 4, the effect of some design variables, such as sheet thickness or rivet row spacing, on the fatigue performance of riveted lap joints is attributed

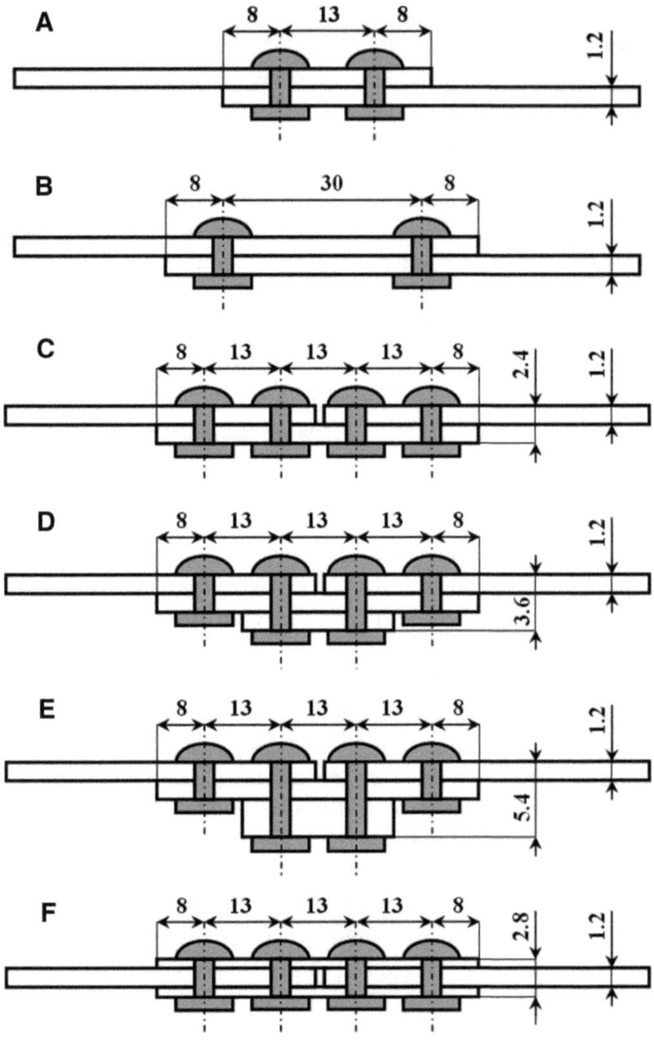

Fig. 6.25 Configurations of riveted joint specimens considered in fatigue tests by Hartman and Schijve (1969)

to variations in SB. In several works a correlation between the magnitude of SB and the fatigue life of joints was sought. An example is the study by Hartman and Schijve (1969) who carried out fatigue tests under CA loading with a mean stress of 69 MPa and under a programme loading on Al 2024-T3 riveted specimens of several configurations shown in Fig. 6.25. Differences in the amount of SB were obtained by varying the rivet row spacing for the lap joint specimens (A and B) and by changing the number and thicknesses of straps for the strap joints (C, D and E). A symmetric double strap joint (F), for which SB does not occur, served as a

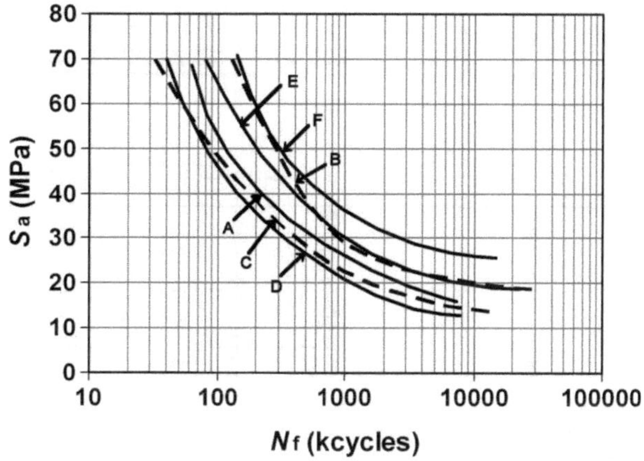

Fig. 6.26 *S-N* curves from fatigue tests of the riveted specimens shown in Fig. 6.25 (Hartman and Schijve 1969)

reference case. The set of stress-life curves for all joints is shown in Fig. 6.26 where S_a is the applied stress amplitude. The bending factor calculated using Schijve's model was found to rate the various joint configurations in a reasonable consistence with the observed fatigue performance. The correlation was, however, upset by the type D joint fatigue behaviour because the order of joints according to increasing fatigue strength was D, C, A, B/E, F, while according to the bending factor value it should be C, A, D/B, E, F. Subsequent fatigue tests by Hartman (1972) under the same type of programme loading performed on 7075-T6 Alclad specimens of type C, D and E confirmed the poor fatigue properties of joint D compared to C. It can be demonstrated that if the fatigue lives observed in the tests by Hartman and Schijve are presented against the amplitude of the combined, i.e. including the SB component, tensile stress $(S + S_b)_a$ with S_b computed from Schijve's model, the scatter of the data is equally large as in the case of the S_a vs. N_f presentation (Skorupa and Korbel 2008). Evidently, the magnitude of SB was not the only factor behind the differences in fatigue performance of the joints considered by Hartman and Schijve. The change in design between lap joints and single shear and double shear strap joints alters not only the severity of SB but also the load transfer. In the case of the strap joints (C, D and E), variations in the number and thicknesses of layers in a stack yielded, in addition to differences in the bending factor, differences in both the load transmission and hole expansion.

If joints differ solely in the magnitude of bending, a good quantitative correlation of calculated or measured SB with fatigue lives can be observed. It can be shown that the $(S + S_b)_a$ vs. N_f results for two lap joints (A and B, see Fig. 6.25) considered by Hartman and Schijve (1969) show an improved consolidation along a single stress-life curve compared to the S_a vs. N_f data (Skorupa and Korbel 2008).

A good consolidation of fatigue test data, when fatigue lives are presented against the combined stress amplitude, has been also reported by Schütz and Lowak (1974) for aluminium alloy lap joints with steel bolts and aluminium alloy rivets fatigued under CA and VA random loading. Their results for specimens with steel bolts of the

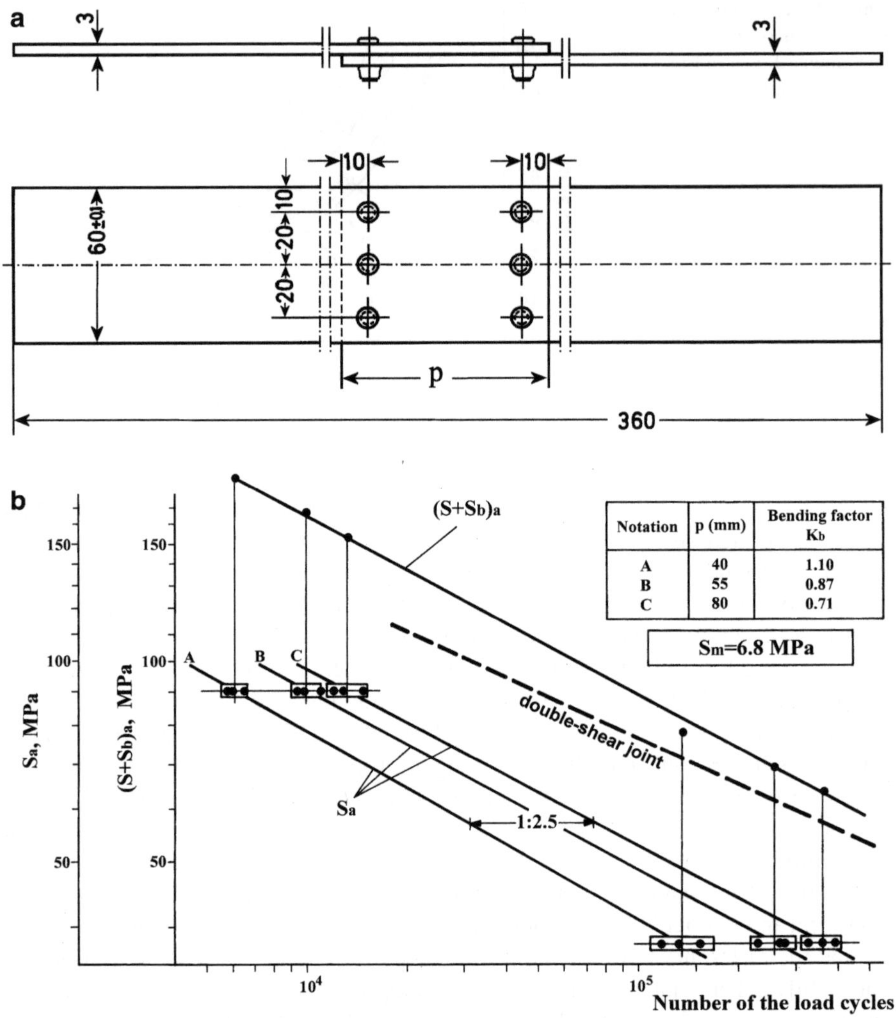

Fig. 6.27 Bolted lap joint geometry (**a**) and fatigue test results presented in terms of the applied and combined tensile stress amplitude (**b**) (Schütz and Lowak 1974)

geometry shown in Fig. 6.27a are plotted in Fig. 6.27b. The bending factor has been determined from double-grid gauge measurements (cf. Fig. 6.15) at locations shown in Fig. 6.16b. The S_a vs. N_f data in Fig. 6.27b indicate, in agreement with many other works (cf. Sect. 4.2), that decreasing the bending factor by increasing the spacing between the rivet rows yields a shift to the right of the corresponding S_a vs. N_f curve. At the same time, minute deviations of the mean $(S + S_b)_a$ vs. N_f data points from the corresponding regression line demonstrate a very good correlation of the observed lives with the amount of SB. Similar to Fig. 6.27 results, but for three-row riveted

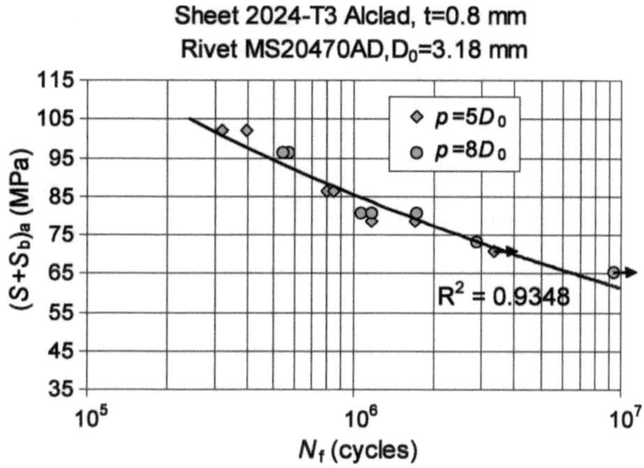

Fig. 6.28 Fatigue lives of riveted lap joints with two different row spacings presented against the combined tensile stress amplitude. Bending stress computed from Schijve's model (Skorupa et al. 2011)

lap joints with two different row spacings, are produced in Fig. 6.28. Note in Fig. 4.9 that when fatigue lives observed in these experiments are presented against the S_a amplitude, the results for either spacing are considerably apart from one another.

Figure 6.29 demonstrates a capability of the $(S + S_b)_a$ vs. N_f presentation to combine within common scatter bands the fatigue test data for three-row riveted lap joints with several sheet thicknesses, different rivet diameters and different row spacings. The correlation coefficients (R^2) quantify a significant consolidation of the results in Fig. 6.29b compared to Fig. 6.29a. It is interesting to note that, for the joint geometries considered in Fig. 6.29, the common D/D_o ratio of 1.5 implies also the common squeeze stress (=squeeze force/rivet shank area).

In Fig. 6.30a and b fatigue test results for lap joint specimens of the standard and staggered thickness configuration with three rows and six columns of rivets are compared. The staggered thickness specimen longitudinal section dimensions are given in Fig. 6.11. The magnitude of SB was estimated using Schijve's model, for details see Sect. 6.2.2.2. It is seen in Fig. 6.30a that the computed K_b does rate the fatigue performance for all three configurations, whereas Fig. 6.30b provides another example of a considerable consolidation of the fatigue lives if the bending stress is accounted for. The results presented in Fig. 6.30 correspond to specimens riveted under the load control to obtain the D/D_o ratio of 1.3. A similar behaviour has been observed for specimens riveted with a higher squeeze force yielding $D/D_o = 1.5$. However, while the results in Fig. 6.30 show a more beneficial influence of the thickness staggering at higher stress levels, the reverse trend has been exhibited in the case of the $D/D_o = 1.5$ specimens. As said earlier (see Sect. 6.2.2), in addition to reducing the level of SB, the thickness staggering is supposed to improve the load transfer through the joint. It is likely that this occurs to a greater extent in the case of the $D/D_o = 1.3$ specimens because for the $D/D_o = 1.5$ specimens a fairly homogenous load transfer through the joint has been observed even for the non-staggered configuration, see Fig. 5.17. The improvement of load transmission should manifest predominantly at higher applied

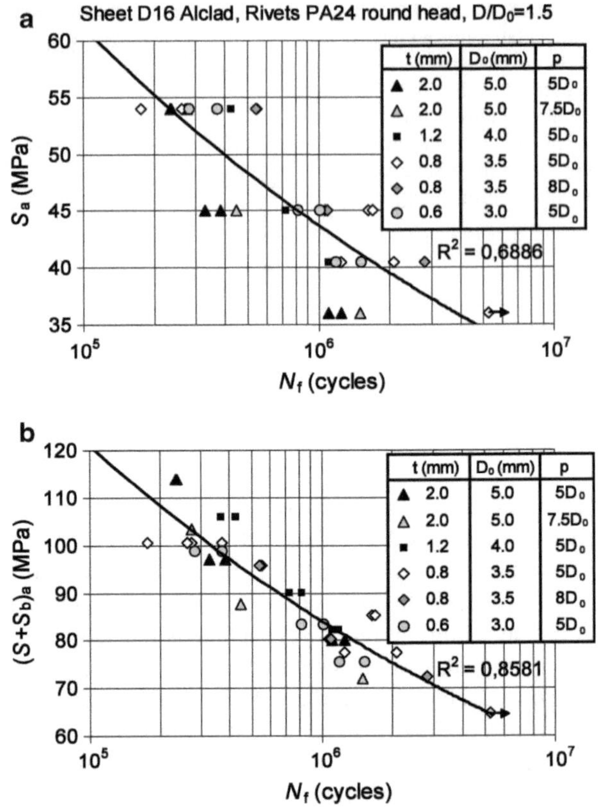

Fig. 6.29 Fatigue lives for riveted lap joints of various geometries presented against applied (**a**) and combined (**b**) stress amplitude. Stress ratio 0.1; bending stress computed from Schijve's model (Skorupa et al. 2011)

stress levels due to the more significant contribution of the bearing force to load transfer, as indeed exhibited in Fig. 6.30a. As detailed in Sect. 3.3.5.3, the staggering operation was found to alter rivet hole expansion in the critical rows as well. A decrease in hole expansion in the 2 mm thick sheet under both the manufactured and driven rivet head was observed for the staggered thickness $D/D_o = 1.3$ specimens (cf. Fig. 3.64), while no change was revealed for the $D/D_o = 1.5$ specimens (cf. Fig. 3.65). Considering the manifold consequences of staggering thickness, some effect of the squeeze force on the joint response to this operation is not surprising.

The point of the literature evidence referred to above is that it unambiguously confirms a deteriorating influence of SB on the fatigue performance of mechanically fastened joints. Reasonably accurate quantitative predictions of the SB effect on fatigue performance utilizing measurements or computations are, however, only possible if SB is a major factor contributing to differences in fatigue behaviour between various joints.

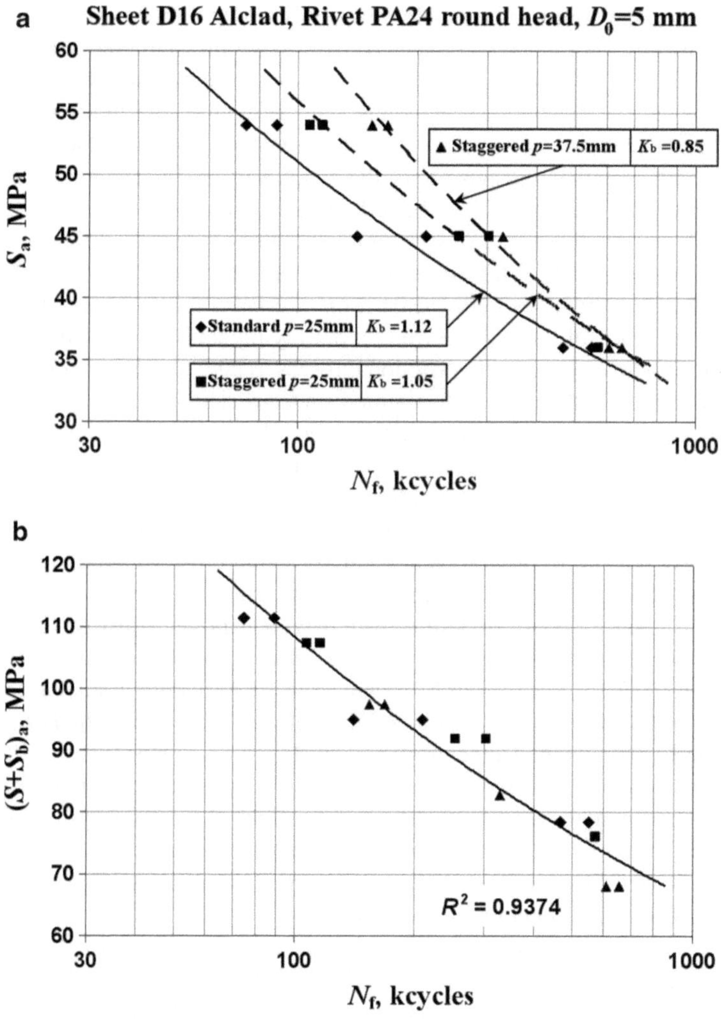

Fig. 6.30 Fatigue lives for standard and staggered thickness riveted specimens sketched in Fig. 6.11 presented against applied (**a**) and combined (**b**) stress amplitude. Stress ratio 0.1; bending stress computed from Schijve's model (Skorupa et al. 2009)

6.5.2 *Effect of Faying Surface Conditions*

In several works, different responses in terms of fatigue life to the change of surface coating for joints exhibiting SB and for non-bending joints have been observed. Schütz and Lowak (1974) report on comparative fatigue tests under a flight simulation loading representative for the transport aircraft wing component performed on lap joints and double strap joints. For either joint type, 2024-T3 Alclad, 1.6 mm thick sheets connected using 4.8 mm diameter universal rivets, and 3.0 mm thick sheets assembled with 4.8 mm diameter steel Huck-Bolts were used. In the case of the double strap joints, a threefold increase in life was exhibited for anodized sheet surfaces compared to alodined and painted surfaces. For the lap joints, anodizing either did not affect the fatigue lives (riveted joints) or yielded nearly three times shorter lives (bolted joints) than in the case of the painted surfaces. To rationalize these observations, the authors invoke different modes of fatigue failure for either of the configurations. Fatigue cracks in double strap joints were nucleated at the holes in the net section of critical fastener rows, while for the lap joints the crack initiation occurred on the faying surface, away from the holes. The brittle anodic coating causes frictional forces to take over a larger portion of the transfer load than in the case of the painted surfaces. This relieves the bearing load, thus decreasing stress concentration at the rivet holes, but at the same time promotes fretting failure on the sheet interface. The anodizing was, therefore, likely to cause less severe stresses at fatigue critical locations in the case of double strap joints, but for lap joints it could create more severe fretting wear conditions.

The above reasoning suggests that the detrimental effect of anodizing on fatigue life, as conditioned by crack initiation apart from the hole, should be more prominent in the case of lap joints riveted with a larger squeeze force. Such a behaviour was indeed observed by Skorupa et al. (2010b) in fatigue tests under CA loading ($R = 0.1$, $S_{max} = 120$ MPa) on 2024-T3 Alclad lap joints of 2 mm thick sheets connected with universal, 4.8 mm diameter AD rivets. For the D/D_o ratios of 1.4, 1.5, and 1.6 they noted a factor of 1.22, 1.30 and 1.36 respectively reductions in life for anodized sheet surfaces compared to the non-anodized condition. Another piece of experimental evidence of the positive effect of the interfaying layer with a small friction coefficient on the riveted lap joint fatigue performance is provided by Müller's (1995) experiments showing a twofold increase in fatigue life for specimens with interfaying teflon foils.

A joint type dependent effect of the interface surface condition on fatigue life has been also observed by Schijve et al. (1977) in comparative tests with and without application of anti-corrosion penetrant on riveted and bolted joints of 2024-T3 Alclad and anodized sheets. The flight simulation loading was of the same type as in the tests by Schütz and Lowak (1974) referred to above. Penetrant is supposed to lessen friction forces at the interfaying sheet surfaces, thus shedding more load on the fastener. The results by Schijve et al. indicate that if the reduction of friction is caused by application of a penetrant, the fatigue life of a symmetric joint (double strap with Hi-lok bolts) slightly increases, while a 40% reduction in life is observed for a joint exhibiting SB (single strap with aluminium rivets). An interesting observation made in these experiments was that fretting patterns did not show systematic differences between similar specimens with or without penetrant. Examples of a deterioration of the riveted lap joint fatigue properties due to the application of penetrant have been presented in Sect. 3.3.4 (cf. Fig. 3.25).

The trends revealed in investigations into the effect of penetrant disagree with observations on the effect of anodizing referred to earlier in this section. Obviously, in addition to the proportion between the frictional and bearing component of the load transfer, also the type and conditions of the mating surface can affect the failure mechanism, and have, therefore, a profound influence on joint fatigue performance. This conclusion can be supported by Wanhill's (1996) investigation addressed in Sect. 3.3.4 (cf. Fig. 3.24).

6.6 Summary of This Chapter

In joints with eccentricities, such as the lap joints, bending moments referred to as secondary bending (SB) occur under nominally tensile loading. At outer fastener rows, stresses induced by SB can attain values comparable to stresses due to axial forces in the sheets. Regardless of the fastener type, geometry parameters, such as the overlap length between the end fastener rows and the sheet thickness, are the primary factors affecting the magnitude of SB in a joint. The literature evidence unambiguously confirms a deteriorating influence of SB on the fatigue performance of mechanically fastened joints.

Stresses due to SB can be determined experimentally in a direct way, for example utilizing strain gauges, or indirectly, from measurements of the joint out-of-plane displacements along lines parallel to the loading direction. For a number of reasons, such measurements are difficult and offer accuracy problems in regions of a higher stress gradient. Whatever experimental technique is applied, only the area beyond the rivet head is accessible, whereas the peak bending stresses in a joint typically occur closer to the rivet hole.

To study the contribution of SB to the local stress state at rivet holes, elaborated elastic plastic FE analyses including simulations of the riveting process are required. Available results suggest that differences between the SB behaviour for the countersunk and universal rivet only occur in the sheet next to the manufactured head, where the peak bending stresses for the countersunk rivet are considerably higher than for the universal rivet. For either rivet type, an increase in the rivet squeeze force causes a migration of the peak bending stress location in the inner sheet from the net section to above the rivet hole.

Nominal bending stress variations along a lap joint can be derived employing simple models based on the elementary theory of beams under bending. The available literature evidence is not conclusive regarding the capability of the beam model to quantitatively estimate the magnitude of SB. However, the beam model was proved to come in useful for studying the dependency of SB in lap joints on some geometric variables and the applied load level.

Examples produced in this chapter demonstrate that fatigue test data for riveted lap joint specimens of various geometries can become consolidated within common scatter bands when the fatigue lives are presented against the combined bending and tensile stress amplitude, with the bending stress computed from the beam model by Schijve. Such a correlation is, however, only possible if SB is a major factor contributing to differences in fatigue behaviour of various joints.

Strain gauge measurements of SB in skin joints of actual fuselage structures under service loading suggest that owing to the presence of lateral reinforcements, bending stresses can be locally considerably lower than in simple specimens without lateral supports. SB in a lap joint becomes reduced by a permanent deformation of the sheets that can occur at higher applied stresses and by the presence of cracks.

Some pieces of experimental evidence reveal different responses in terms of fatigue life to the change of surface condition for joints exhibiting SB and for non-bending joints. The corresponding observations are, however, difficult to interpret.

Chapter 7
Crack Initiation Location and Crack Shape Development in Riveted Lap Joints – Experimental Trends

In this chapter the nucleation and shape development of cracks in longitudinal lap joints under static and fatigue loading will be considered. The issues of special attention are the influence of a squeeze force on the mode of fatigue failure and the significance of fretting on a sheet interface for fatigue crack initiation.

7.1 Crack Initiation Site

7.1.1 Static Loading

Either sheet or rivet failure can occur under static loading applied on a riveted lap joint. Possible modes of rivet failure are illustrated in Fig. 7.1 in the form of a schematic diagram of the rivet strength observed in tests P_u/D_o^2, where P_u is the load per rivet at failure, presented as a function of sheet thickness to rivet diameter ratio (t/D_o). In general, three regions can be distinguished:

- failure by sheet bearing,
- transition zone, which may be failure by other means such as rivet head pulling through sheet, rivet head break out or burst off, etc.,
- failure by rivet shear.

Test methods and evaluation of rivet strength are given in MIL-HDBK-5 (2001). The test results' evaluation leads to fastener strength tables giving static strength for a combination of sheet and fastener material, sheet thickness and fastener diameter. It seems that failure by rivet shearing is not significantly influenced by sheet thickness, Fig. 7.2. It is evident that the sheet net section along the fastener row must be able to carry the load as well.

The mode of joint failure is in the first place dependent on the rivet pitch in a row to rivet diameter ratio (s/D_o). This conclusion can be backed up by many test results, e.g. of Hartman and Klaassen (1956). Their specimens made of 0.8 mm

A. Skorupa and M. Skorupa, *Riveted Lap Joints in Aircraft Fuselage*, Solid Mechanics and Its Applications 189, DOI 10.1007/978-94-007-4282-6_7,

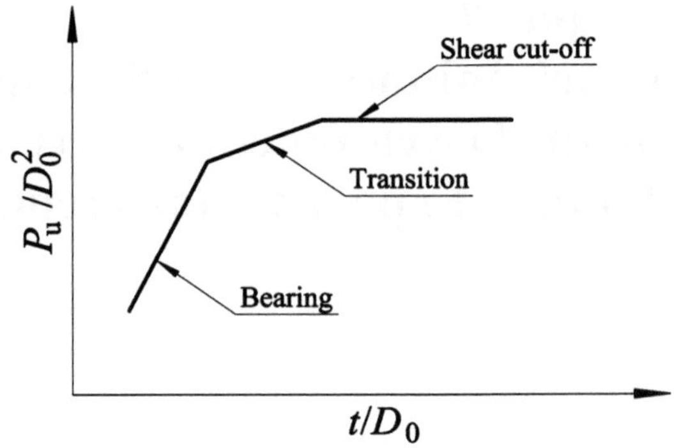

Fig. 7.1 Schematic diagram of rivet static strength versus sheet thickness to rivet diameter ratio (Homan 2008)

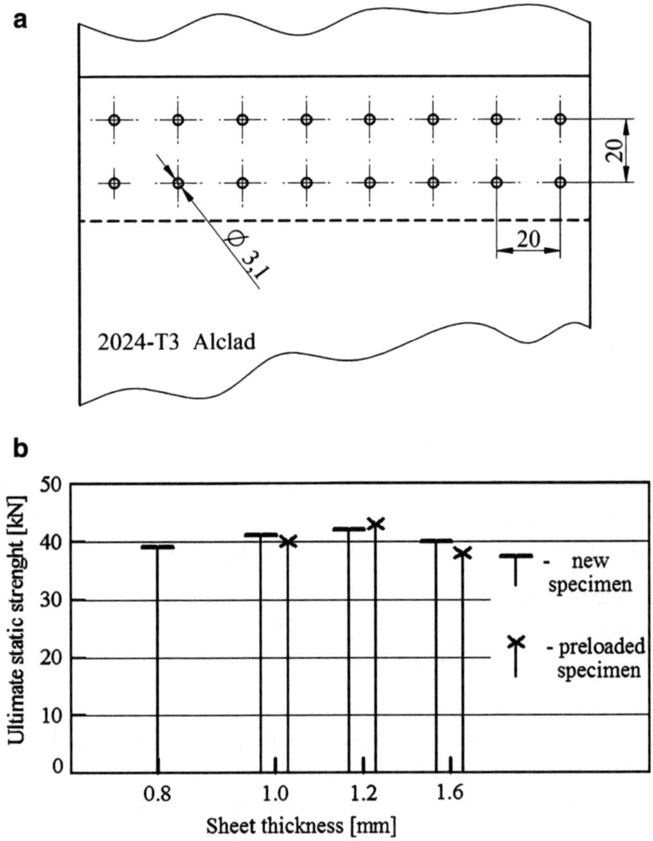

Fig. 7.2 Static tests of lap joints with two snap rivet rows: (**a**) specimen; (**b**) test results (Hartman 1954)

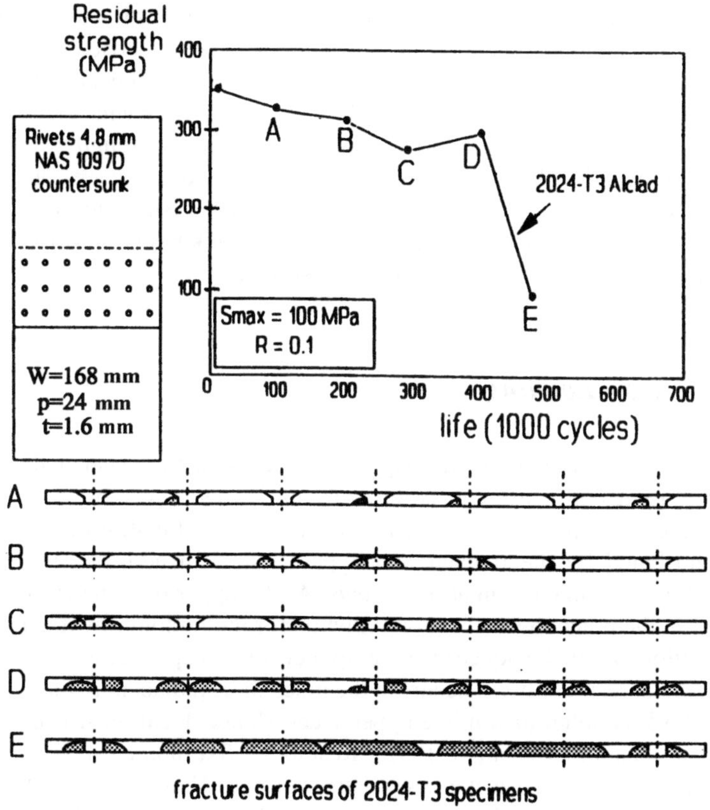

Fig. 7.3 Reduction of static strength of a riveted lap joint due to fatigue (Schijve 1994)

thick, 2024-T3 Alclad sheets with two rows of protruding head AD rivets 3.1 mm in diameter failed by rivet shearing for s/D_o of 6.5. For lower s/D_o-values (4.3 and 3.2), sheet failure in the net section occurred.

It is evident that the failure mode is also determined by the sheet material. Double-row riveted specimens (AD and DD rivets, $s/D_o = 4.2$) made of the 2024-T3 Alclad alloy failed in the sheet, while those made of the 7075-T6 Clad material failed by rivet shearing (Hartman et al. 1965). Sheet thickness and rolling direction relative to load direction had no influence on the failure mode. The above results reflect a higher static strength of the 7075 aluminium alloy compared to the 2024 material (see Sect. 3.1).

Hartman (1954) observed that the residual static strength of riveted lap joints was not significantly influenced by fatigue loading if the latter did not cause any visible cracking in the sheet, Fig. 7.2. In experiments described by Schijve (1994), Soetikno measured a reduction in static strength for lap joint specimens with countersunk rivets, Fig. 7.3. Fatigue tests were interrupted after $n \times 10^5$ cycles ($n = 1,2,3\ldots$; fatigue life about 5×10^5 cycles) for a static test until failure. If fatigue cracks were not visible (below 2×10^5 cycles), the residual strength was about 90% of that for the as-manufactured specimens. Roebroeks (1994) carried out tests under CA loading on three-row riveted lap joints of 2024-T3 alloy sheets with NAS 1097 rivets. Initially, the joint residual strength decreased slowly with increases in the number of cycles. A rapid, threefold decrease of the residual strength was associated with MSD at most rivet holes. The observations addressed above are consistent with results by Schra et al. (1995) who observed that during static tests on lap joints presented in Fig. 3.38a, shear failure of the countersunk rivets occurred as long as no cracks were visible on the outer specimen surface.

7.1.2 *Fatigue Loading*

In fatigue tests, failure of riveted lap joints occurs in the sheet material. For a double-row joint both rows can be critical, but if there are more than two rows (Fig. 7.4), the outer rows are critical due to the largest combination of tensile stresses being contributed by the transfer and bypass loads. In addition, the secondary bending obtains its maximum at these rows. At the upper row, the critical location is situated in the outer sheet (location A) while for the bottom row it is in the inner sheet (location B). If the squeeze force applied to a snap rivet is large enough, a more beneficial residual stress occurs in the inner sheet (adjacent to the driven head, see Sect. 3.3.5.4) rather than in the upper sheet. Hence, location A is usually more critical than location B (Schijve 1992). Results by Skorupa et al. (2011) indicate that this may not be the case at a lower squeeze force. They performed fatigue tests on lap joint specimens similar to those used by Skorupa et al. (2010a) for the load transfer measurements (cf. Fig. 5.15a) except that six rivets in a row were installed. For all of a total of four specimens riveted with squeeze forces leading to the ratios of the rivet driven head diameter to shank diameter (D/D_0) ranging from 1.3 to 1.4, the fatigue crack nucleation and failure occurred in the sheet adjacent to the

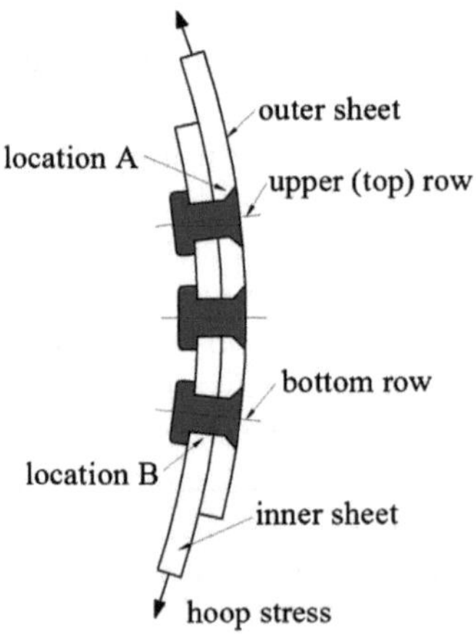

Fig. 7.4 Longitudinal lap splice joint, nomenclature

rivet driven head. However, the $D/D_0 = 1.5$ specimens fractured in the sheet next to the manufactured head. The above behaviour can be explained by the load transfer measurement results shown in Fig. 5.17 that were obtained for the same cyclic load level as applied in the aforementioned fatigue tests. Though the measured hole expansion was always larger in the sheet under the rivet driven head than under the manufactured head, for the specimens riveted with the lower squeeze forces this beneficial effect was evidently overwhelmed by the negative influence of the higher transfer load in the sheet adjacent to the driven head. For the $D/D_0 = 1.5$ specimens, load transfer through the end rivet rows is more symmetrical (cf. Fig. 5.17), which causes the effect of hole expansion to dominate.

When a countersunk rivet is installed, the outer sheet hole is countersink, which additionally increases stress concentration. Therefore, again the upper row is critical in most cases. The only exception is the NACA rivet for which cracks are observed in the inner sheet (adjacent to the manufactured head). For dimpled joints, cracks occur more frequently in the inner sheet in the upper row than in the outer sheet (Hartman and Duyn 1952).

The above consideration implies that the fatigue behaviour of lap joints should depend on the location of failure, as indeed backed up by fatigue tests of Rans (2007) on riveted specimens of two configurations shown in Fig. 7.5a. For specimen type 1, the crack nucleation is imposed to occur next to the manufactured head (countersunk or protruding) in order to simulate failure in the outer sheet. With a specimen of type 2, inner sheet failure is simulated because the crack initiation next to the rivet driven head is ensured. Figure 7.5b and c indicate that for both types of rivet, snap and countersunk, the fatigue life is longer if failure occurs in the inner sheet. The only exception are specimens with protruding head rivets formed with a relatively low squeeze force, Fig. 7.5b, in accordance with the results by Skorupa et al. (2011) cited earlier in this section. Fatigue tests on specimens of the same configurations were carried out earlier by Müller (1995). He also noted that in the case of countersunk rivets, the fatigue lives were significantly shorter if outer sheet failure was observed, Fig. 7.6.

It is possible that the sheet material can exert an influence on the site of cracking. Schijve and Jacobs (1960) observed the 7075 Al alloy specimens with snap rivets to exhibit some preference for failure in the sheet adjacent to the rivet working head, but for the 2024 alloy specimens, cracking in the outer sheet prevailed.

Fatigue cracks in the sheet can start at different locations depending on the rivet type and the clamping force. A particularly large number of works document the influence of the squeeze force on the crack initiation site, e.g. Hartman (1966), Schijve (1992), Müller (1995), Harish et al. (1999), Skorupa et al. (2009). Common locations for specimens with snap and countersunk rivets are the following:

- fatigue cracks initiated and propagating in the net cross section (rivet row line) at the edge of the hole on the faying surface, Fig. 7.7a,
- fatigue cracks initiated outside the rivet hole, but propagating through the hole, usually shifted above the net cross section, Fig. 7.7b,
- fatigue cracks initiated above the hole (near the edge of the clamping area) and propagating outside the hole, Fig. 7.7c.

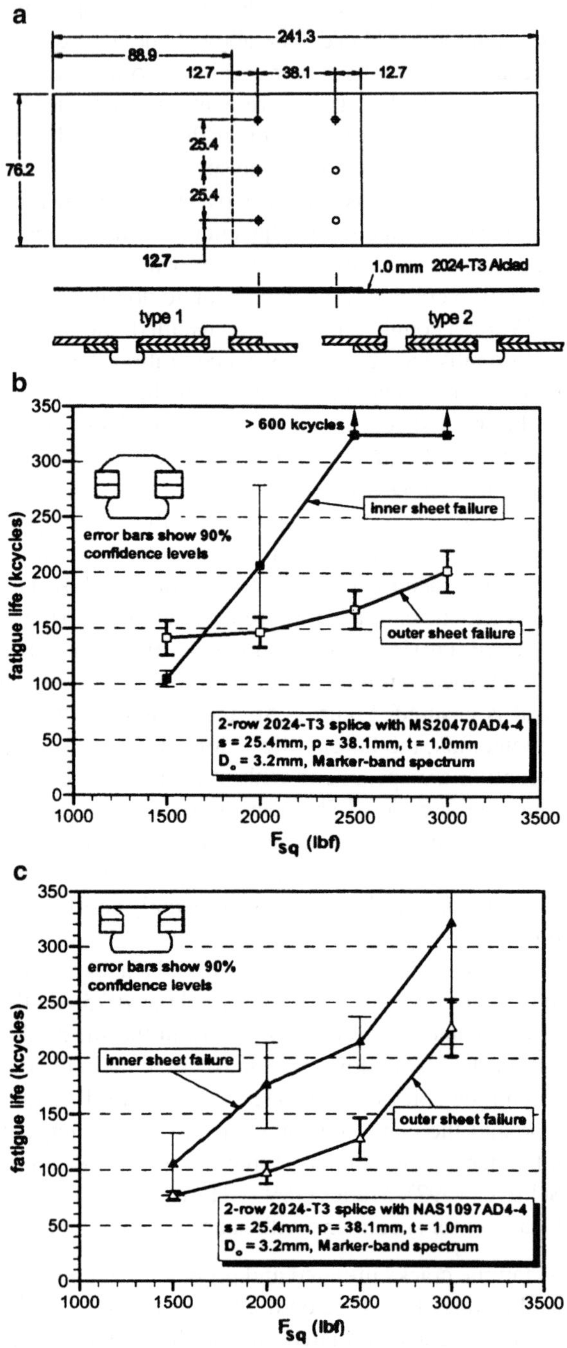

Fig. 7.5 Fatigue lives for inner and outer sheet failure in a riveted lap joint: (**a**) specimens used in the tests; (**b**) fatigue lives for universal rivet; (**c**) fatigue lives for countersunk rivet (Rans 2007)

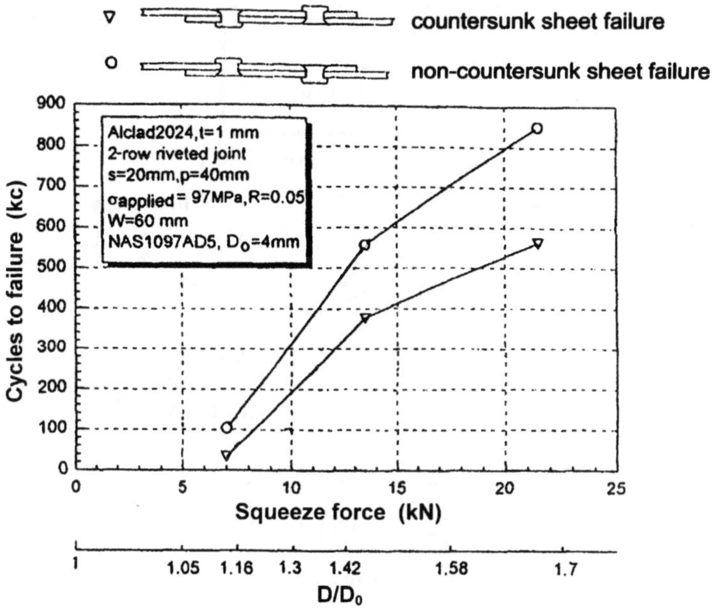

Fig. 7.6 Fatigue lives of lap joints for outer and inner sheet failure (Müller 1995)

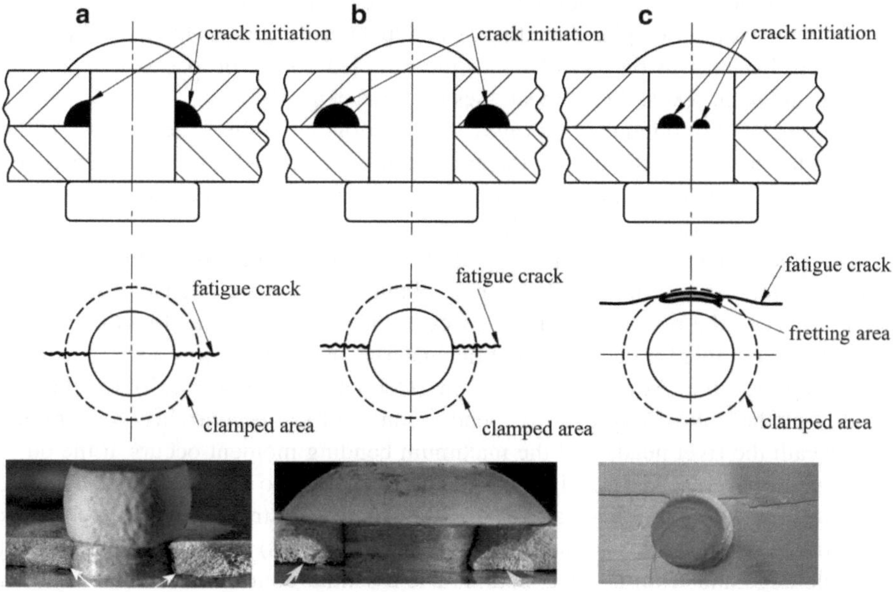

Fig. 7.7 Common fatigue crack initiation sites in lap joints: (**a**) at the hole edge; (**b**) outside the hole; (**c**) above the hole

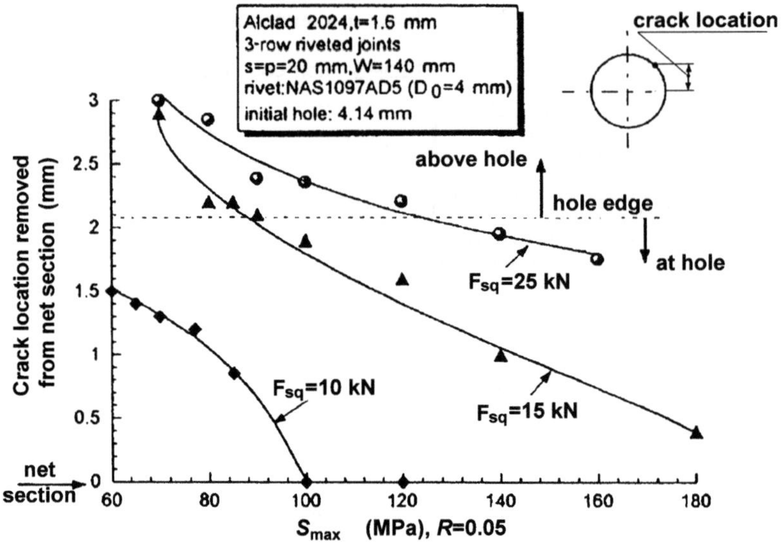

Fig. 7.8 Dependence of the fatigue crack initiation sites in riveted lap joints on the applied stress for different rivet squeeze forces (Müller 1995)

A special case is the crack initiation in dimpled lap joints. The cracks start at the edges of the dimpling cones, which may be due to local stress concentration and reinforcement of out-of-plane bending by residual tensile stresses from the dimpling operation, but also in the periphery of the dimpled zones, probably activated by fretting and local residual tensile stresses (Schijve et al. 1977).

The crack initiation in the minimum cross section of the sheet (net section) at the edge of the hole or at the end of the countersink, Fig. 7.7a, is found for snap and countersunk rivets if the clamping force is limited and a very small part of the load is transmitted by friction between the sheets.

The crack nucleation shown in Fig. 7.7b is a consequence of modified residual stress distribution and improved clamping obtained by a more intense squeezing of the rivet. Then the residual tensile stresses in the circumferential direction around the hole, favourable for decohesion of the material, are moved away a short distance from the hole edge. At the same time, the rivet heads increase locally the flexural rigidity of the joint. Consequently, bending due to local eccentricity is restrained underneath the rivet heads and the maximum bending moment occurs at the outer edge of the rivet head (cf. Figs. 6.13 and 6.14). Due to the above phenomena, the site of the maximum local tensile stress (sum of stress due to axial force, appropriate residual stress component and bending stress) moves away from both the hole edge and from the net section. The essential contribution of the squeeze force to migration of the crack initiation location from the net section is evident in the experimental results by Müller (1995) shown in Fig. 7.8. Here, the data in the 'at hole' region (except those on the abscissa) correspond to the crack location according to Fig. 7.7b. Similar observations have also been made by Harish et al. (1999), Szolwinski et al. (1997) and Rans (2007). For example, Rans noted initiation of cracks at the hole edge in the countersunk sheet (AD rivet 3.2 mm dia, 2024-T3 Alclad sheet 1 mm thick) if the squeeze force was equal to 8.9 kN ($D/D_o = 1.5$). At a higher squeeze force of 13.3 kN ($D/D_o = 1.7$) the crack nucleation location was shifted approximately 0.5–1 mm away from the hole edge. It should be emphasized that this type of crack initiation is predominantly controlled by fretting corrosion.

If for well-driven rivets the clamping is very high, the crack initiation and propagation occur above the hole (Fig. 7.7c). Such a case is also illustrated by the data points in Fig. 7.8 falling in the "above hole" region. Apparently, the material around the rivet hole, together with the rivet itself, acts as a single continuous piece of material. The nature of the load transmission has changed from a pin loaded hole type to a kind of contact-area load transmission. It is partly similar to spot welded joints. The maximum stress concentration then occurs ahead of the rivet hole where secondary bending and fretting corrosion cooperate to initiate fatigue crack nuclei. The crack starts at the top of the fretted area (shaded area in Fig. 7.7c), usually at some neighbouring locations simultaneously, and no longer grows through the hole (Schijve 1992, 1994). The phenomenon of fretting and its implications for fatigue crack initiation are considered in more detail in the next section.

Figure 7.8 indicates that the applied fatigue stress level also affects the location of the crack nuclei. At a higher applied stress, maximum secondary bending occurs closer to the net section due to reduced bending restraint of the rivet head which, in addition to higher bearing load, leads to crack initiation closer to the net section than at a lower applied stress.

Results of Müller's (1995) parametric experimental study of crack initiation location are summarized in Fig. 7.9 where f is defined as the ratio of the crack initiation location distance from the net section to the expanded hole diameter. It is seen that, in addition to the squeeze force (represented in this plot by the D/D_o ratio) and the applied stress, crack initiation location is also affected by the t/D_o ratio. For larger t/D_o-values the constraint of the rivet head on bending in the net section is smaller and the bearing load becomes higher, which leads to crack nuclei closer to the rivet hole. The t/D_o ratio has a small effect as long as the driven head is still small (low squeeze force), but its influence increases significantly for larger driven heads (higher squeeze force).

Sealing can change the crack initiation and propagation location, as observed by Schütz (1963) for lap joint specimens with protruding head rivets with and without sealing. Cracks in the specimens without sealing started from the edge of the rivet hole. For the joints with sealing the fracture surface was outside the rivet row net cross section. In both cases the outer sheet failed.

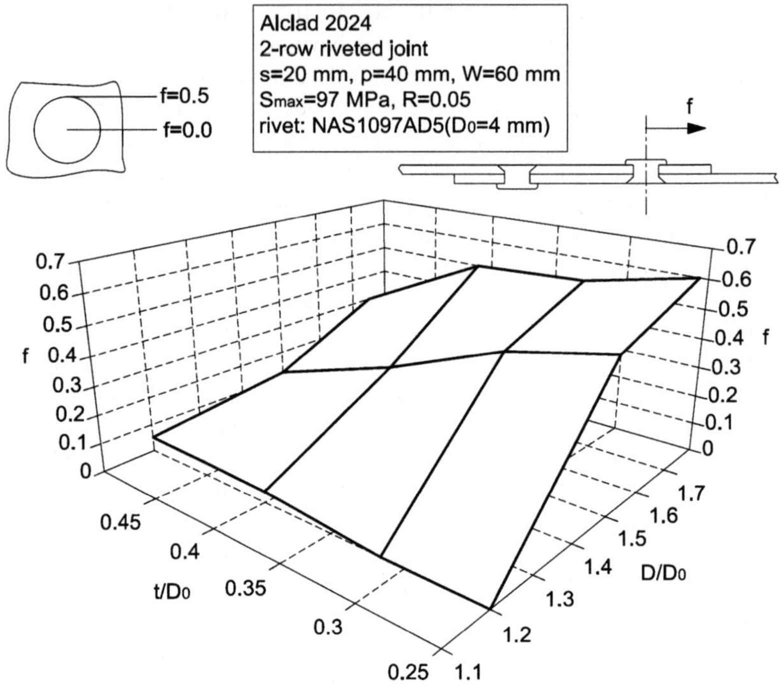

Fig. 7.9 Fatigue crack initiation sites in riveted lap joints for different sheet thicknesses and rivet squeeze forces expressed by *D/D*$_0$ (Müller 1995)

7.2 The Role of Fretting

7.2.1 The Phenomenon of Fretting

Fretting is a contact damage process arising from microslip associated with small-scale oscillatory motion of clamped structural members. The damage formation due to fretting can be divided into three stages. During the first stage the thin clad or oxide layer is removed from the sheet surface due to mechanical wear process. This process results in the initial formation of wear debris between the contact surfaces. For aluminium alloys, the wear debris is easily identified as a black powder surrounding the clamped area, Fig. 7.10. The next stage is characterized by the appearance of near-surface plastic deformation leading to high shear stresses which cause grain-sized microcracks, Fig. 7.11a. Also, additional wear and new debris formation occurs. In the presence of a tensile stress field the microcracks propagate into bulk material, Fig. 7.11b. At this third stage of fretting damage the influence of contact stresses becomes less dominant and crack growth is governed by global stresses. Fretting contact differs from sliding contact in that the applied tangential force does not result in global relative motion of the contacting surfaces. The contact

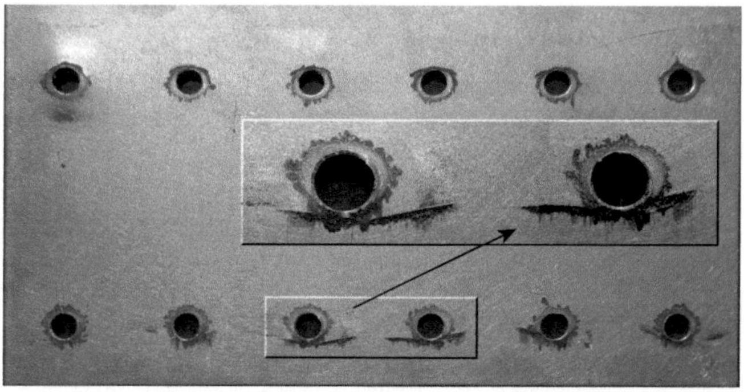

Fig. 7.10 View of a surface with fretting

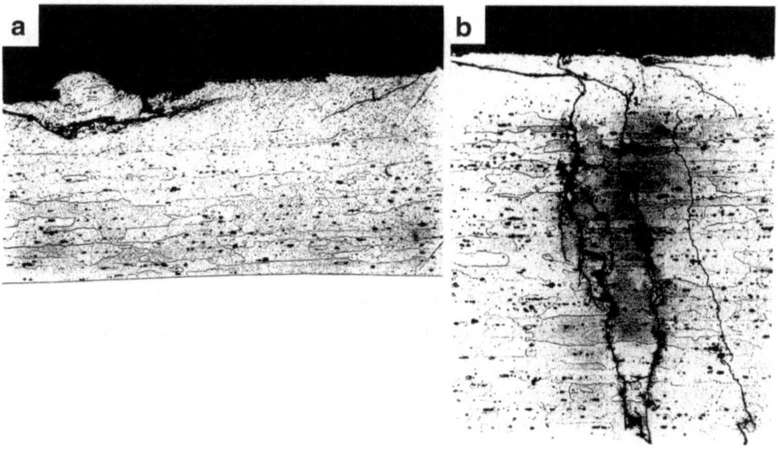

Fig. 7.11 Damage formation due to fretting for the 7075-T6 clad Al alloy: (**a**) debris and cracks in a clad layer; (**b**) cracks propagating into the bulk material (Hartman et al. 1963)

surface is divided instead into a zone of microslip (also referred to as the stick-slip zone), where the surfaces experience relative motion in the order of 25–100 μm, and a stick region where no relative movements occur.

The accumulation of wear debris between the contact surfaces leads to a dramatic increase of the friction coefficient. For example, in fretting tests for 2024-T351 by Szolwinski (1998) the initial value of the coefficient of friction (μ) in the absence of fretting damage has been measured to be around 0.15. Already after a few thousand cycles a quick rise in μ corresponding to the formation of fretting damage was noted. When a value of 0.65 was reached, no further increase in μ occurred. It is worth noting that the behaviour of the friction coefficient in these tests, Fig. 7.12, is in a qualitative and even a quantitative agreement with the friction force evolution during the fatigue cycling in the experiments of Hartman (1961) addressed in Sect. 5.4.

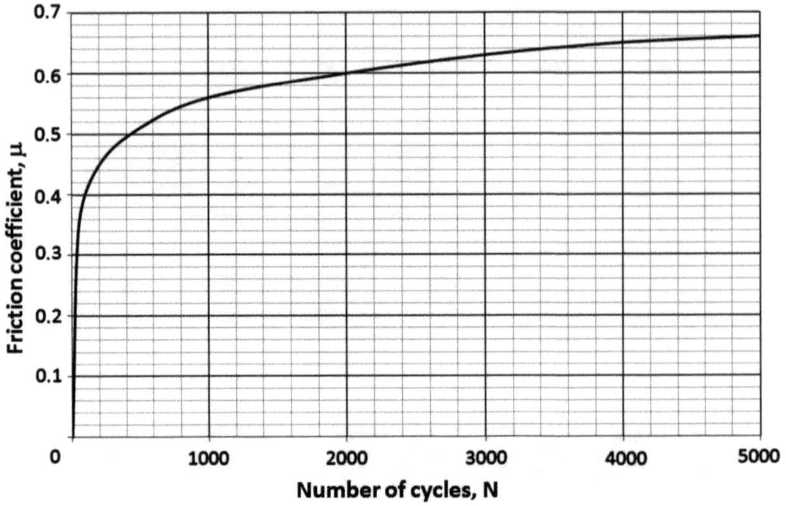

Fig. 7.12 Friction coefficient behaviour in a fretting test on the 2024-T351 Al alloy (Szolwinski 1998)

In view of a synergistic contribution of the complex mechanical and chemical phenomena, fretting damage can be affected by a number of variables, like clamping pressure between contact surfaces, magnitude of rubbing movements, the material and its surface condition, loading parameters, and environmental conditions (Schijve 2009a).

In the published literature, examinations of the sheet interface in the overlap region carried out after dismantling a riveted joint show a black oxide deposit around the rivet holes. According to Brown and Straznicky (2009), contact areas predicted by means of the FE method for the countersunk rivet agreed reasonably with fretted regions observed after dismantling fatigued riveted specimens conforming to the geometry and loading conditions assumed in their FE simulations referred to in Sect. 5.4. From the pictorial documentation produced in their paper, for universal rivets the agreement was less satisfactory. In the work addressed in Sect. 5.4, Hartman (1961) reported that the friction area indicated by black fretting debris around the holes after dismantling the specimens showed small and larger black spots in an area approximately equal to that beneath the rivet driven head which, as said previously, was extra large in this investigation. The latter observation is in disagreement with the FE results by Brown and Straznicky that indicate much smaller contact areas, cf. Fig. 5.26.

Fretting is a problem associated with aging aircraft. Bakuckas et al. (2007) inspected the faying surface quality of longitudinal lap joints of the fuselage panel removed from a retired Boeing 727–232 passenger aircraft. While in service, the airplane accumulated 59,497 flights, which was near its design service goal of 60,000 flight cycles. Overall, 66% of the faying surfaces exhibited some degree of fretting in the regions surrounding rivet holes.

7.2.2 *Cracking in the Presence of Fretting*

Fretting, which is recognized as one of the major factors in the initiation of fatigue cracks in riveted joints, often leads to their multi-site damage (Hurricks 1970; Schijve 1992; Szolwinski and Farris 1996; Hoeppner et al. 1996). For example, in fatigue tests of Liao et al. (2001) performed on a fuselage splice with three rivet rows (2024-T3 Alclad sheet 1 mm thick, countersunk rivet MS20426AD5-5) the cracks were observed to typically originate at heavily fretted area around the rivet holes a short distance away from the hole. However, investigation results on the dependency between the occurrence of crack initiation and the amount of fretting are not fully conclusive.

Hartman's (1961) experiments referred to in the previous section indicated that if the cracks were found, they did not always initiate at the hole with severe fretting. He concluded that "in riveted joints, coinciding of the initiation point of the fatigue crack with a point of severe fretting between the sheets is not a fixed rule".

Piascik et al. (1994) attempted to find a correlation between the amount of fretting and the fatigue crack initiation site. They performed a full scale testing of a riveted lap joint in the fuselage skin (2024-T3 Alclad sheet material, four rows of Briles rivets, sealant applied to the mating surfaces), Fig. 7.13a. The fatigue test was conducted up to total failure formed along upper row J and terminating at the adjacent tear straps. A detailed examination of the fracture surface from the long crack revealed that outer skin fatigue cracks were initiated at all rivets. All remaining rivet locations contained in the lower three rows I, H, G were individually removed from the panel. Next, each location was sawed exposing the aft and forward section of the rivet hole. Due to this operation even small fatigue cracks located on the rivet hole surfaces were visible. Figure 7.13b details the location and initiation site (arrow) of each fatigue crack in the considered rivet rows. As can be seen, the rows contained cracks that initiated at the rivet holes and at the faying surface. The amount of fretting was estimated by determining an arithmetic average of the fretted surface area observed on the faying surface of the outer sheet, Fig. 7.13c. A scaling factor was assigned to estimate the amount of fretting in each quadrant shown in Fig. 7.13d. The value of 0 was assigned when no fretting was observed and the value of 5 represented the highest degree of fretting. A fretting average (FA) was then calculated for each rivet. As seen in Fig. 7.13, all holes of the critical rivet row J except hole 11 exhibited a relatively high $FA > 3$. With the remaining rows, however, some holes that exhibited an $FA > 3$ did not contain fretting initiated cracks (e.g. hole 3 in row I, hole 2 in row H). At the same time, fretting initiated cracks were found in rivet holes with $FA < 3$ (e.g. holes 7 and 10 in row I). It is seen that a direct correlation of the FA with the rivet hole cracking is not possible, in agreement with the conclusion of Hartman (1961) cited above.

It is interesting to note that fretting is not a prerequisite for the onset of fatigue failure away from the hole edge. Müller (1995) observed fatigue crack nuclei in 2024-T3 Alclad riveted lap joints at locations where the clamping contact beneath the rivet heads terminates and large fretting movements between the sheets occur.

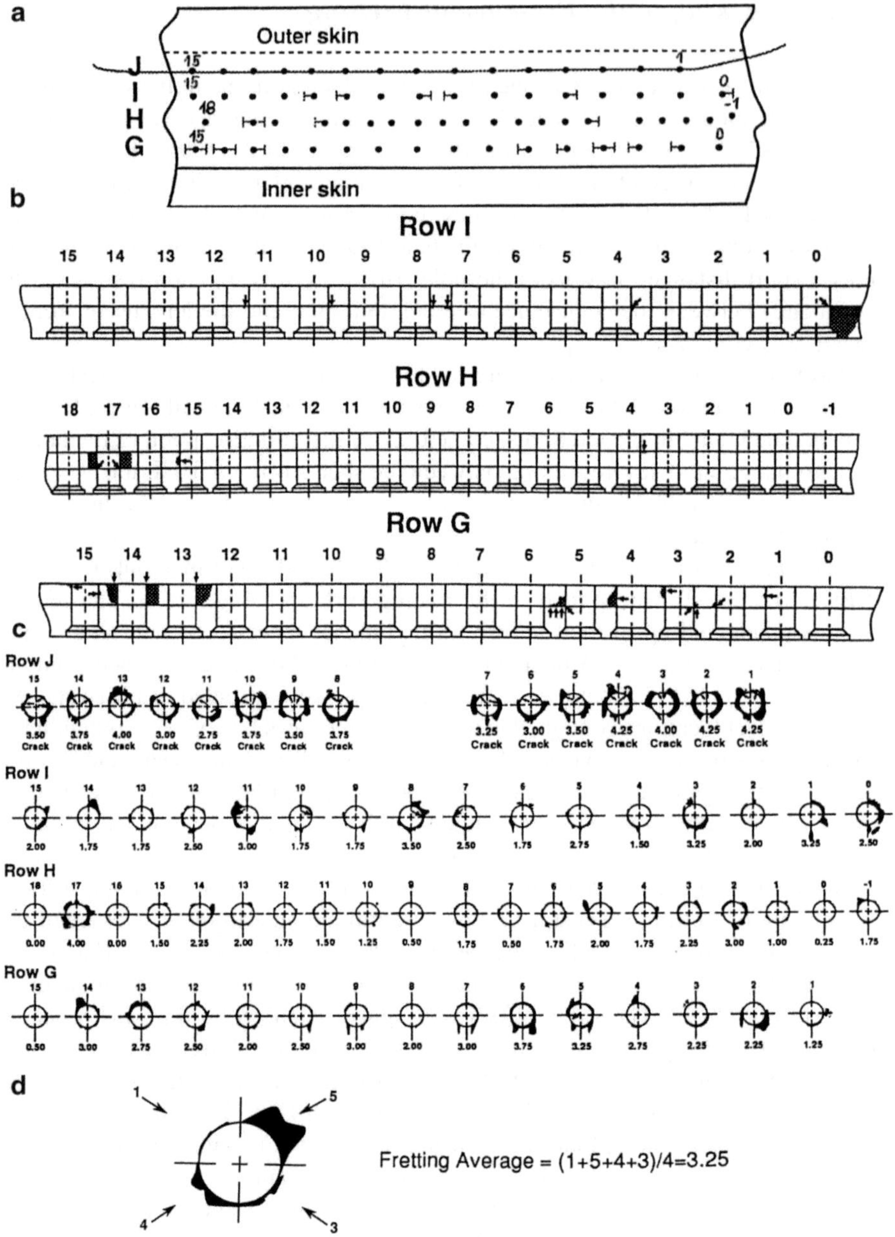

Fig. 7.13 Characterization of fretting in a fuselage riveted lap joint: (**a**) scheme of the joint; (**b**) location of fatigue cracks (*arrows* show the initiation sites); (**c**) interior surface of the lap splice outer skin (shown are locations of fretting and fretting average values); (**d**) estimation of the fretting average value (Piascik et al. 1994)

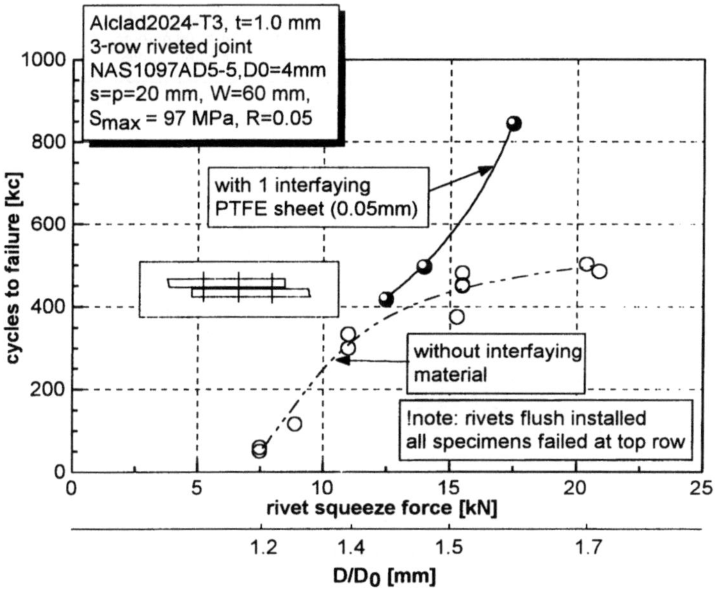

Fig. 7.14 Fatigue lives of riveted lap joints with and without an interfaying PTFE sheet for different rivet squeeze forces (Müller 1995)

However, the cracks initiated beyond the rivet hole edge also in specimens with a thin Teflon foil between the overlap of the riveted joint, though the fracture surface did not show any signs of fretting. Fatigue lives of the Teflon foil equipped specimens were higher than for specimens without the Teflon layers, the effect becoming more pronounced with increasing squeeze force level, Fig. 7.14.

7.3 Fatigue Crack Shape Development

First fatigue small cracks are initiated at the faying surface of the sheets. As a consequence, these crack nuclei initially grow as part through cracks. The secondary bending also stimulates the part-through character. Depending on the proximity of a rivet hole, these cracks can have a quarter elliptical shape (crack starts near the rivet hole edge) or a semi-elliptical shape (crack nucleated away from the hole), as depicted schematically in Fig. 7.7a and b respectively and observed experimentally by Rans (2007), Fig. 7.15a and b respectively. It should be noted that the quarter elliptical or semi-elliptical shape is a rough approximation only. Fractographic examinations, e.g. by Piascik et al.(1994), reveal that two diametrically opposed corner cracks at a rivet hole rarely propagate in equal increments and, therefore, are symmetric only at the beginning of the fatigue endurance. There is a strong tendency for the cracks to grow faster in the direction of the row of rivets than

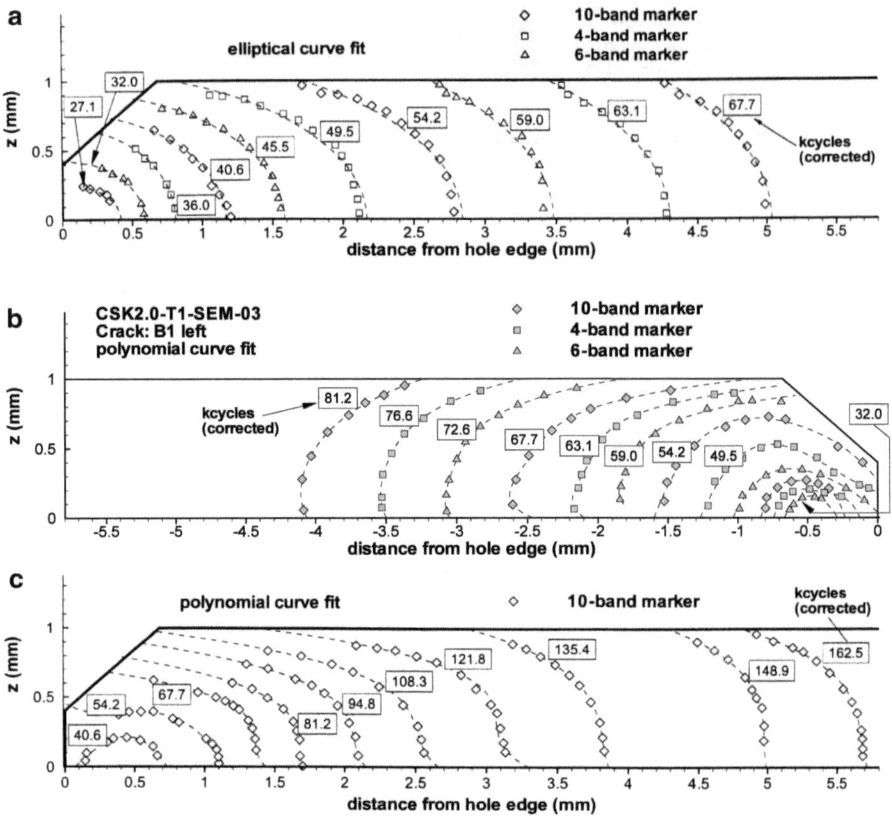

Fig. 7.15 Crack front reconstruction for: (**a**) quarter elliptical crack; (**b**) semi-elliptical crack; (**c**) intact ligament region next to the countersink surface. Marker load spectrum used to determine the crack front shape (Rans 2007)

towards the outer sheet surface. Hence, many semi-elliptical cracks intersect the rivet hole and become quarter elliptical before they reach the outer surface, as exemplified in Fig. 7.15b. Sometimes no marker bands or striations are observed in the region next to the countersink surface of the hole indicating a small intact ligament region which eventually ruptures due to overloading, Fig. 7.15c. Such a phenomenon was also observed by Ahmed et al. (2005) during a fuselage panel test described in Sect. 8.2.4. It is also seen in Fig. 7.15 that after penetration of the whole sheet thickness, the crack becomes a through crack and then it can be detected by a visual examination. However, the crack front remains slanted to some extent due to secondary bending. Figure 7.16 presents the development of cracks initiated according to Fig. 7.7c. It is seen that the cracks finally link up by failure of the remaining ligaments, as also observed by Schijve (1992) and Wanhill (1996).

Figure 7.17 presents crack depth and crack length along the specimen faying surface plotted against the percentage of failure life (Schra et al. 1995). Several lap

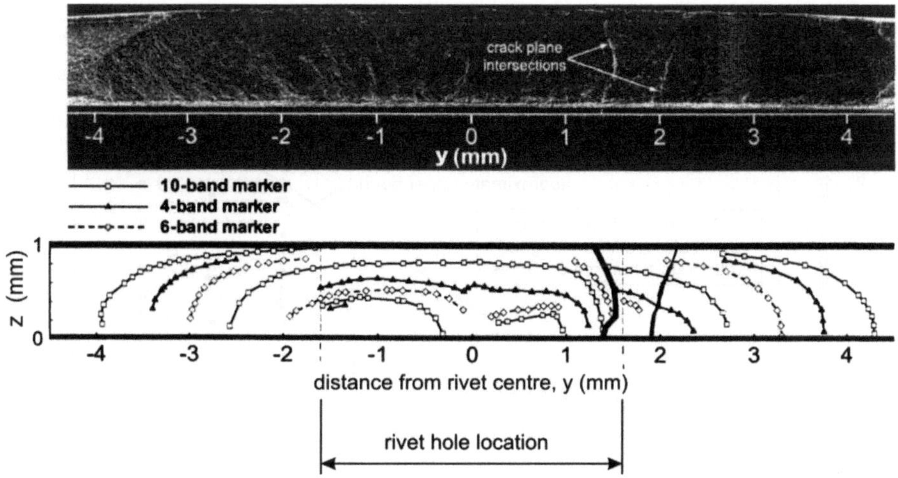

Fig. 7.16 Example of crack front reconstruction for a crack initiated and propagating outside a rivet hole. Marker load spectrum used to determine crack front shape (Rans 2007)

joint specimens with countersunk rivets (see Fig. 3.38a) were fatigued to various percentages of the fatigue life and then statically loaded to failure. The crack dimensions were measured at the fracture surfaces of the non-countersunk sheet. The curves in Fig. 7.17 represent growth of the most advanced cracks. At 30% of fatigue life there were already about 0.4 mm deep and 0.5 mm long cracks along the faying surface. The data indicate, therefore, that the initiation of a first crack will occur shortly after the start of fatigue loading. Based on the plots in Fig. 7.17, the crack growth rate (CGR) was calculated. Figure 7.18 indicates that the CGRs in the thickness direction were low and independent of the crack depth. The CGRs along the faying surface were about a factor of three higher and increasing slowly with increasing crack length.

The same trends were observed by Wanhill and Koolloos (2001) who analyzed crack growth data obtained fractographically from longitudinal lap splices made of the 2024-T3 Alclad alloy. The lap splices came from several service aircraft types (Fokker, British Aerospace, cf. Fig. 2.10). It was noted that the CGRs in the thickness direction were, on average, fairly constant and similar for all aircraft types, Fig. 7.19. An estimate of the CGR for the F100 full scale test at a 100% design load was derived from the factor $(100/110)^{7.1}$, where 7.1 is the Paris law exponent for Al 2024-T3 in the same range of CGRs. The CGRs along the faying surface were also similar, but exhibiting an increasing tendency when the crack length was increasing.

An increasing CGR along the faying surface with increasing crack length was observed by Piascik and Willard (1997) for the longitudinal lap joint coming from a Boeing 747–400 fuselage panel. The panel consisted of four full bays. Four rows of Briles rivets were used for joining two sections of the fuselage skin made of the 2024-T3 Alclad alloy. Figure 7.20 gives the results obtained based on marker bands

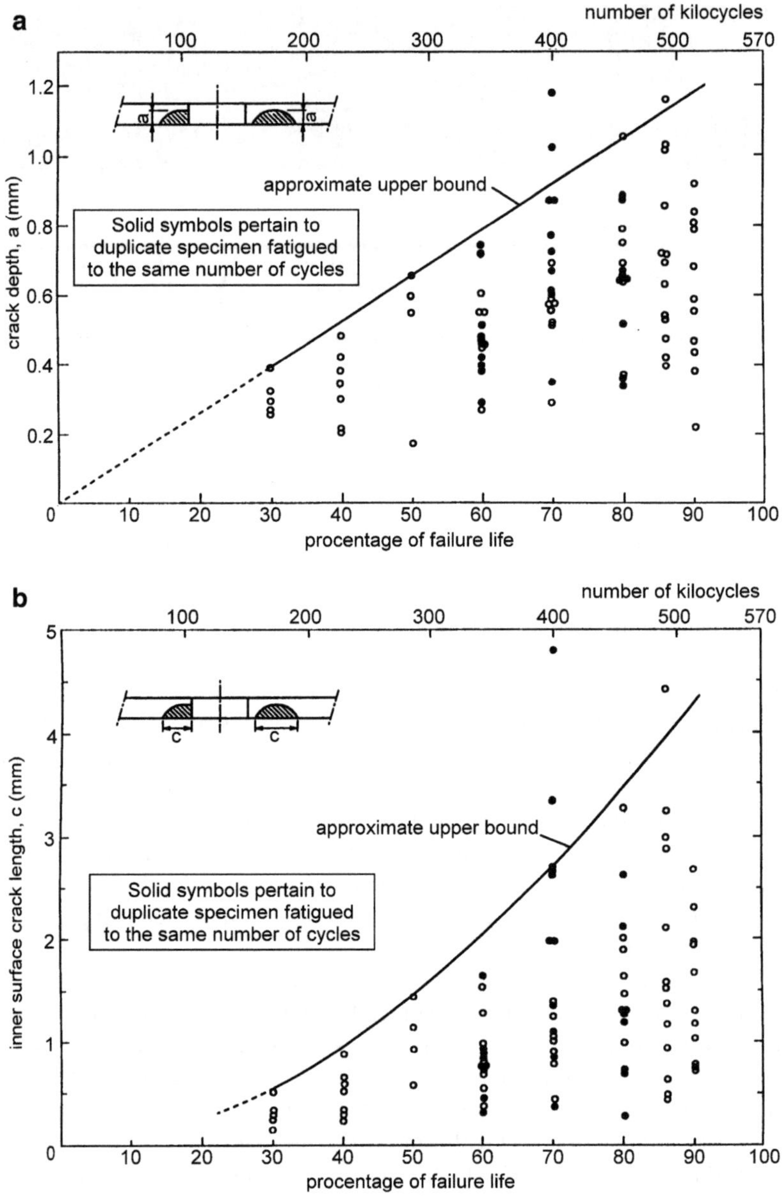

Fig. 7.17 Crack depth (**a**) and crack length (**b**) as a function of the number of fatigue cycles (Schra et al. 1995)

for the upper rivet row in two bays. The linear double logarithmic CGR versus crack length plot may suggest that local stresses, like those resulting from rivet/hole interference, have little effect on the CGRs when the cracks grow out of the local stress region into the remote stress dominated regime.

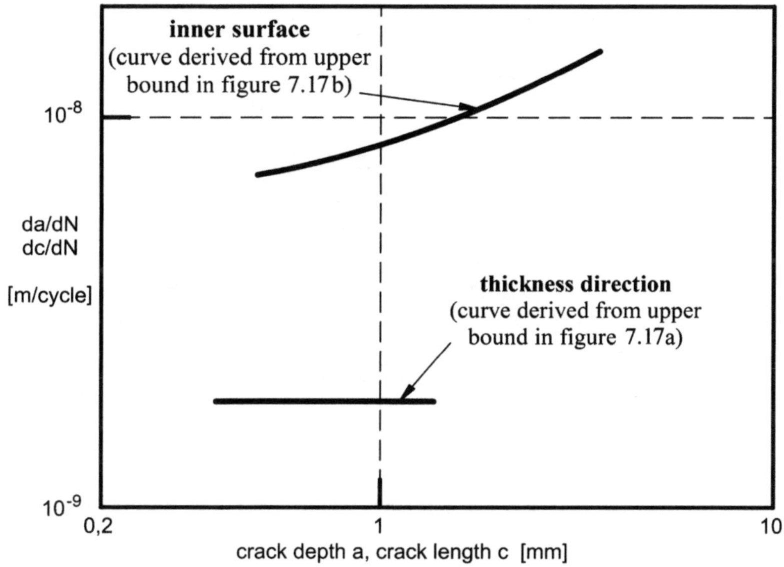

Fig. 7.18 Crack depth and crack length growth rate as functions of crack size Schra et al. (1995)

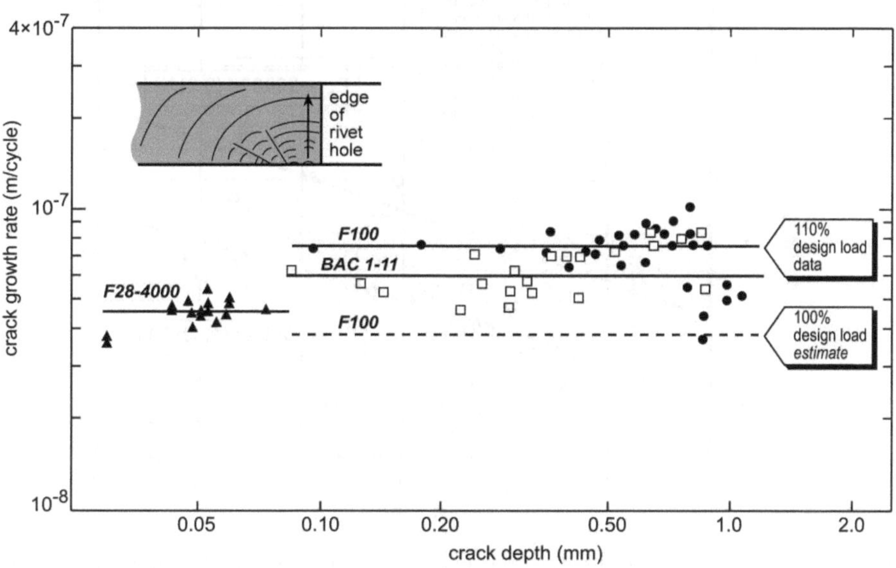

Fig. 7.19 Summary of transverse CGRs obtained from fractographic measurements for different lap splices (Wanhill and Koolloos 2001)

As could be expected, the rivet squeeze force influences the crack growth behaviour. Rans et al. (2007) observed a distinct difference in the CGRs along the faying surface for two squeeze forces, Fig. 7.21. A threefold reduction in the CGRs was found for the higher squeeze force due to a more favourable residual

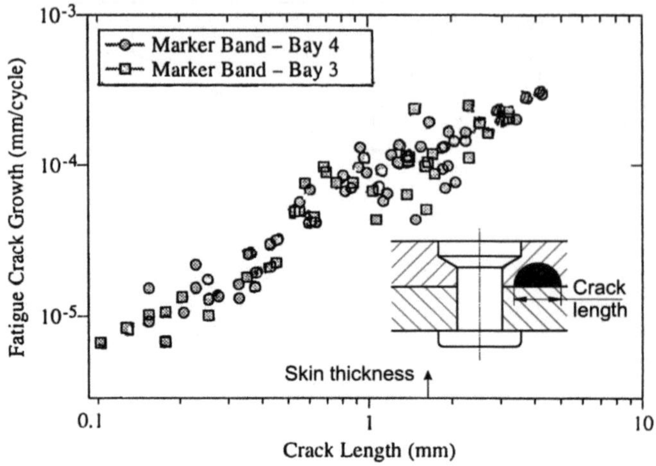

Fig. 7.20 Results on fatigue crack growth rates in lap splice joints of the Boeing 747–400 fuselage (Piascik and Willard 1997)

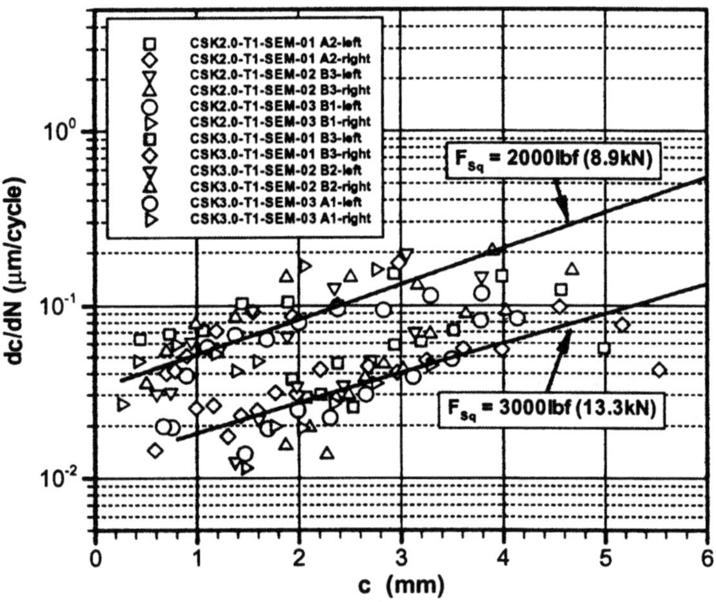

Fig. 7.21 Longitudinal crack growth rates in riveted lap joints from a SEM reconstruction (1 mm thick 2024-T3 Alclad sheet, 3.2 mm dia NAS1097AD4 rivet). For specimen geometry see Fig. 7.5 (Rans 2007)

stress field induced during the riveting process. A similar effect was also reported by Cavallini et al. (2001). The crack length as a function of the number of fatigue cycles noticeably decreased with increasing rivet driven head diameter.

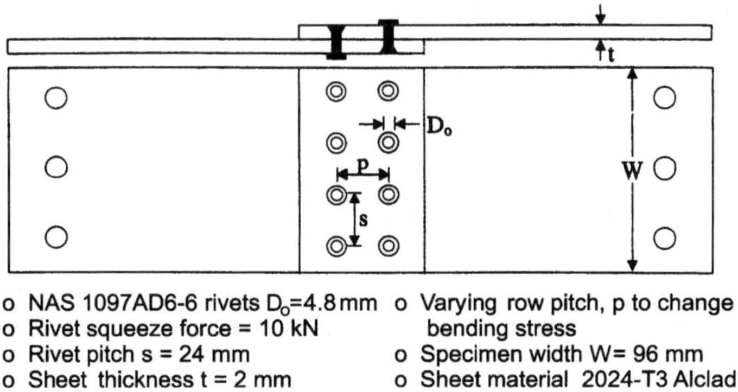

Fig. 7.22 Lap splice joint tested by Fawaz (1997)

Table 7.1 Effect of secondary bending on the crack shape in asymmetric lap splice joint

		60% N_f		90% N_f	
p (mm)	$K_b = S_b/S$	Mean a/t	Mean a/c	Mean a/t	Mean a/c
16	1.998	0.509	1.443	0.771	0.787
24	1.682	0.398	1.472	0.829	0.918
32	1.510	0.508	1.473	0.837	0.992
48	1.246	0.592	1.442	0.692[a]	1.261[a]

[a]Fatigued to 80% N_f

The effect of secondary bending on the crack shape development was investigated by Fawaz (1997) who used an asymmetric riveted lap splice joint shown in Fig. 7.22. The row pitch p was varied in order to control the degree of the secondary bending. The bending stress S_b was calculated using the analytical model by Schijve (cf. Sect. 6.2) for the applied stress S of 100 MPa. Dimensions of quarter elliptical cracks in the non-countersunk sheet were measured at 60% and 90% of the total fatigue lives N_f of the specimens. A marker-band spectrum was used for crack growth reconstruction. The ratios of crack depth to sheet thickness (a/t) and of crack depth to crack length (a/c) calculated from the Fawaz measurements are listed in Table 7.1.

A first conclusion that can be drawn from Fawaz's experiments is that the crack shapes are different at different percentages of fatigue life. While at 60% N_f the a/t and a/c ratios are independent of the bending coefficient K_b, at 90% N_f both ratios increase when the bending stress decreases. The increase in the a/c ratio is expected since the stress distribution over the thickness becomes more tensile with decreasing bending stress. In accordance with results obtained in other works referred to above, the crack advance along the sheet surface significantly exceeds that in the thickness direction, as indicated in Table 7.1 by a/c-values lower at 90% N_f than at 60% N_f.

In addition, Fawaz noted that the cracks became more and more shallow with increasing sheet thickness t. This implied that the a/c ratio was decreasing with t, but the development of the *a/t* ratio throughout fatigue life was not affected by thickness. Similarly, de Rijck (2005) did not observe any dependency between the *a/t* ratio evolution with the number of cycles and *t*.

7.4 Summary of This Chapter

A mode of lap joint failure under static loading depends on the rivet pitch in row to rivet diameter ratio (s/D_0). Lower s/D_0 ratios are favourable for sheet failure in the net section, while at higher s/D_0 ratios failure by rivet shear is observed. The failure mode under static loading is also determined by the sheet material. It seems that failure by rivet shear is not significantly influenced by sheet thickness.

Under fatigue loading, failure of riveted lap joints occurs in the sheet. Outer rivet rows are critical due to the largest load transmission and secondary bending. For the upper row the critical location is situated in the outer sheet, especially if a countersunk rivet is installed (additional stress concentration due to the countersink). For the bottom row the inner sheet is critical. The fatigue life of lap joints is usually longer if failure occurs in the inner sheet.

Fatigue cracks in the sheet can start at different locations depending on the rivet type and the squeeze force. If the squeeze force is relatively low, the fatigue cracks initiate and propagate in the net cross section at the edge of the hole on the faying surface. Fatigue cracks nucleated outside the rivet hole on the faying surface, but propagating through the hole, are a consequence of a more intense rivet squeezing. The crack initiation and propagation above the hole is observed for very tight rivets. In that case fretting between the mating surfaces is recognized as one of the factors influencing the fatigue crack initiation. However, a direct correlation between the amount of fretting and the crack initiation is not possible. Fretting is a problem associated mainly with aging aircraft. Sealing and the applied fatigue stress level can also affect the crack initiation and propagation location.

Fatigue cracks initially grow as part-through cracks. Depending on the proximity of a rivet hole, these cracks can adopt a quarter elliptical shape (crack starts at the hole edge) or a semi-elliptical shape (crack initiated away from the hole). The crack shape development of the part-through cracks is secondary bending dependent. Through cracks typically have an oblique front.

There is a strong tendency for cracks to grow faster in the rivet row direction than in the sheet thickness direction. The crack growth rate along the faying surface increases with crack length, while the crack growth rate in the thickness direction is fairly constant for lap splices of different aircraft types.

Chapter 8
Multiple-Site Damage in Riveted Lap Joints – Experimental Observations

The present chapter is divided into two sections. In Sect. 8.1, two passenger aircraft accidents due to multiple-site damage (MSD) of riveted lap joints are described in detail. Section 8.2 presents some results of experimental investigations into MSD performed on full-scale fuselage panels and lap joint specimens.

8.1 Examples of Aircraft Catastrophic Failure Due to MSD

A situation when many cracks are initiated at about the same time in a row of rivets, grow in a similar manner and suddenly coalesce to form a single long crack which can lead to catastrophic failure was recognized a long time ago. A few warnings had been obtained in full-scale fatigue tests before the Japan Air Lines Boeing 747 and the Aloha Boeing 737 accidents that occurred in August 1985 and April 1988 respectively (Schijve 1992; Schra et al. 1995).

The JAL Boeing 747 SR-100 experienced a violent decompression a dozen minutes after take-off at an elevation of 7,200 m (Terada 1995; Blom 2001). A sudden fracture started in bay 2 of the aft pressure bulkhead and subsequently a total fracture of the L18 splice occurred, Fig. 8.1. Within a few seconds, a portion of the aft pressure bulkhead was blown off and the passenger cabin pressure increased the inner pressure in the tail section, Fig. 8.2a. The air blast escaped through an opening in the fuselage skin between the tail cone and the front section of the stabilizer. This resulted in loss of the auxiliary power unit (APU), vertical fin, rudder and hydraulic line system. Due to this uncontrollable condition, the aircraft crashed in the mountains. Of 524 passengers only four survived. It was the largest-ever loss of life from a single aircraft accident. In the case of a blast of air escaping through the tail cone, the consequences of such an accident should not have necessarily led to a catastrophic end. But this did not occur due to the high strength of the tail cone.

A. Skorupa and M. Skorupa, *Riveted Lap Joints in Aircraft Fuselage*, Solid Mechanics and Its Applications 189, DOI 10.1007/978-94-007-4282-6_8,

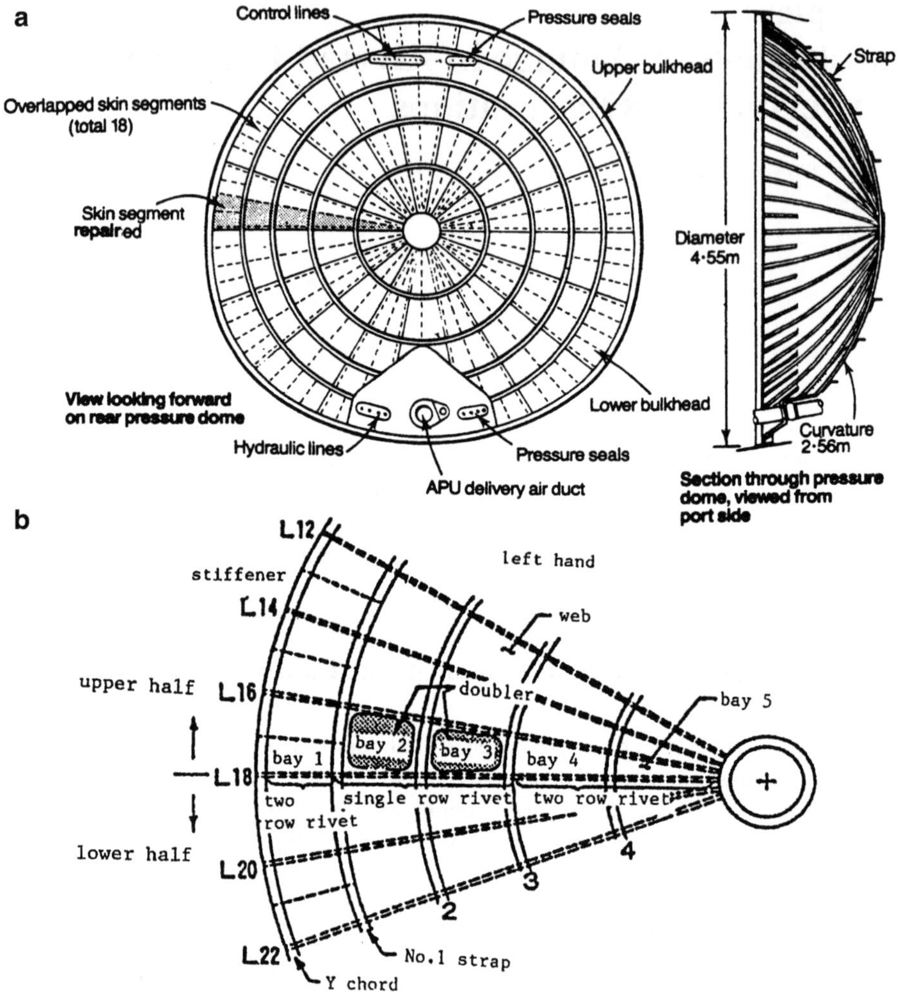

Fig. 8.1 JAL Boeing 747 SR-100 accident: (**a**) rear bulkhead design; (**b**) repaired bulkhead area (Terada 1995; Blom 2001)

Prior to the accident, this aircraft fuselage had been seriously damaged during a bad landing in June 1978. The aircraft was then repaired. The lower half of the aft bulkhead was to be replaced and a splice plate was to be inserted between the webs of the upper and lower halves of the bulkhead. However, the repair was not done in the correct way, Fig. 8.2b. As can be seen, the repair was carried out by using two splice plates instead of a single one. In consequence of this repair, the whole load was carried through the centre rivet row only. The strength of the L18 splice

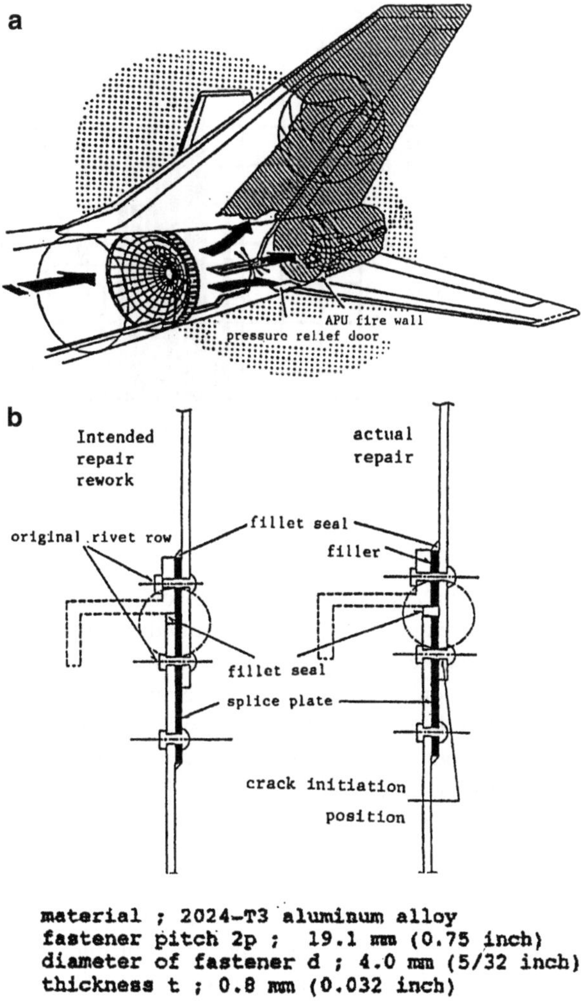

Fig. 8.2 JAL Boeing 747 SR-100 accident: (**a**) estimated fracture of the aircraft rear section; (**b**) lap joint intended and actual repair (Terada 1995; Blom 2001)

plate with a single rivet row was estimated to be below 70% of the intended double-row splice design. The airplane crash occurred 12,319 flight cycles after the repair (totally the aircraft had accumulated 25,030 flight hours and 18,835 flight cycles).

MSD was observed in the spliced portion of the bulkhead webs. Figure 8.3 shows the distribution of fatigue cracks emanating from all rivet holes of the splice obtained from the post-accident investigation. The ordinate and the abscissa present the normalized crack length (rivet pitch 19.1 mm) and the location of the rivets counted from the left to the centre of the bulkhead respectively. It can be seen that in the repair area some link-up cracks occurred.

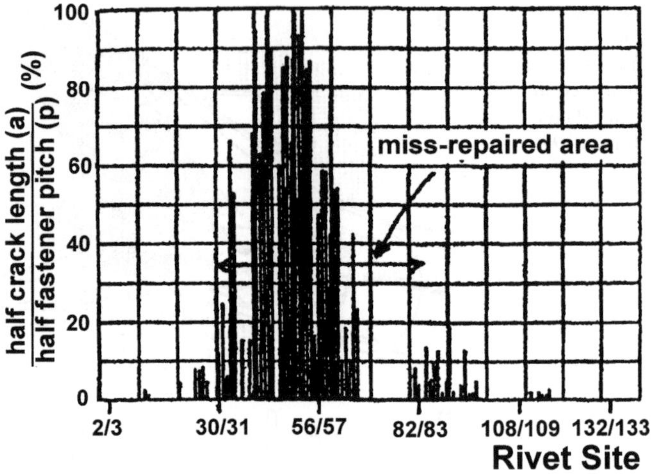

Fig. 8.3 Crack distribution in the mis-repaired joint of the B 747 SR-100 (Terada 1995)

The Aloha Airlines early model Boeing 737 suffered catastrophic failure of the forward fuselage. Despite uncontrolled decompression, the aircraft landed safely with 94 people on board (only one member of the crew was a fatality). Some information on this accident is presented in Fig. 8.4. A major portion of the upper skin and of the section 43 structure was separated. Damage extended from the main door aft about 18 ft. Perhaps this large opening could have been due in part to the high velocity air flow along the fuselage which suppressed the development of a crack flap necessary for crack curving and, hence, crack arrest (Kosai et al. 1992).

A post-failure investigation of the aircraft by the U.S. National Transportation Safety Board (Aircraft Accident Report 1989) revealed the cause of the mishap: "The failure mechanism was a result of multiple site fatigue cracking of the skin adjacent to rivet holes along the lap joint upper rivet row and tear strap disbond which negated the fail-safe characteristics of the fuselage". This mechanism was promoted by corrosion due to the disbonding of epoxy adhesive used to bond the skin sheets, see Fig. 8.4a. Water was able to enter the crevices and started the corrosion process. The aircraft operated in a coastal environment with an exposure to salt and humidity. Examination of several pieces of the aircraft skin showed a moderate corrosion along most of the joints. In that early version of the B 737 joint, countersunk rivet heads created the "knife-edge effect" condition in the outer skin that increased the risk of fatigue crack initiation (cf. Fig. 3.7).

The accident occurred after 89,681 flights (most Aloha flights were of short distance), whereas the design service goal (DSG) was 75,000 flights. The design life of 20 years was, however, not exceeded because the aircraft had been in commercial operation for 19 years.

After the Aloha accident the problem got its name: Multiple-Site Damage. This particular type of failure led to awareness of the so-called aging aircraft problem which has become the major focus of airframe fatigue related research. Also, many conferences have been devoted to this topic.

Eastin and Mowery (2007) claimed that MSD was the reason for 7 out of 11 aircraft accidents that they analyzed in detail within the years 1968–2002.

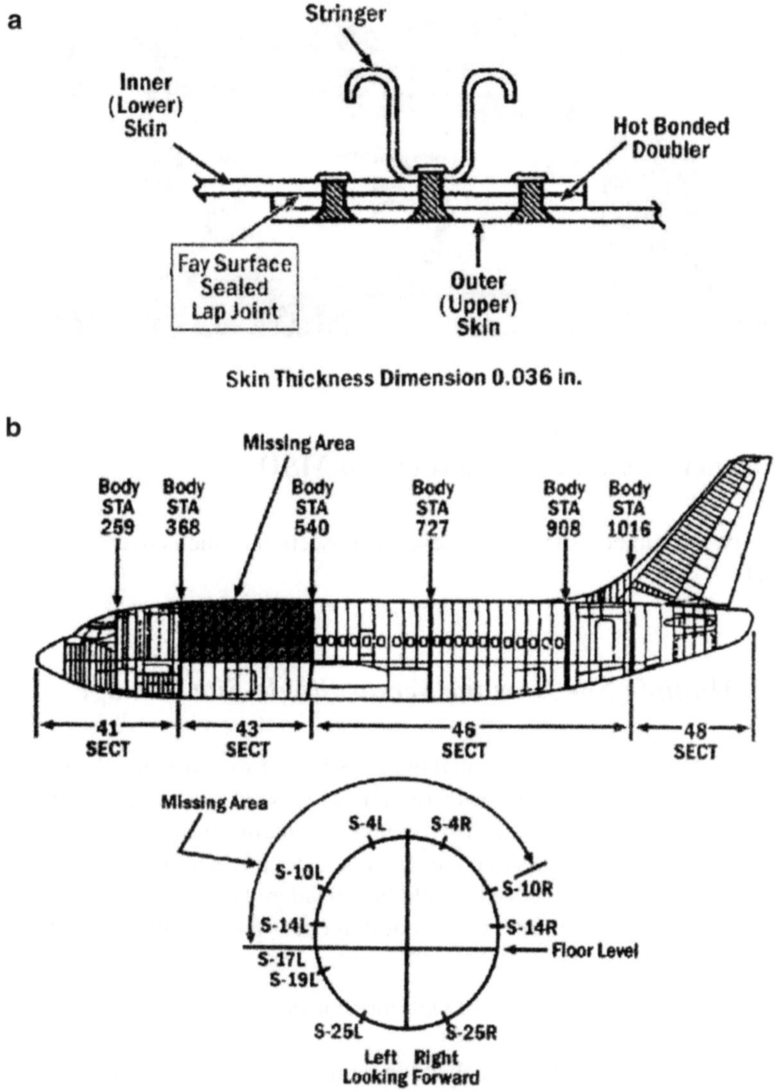

Fig. 8.4 Aloha Boeing 737 accident: (**a**) longitudinal lap joint design; (**b**) fuselage damage area; (**c**) aircraft after landing (Aircraft Accident Report NTSB/AAR-89/03 1989)

c

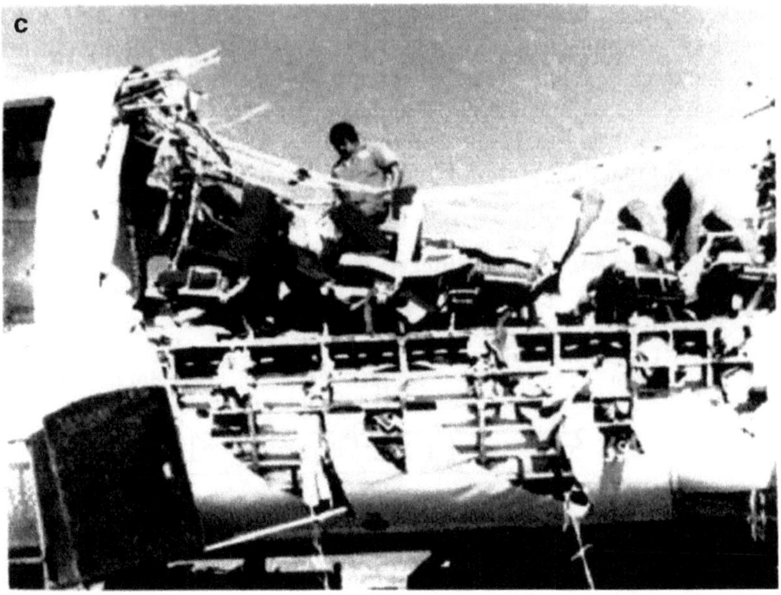

Fig. 8.4 (continued)

8.2 Experimental Investigations of MSD

For clarity, the subject is treated in several subsections which cover various aspects of MSD.

8.2.1 Multiple-Site Damage Versus Single-Site Damage

Besides the MSD situation illustrated in Fig. 8.5a, the so-called Single-Site Damage (SSD) may occur if a crack grows from a rivet to a substantial length until failure and this growth is not influenced by cracks emanating from neighbouring rivets, Fig. 8.5b. SSD may be assumed only if accidental manufacturing defects are probable. In all other cases MSD should be considered (Schra et al. 1995).

Viewing the damage tolerance requirements, the significance of MSD is twofold:

- growth of medium-to-long MSD fatigue cracks is faster than that of an isolated crack of the same size,
- the critical size of an MSD crack at the required residual strength level is smaller than that of an isolated crack.

The first statement is confirmed by the Schra et al. (1995) experiments referred to earlier (cf. Sect. 3.3.5.1, specimen shown in Fig. 3.38a). Figure 8.6a and b present

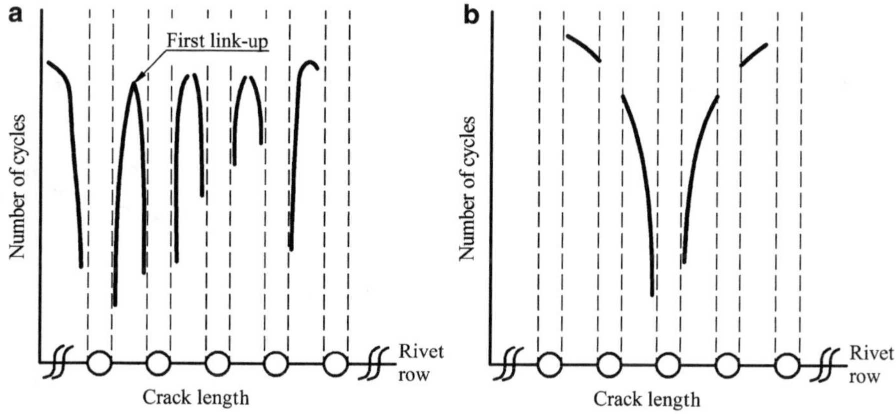

Fig. 8.5 Schematic of: (**a**) MSD situation; (**b**) SSD situation

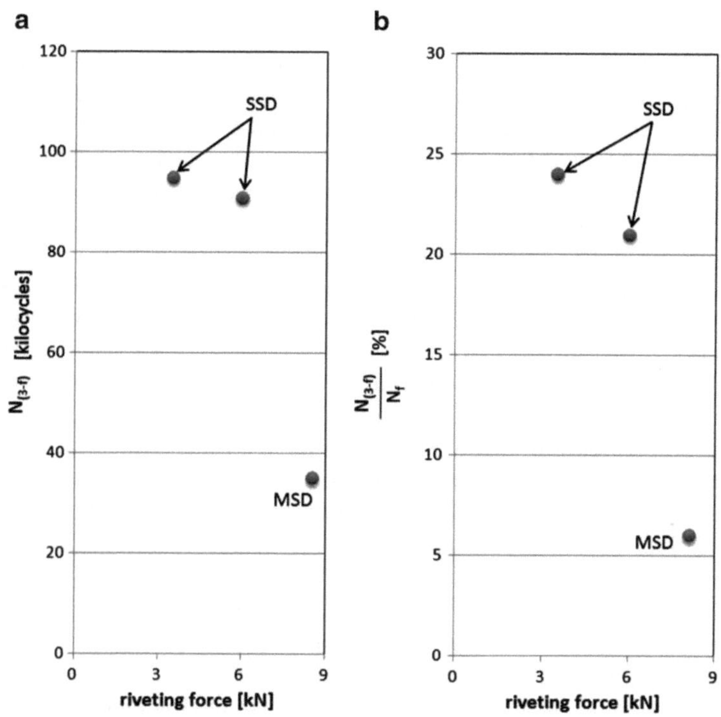

Fig. 8.6 Absolute (**a**) and relative (**b**) fatigue crack growth lives for visible crack growth at the outer specimen surface as a function of the riveting force of the central rivet in the outer row (Schra et al. 1995)

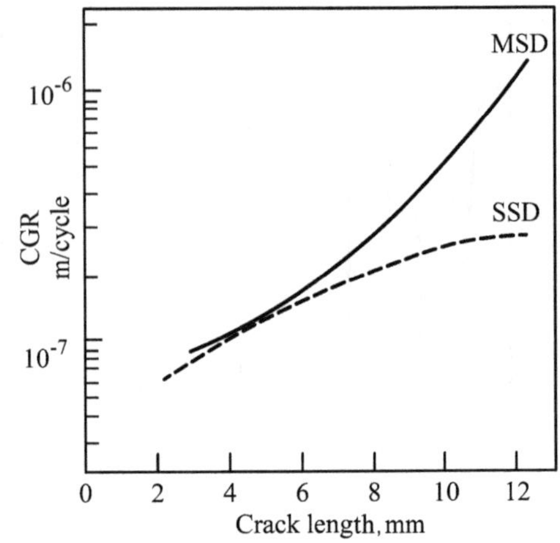

Fig. 8.7 Comparison of crack growth rates for the most advanced visible cracks in specimens representing an SSD and MSD situation (Schra et al. 1995)

the absolute and relative (percentage of the total fatigue life N_f) respectively fatigue crack growth lives from a surface, 3 mm long crack to failure ($N_{(3\text{-}f)}$) as a function of the riveting force. An SSD situation was created by installing the central rivet in both critical outer rows with a smaller force, e.g., 4 and 6 kN, than the normally used F_{sq} of 8.5 kN. For the latter force the MSD situation exists. It is seen that the visible crack growth is much slower for the SSD situation than under the MSD condition. In the case of SSD the fatigue crack growth life takes about 20–30% of the total life, while for the MSD situation it is only equal to 0.06 N_f. Under the SSD condition, the first fatigue crack was visible at the outer specimen surface during approximately the last 20% of the fatigue life, and for the MSD situation the visible crack growth took the last 6–9% of the total life. Figure 8.7 shows the crack growth rate (CGR) as a function of the surface crack length. Only for small crack lengths is the CGR for the specimen with one badly driven central rivet ($F_{sq} = 6$ kN) representing an SSD situation similar to that for the reference specimen (all rivets driven with $F_{sq} = 8.5$ kN) representing an MSD condition. With increasing crack length, the CGR for the SSD situation becomes much lower than for the MSD situation.

Tests on Boeing 747–400 fuselage panels conducted in order to compare the MSD and SSD situations are reported by Gruber et al. (1997). The test facility used in the experiments and shown in Fig. 8.8 consists of two cylindrical fixtures to match Boeing's typical narrow- and wide-body airplanes. The overall geometry of the fixtures is consistent with typical fuselage design. One end bulkhead is fixed and the other is on rollers to permit axial expansion during pressurization. The fixtures have a rectangular cutout designed to accept the test panel. The test panels are attached to the fixture at the skin, frames and stringers. Compressed air is used as the pressurizing medium. Polystyrene blocks are placed within the fixtures to reduce the required air volume. The six-bay panel consisted of bonded tear straps and floating frames connected to hat section stringers via stringer clips, Fig. 8.9a. The lap joints of a typical three-row configuration were assembled using 3/16 in. diameter, 100° countersunk rivets and faying surface sealant. The rolling direction

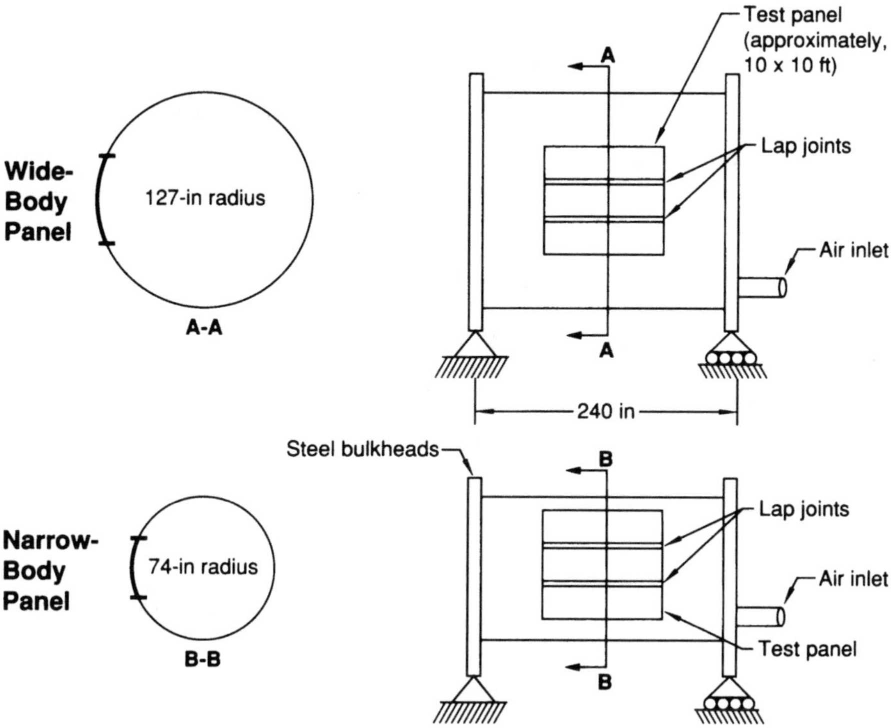

Fig. 8.8 Schematic of pressure test fixtures applied by Boeing (Gruber et al. 1997)

of the skin (2024-T3) was oriented along the fuselage axis. The stringers and the frames were made from 7075 aluminium clad sheet. Simulated MSD was introduced in the upper rivet row by installing saw-cuts in the outer skin at the faying surface (handsaw producing a 0.006–0.008 in. wide cut). The saw-cuts on both sides of the rivets were introduced after the fastener holes had been drilled and countersunk to size, but prior to application of the sealant and rivet installation. The MSD saw-cuts were introduced in four bay regions of the panel.

Test 1 on panel FAA 1 was conducted on the lap joint with 0.05 in. long cuts simulating MSD, as shown in Fig. 8.9b. A 5 in. saw-cut centred at frame station 120 was also introduced in the outer skin only. The panel was pressure cycled at 8.6 psi. No crack initiations were visible after 281 pressure cycles. At that point, the station 120 tear strap bridging the saw-cut was cut. As the cycling continued, the main crack steadily grew to approximately 6.5 in., at which point the 0.05 in. MSD saw-cuts one rivet ahead of the tips of the main crack had grown from under the rivet and became visible. The main crack extended dynamically after reaching a length of 17.2 in. and arrested at a length of 37.4 in., as can be seen in Fig. 8.10. The tear straps contained the dynamic crack, but visible cracks did develop in the

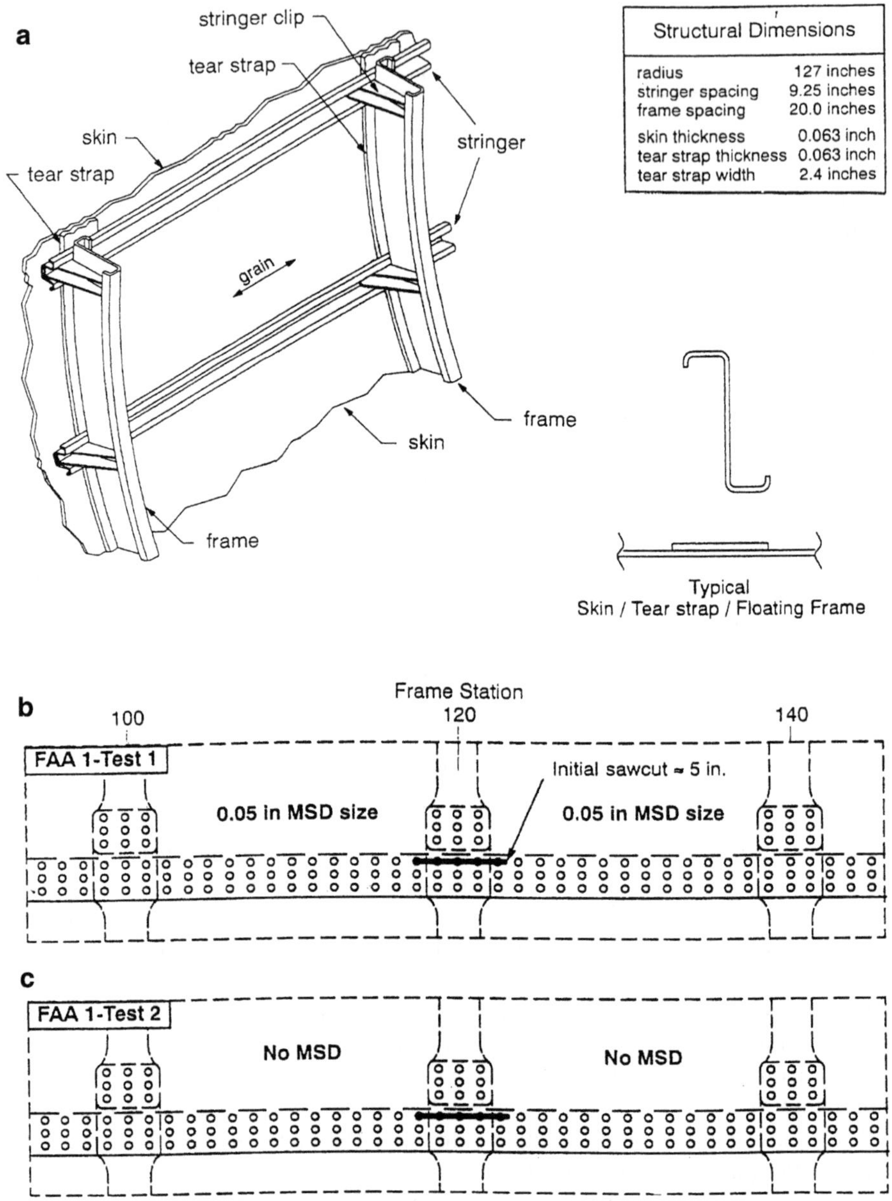

Fig. 8.9 Structural details of a B 747–400 fuselage test panel (**a**) and initial sawcut installations for a lap joint with MSD (**b**) and without MSD (**c**) (Gruber et al. 1997)

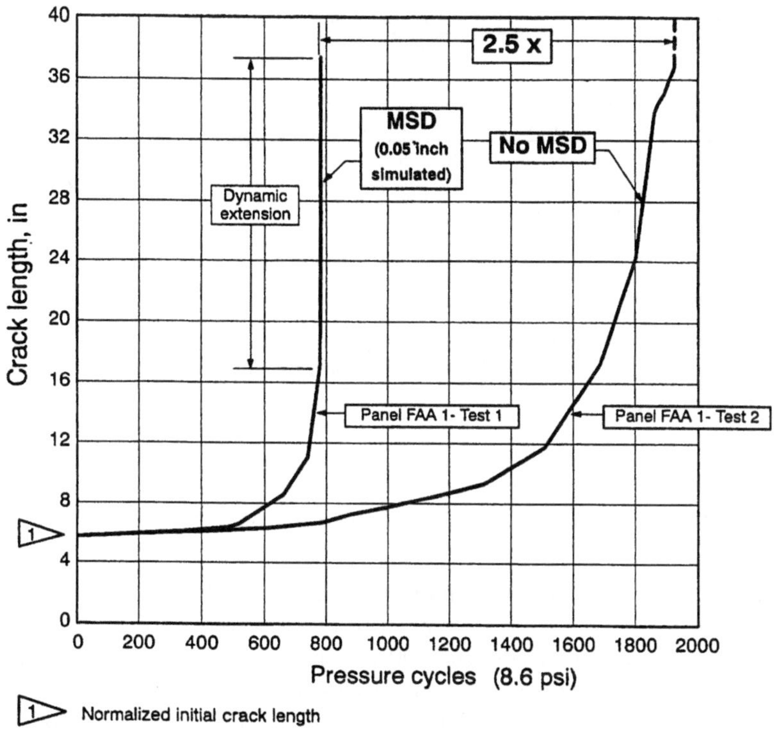

Fig. 8.10 Crack growth comparison: MSD vs. non-MSD (Gruber et al. 1997)

fasteners common to the tear straps at the next adjacent frames (stations 100 and 140). Residual strength[1] of the panel was not determined in this experiment.

Test 2 of panel FAA 1 was conducted on the lap joint without MSD, Fig. 8.9c. A 5 in. saw-cut was installed in the outer skin and tear strap centred on frame station 120. After 1,172 cycles no crack growth had been detected from the saw-cut. Subsequently the saw-cut was extended to 5.72 in., terminating approximately 0.25 in. beyond the fastener hole at each end of the saw-cut. As the cycling continued, the crack steadily grew to a length of 36.91 in. with the crack tips adjacent to the edges of the tear straps, Fig. 8.10. When the pressure was increased up to 9.4 psi, the crack grew stably to a length of approximately 38.2 in. and then it extended dynamically to the ends of the panel, resulting in catastrophic failure. As can be seen in Fig. 8.10, the impact of the 0.05 in. MSD on the test panel fatigue behaviour was most noticeable in terms of crack growth life. The crack growth life for the panel without MSD was 2.5 times longer than for the panel with MSD.

[1]Research into residual strength of fuselage panels will be considered in detail in Chap. 10.

Table 8.1 Test results on the influence of the squeeze force on MSD (Terada 1995)

Specimen	Number of initial flaws	Squeezing stress, MPa Rivet row Top	Bottom	N_{link}, cycles	N_f, cycles
A	1	936	936	364,700	368,300
B	15	936	936	273,400	276,580
C	1	624	936	72,000	76,880
D	15	624	936	24,000	24,540

Altogether, the results considered in the present section suggest that crack growth life of longitudinal lap joints under the SSD condition is a factor of 2.5–3 longer than for joints with MSD.

8.2.2 *Influence of the Riveting Force on MSD*

An investigation by Terada (1995) into circumstances of the JAL accident described in Sect. 8.1 involved tests on the influence of the squeeze force on MSD in lap joints. Double-rivet-row specimens 300 mm in width consisting of 1 mm thick, 2024-T3 sheets joined with countersunk MS20426AD5 rivets (4 mm dia) were used. The initial flaws were introduced at the top row rivet holes (in 1 or all 15 holes) in the sheet adjacent to the manufactured rivet heads by the electric discharging method before the riveting. The length and the tip radius of the notches which were cut at both sides of the holes were 1 and 0.1 mm respectively. CA loading of (54 ± 42) MPa at a frequency of 5 Hz was applied in the test. Rivets in either row were installed with a different squeeze force corresponding to the squeeze stress of 936 MPa (standard force) and 624 MPa (2/3 of the standard force). In Table 8.1, the number of cycles when the first link-up occurred (N_{link}) and the total fatigue life (N_f) are given. The number of specimens was 2 or 3 for each case. Figure 8.11 presents crack propagation curves for the considered cases.

The data in Table 8.1 indicate that when the squeeze stress at the top rivet row is 2/3 of the nominal value, the number of initial flaws significantly affects fatigue life. The fatigue life of specimens with flaws in all top row holes was about 1/3 of that with one flaw only. If the specimens were riveted correctly, the number of initial flaws was of a lower importance for total life. In that case, the specimens with flaws in all holes showed fatigue life as high as 75% of that observed for the specimens with one flaw. Terada concludes that MSD in aging aircraft structures stems mainly from clamping force deterioration, especially at the top rivet row of the joint. Such a deterioration in service conditions can result from corrosion or fretting under cyclic loading and excessive out-of-plane deformations during service operations (Terada 2009, private communication).

Another interesting experiment was carried out by Bakuckas et al. (2007) who tested a panel removed from the crown of a Boeing 727 airplane near its design

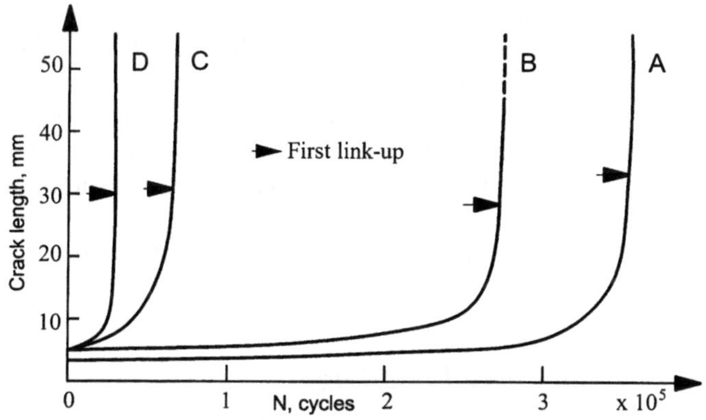

Fig. 8.11 Effect of the squeeze force on fatigue crack propagation (Terada 1995)

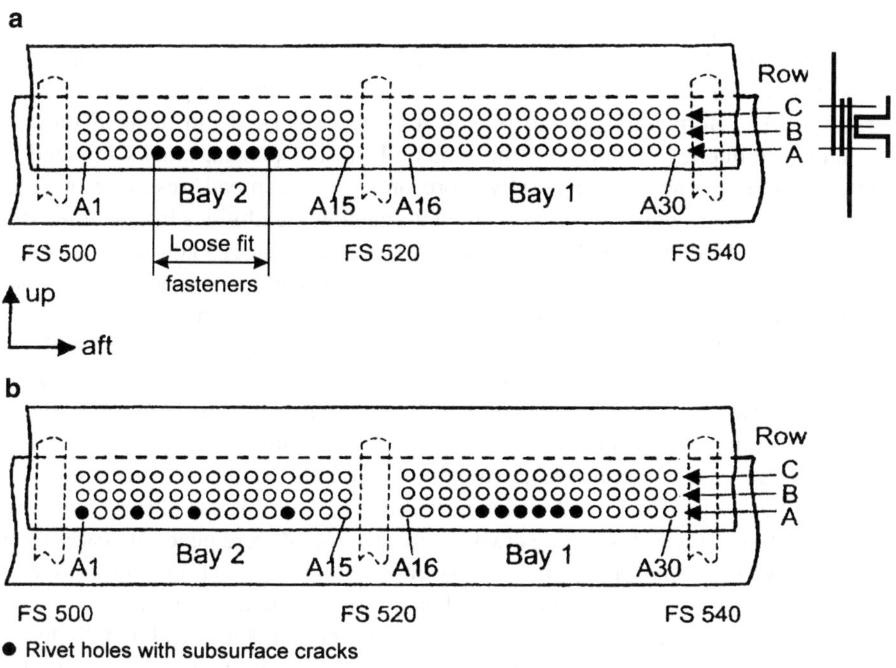

Fig. 8.12 Damage scenario in a lap joint of the test panel (**a**) and rivet holes with subsurface fatigue cracks (**b**) (Bakuckas et al. 2007)

service goal (59,197 flight cycles). After additional 12,000 cycles which simulated in-service loading conditions, seven middle countersunk rivets in Bay 2 (A5-A11, see Fig. 8.12a) were replaced with snug tightened fasteners (Hi-Loks) to simulate rivets with a poor clamping. Rivets in Bay 1 were left unchanged. At the end of the test, six mid-bay rivets (A20-A25) in Bay 1 had subsurface cracks in the inner skin, Fig. 8.12b. This could be expected because the hoop stress adopts the highest values at the mid-bay location and the lowest at the frames (cf. Fig. 2.2). None of the cracks had grown to become a through-the-thickness crack and, hence, they were not visible from the inner surface of the skin. For Bay 2, the fractographic examination revealed cracks at rivets A1, A4, A7 and A12. Unlike in the case of Bay 1, these cracks were dispersed along the bay and consequently no link-up occurred. These observations are not consistent with the earlier presented results by Terada and lead to the conclusion that poorly driven rivets do not necessarily contribute to MSD.

8.2.3 MSD Under Biaxial Loading

Vlieger (1994) compared results for uniaxially and biaxially tested lap joint specimens with regard to crack initiation and propagation in the critical rivet row, Fig. 8.13 (for specimen details see Fig. 2.8a). A hoop stress of 14 and 16 ksi represented fuselage pressurisation. During the biaxial test the specimens were subjected to "in phase" stresses with the biaxiality ratio (longitudinal to hoop stress) of 0.5. Both uniaxial and biaxial tests were carried out at an *R*-ratio of 0.1 and a frequency of 4 Hz. In the uniaxially tested panels, cracks initiated in the critical rivet row were generally more or less widespread, resulting in more separate cracks growing simultaneously without showing obvious interactions. It is seen in Fig. 8.13 that just before the panel failure, i.e. after 213 kcycles, the cracks initiated at various locations along the rivet row still had not linked up. The inset in the figure shows that the cracks had propagated from adjacent rivets in opposite directions, more or less parallel to each other without showing a link-up. With the biaxially tested specimens, the crack initiation in the critical rivet row generally occurred at two or more rivets positioned close together, resulting in more cracks growing simultaneously and linking up of these cracks during the test continuation. In these panels a continuous crack extending over a large number of rivet spacings occurred before final failure. As exemplified in Fig. 8.13, it was observed that the last inspections at which separate cracks were still found were generally carried out after 95% of the panel fatigue lives. This implies that the crack growth ultimately leading to panel failure occurred in less than 5% of the life until failure.

Recalling the notation from Fig. 2.8b, it should be noted that all biaxially tested panels failed across the critical rivet row (mode A, 67%) or due to cracks in the skin at the tear strap ends and in the critical rivet row (mode C, 33%). On the contrary, the uniaxially tested panels generally failed when cracks in the skin at the tear strap ends became unstable because no cracks had initiated and grew in the critical rivet row during the test (mode B, 60%). However, some uniaxially loaded panels failed in mode A (20%) or in mode C (20%). A conclusion that can be reached from the work by Vlieger (1994) is that the biaxially loaded panels show a much longer fatigue life (295–1,195 kcycles) compared to the panels fatigued under uniaxial loading (204–416 kcycles).

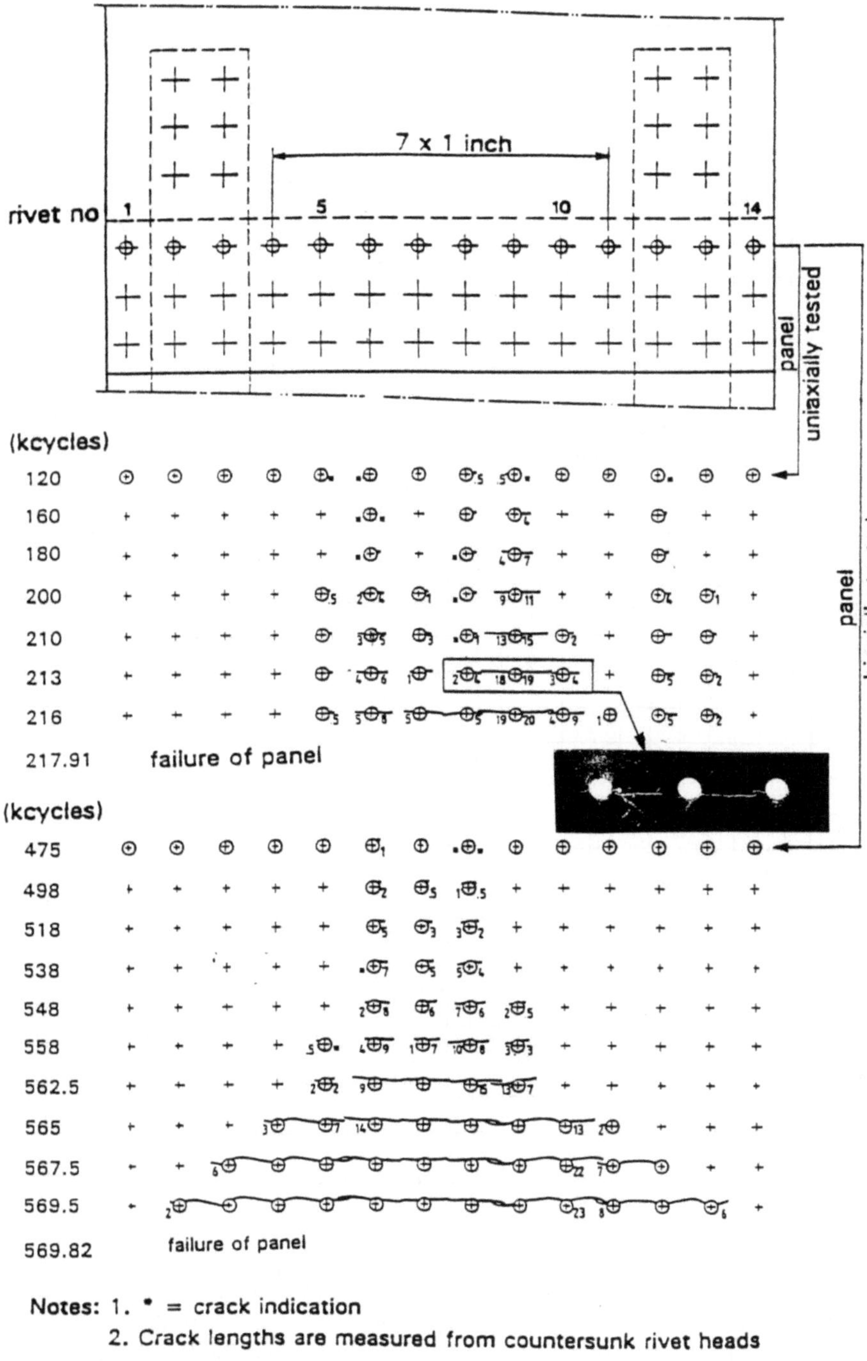

Fig. 8.13 Typical crack initiation and crack propagation in the critical rivet row for uniaxial and biaxial lap joint specimens. Hoop stress of 14 ksi (Vlieger 1994)

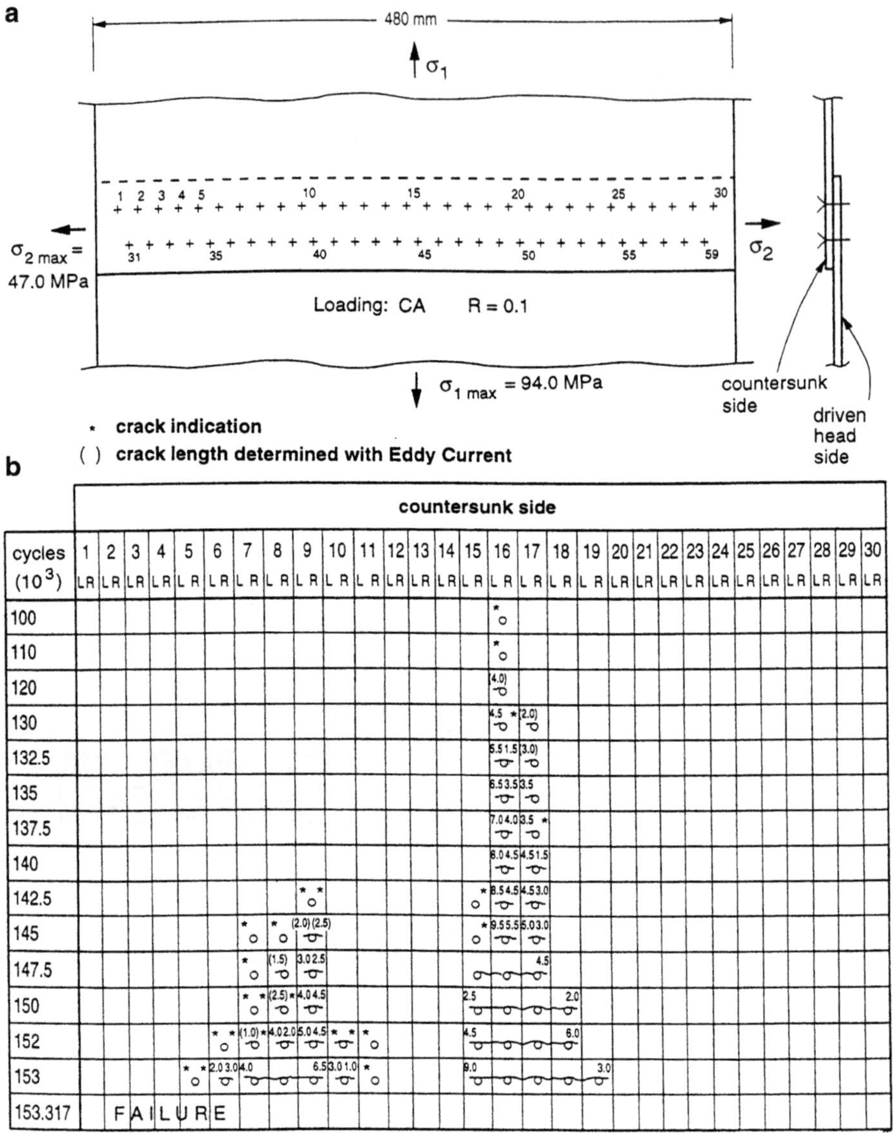

Fig. 8.14 Dimpled lap joint biaxially tested: (**a**) specimen; (**b**) results for specimen D3a (Ottens 1995)

Ottens (1995) conducted fatigue tests under biaxial loading on specimens shown in Fig. 8.14a. This type of longitudinal lap joints with two rows of dimpled rivets is used in the Fokker F28 fuselage. The test results are given in Table 8.2 and the MSD development for one of the specimens is presented in Fig. 8.14b. It can be seen that crack growth from a detected crack to failure covers only about 10–20% of the total

Table 8.2 Summary of the test results for dimpled lap joint specimens (Ottens 1995)

	Fatigue life, kcycles		
Specimen no.	First crack	First link-up	Failure
D3a	120	148	153
D4a	112	135	137
D5a	112	124	127

fatigue life. The part of the fatigue life between the first crack link-up and failure is less than 3% of the total fatigue life. It should be noted that the dimpling process causes a large plastic deformation in the vicinity of the rivet hole. This can lead to relatively early crack initiation in the form of circumferential cracks at the edge of the dimpled countersink. Additional information offered by Ottens is that in-service cracks were found at 35,000–45,000 flights.

8.2.4 MSD Tests on Fuselage Panels

A Boeing 747–400 forward fuselage which came off the production line (April, 1988) was tested up to 60,000 full pressure cycles by Piascik et al. (1994) and Piascik and Willard (1997). One pressure cycle from 0 to 9.0 psi took 2 min. A panel removed from the fuselage after the test and subsequently examined in detail consists of four lap splice joint frame bays (2, 3, 4 and 5) and two partial bays (1 and 6), Fig. 8.15a. Each bay is separated by tear straps (TS) bonded and riveted to the inner surface of the skin. The lap splice joint is formed by two sections of the skin (Alclad 2024-T3) using a four-row riveted (Briles design) construction, Fig. 8.15b. A horizontal stiffener, not depicted in the figure, is riveted (third row from the top) along the length of the lap splice joint inner surface. A sealant was applied to the lap splice mating surfaces to prevent internal pressure loss during flight and to inhibit the joint corrosion.

After 20,000 pressure cycles, *in situ* visual inspections of the outer skin surface were conducted at intervals of approximately 1,700 cycles. More frequent visual inspections were performed for local regions that exhibited evidence of cracking. The first evidence of outer skin cracking was observed along the upper rivet row near the middle of Bay 5 after 33,570 pressurization cycles. After 35,000 and 38,333 cycles, visible cracks were observed also in Bays 4 and 2 respectively, Fig. 8.15c (cracks marked by the arrows). No further cracking was observed in Bay 2 until 50,250 pressurisations when three mid-bay cracks were noted, Fig. 8.15d. At 55,500 cycles, ten rivet locations contained cracks, Fig. 8.15e. The first evidence of fatigue crack link-up between the adjacent rivets was detected in Bay 2 at 58,200 cycles, Fig. 8.15f. Additional pressurizations totalling 58,500 cycles had been performed before the cracks at all upper rivet row locations in Bay 2 linked-up forming a single 18.85 in. long crack terminating at the adjacent tear straps, Fig. 8.15g and h. At 59,900 cycles, Bay 4 exhibited a crack link-up between two rivets in the upper row.

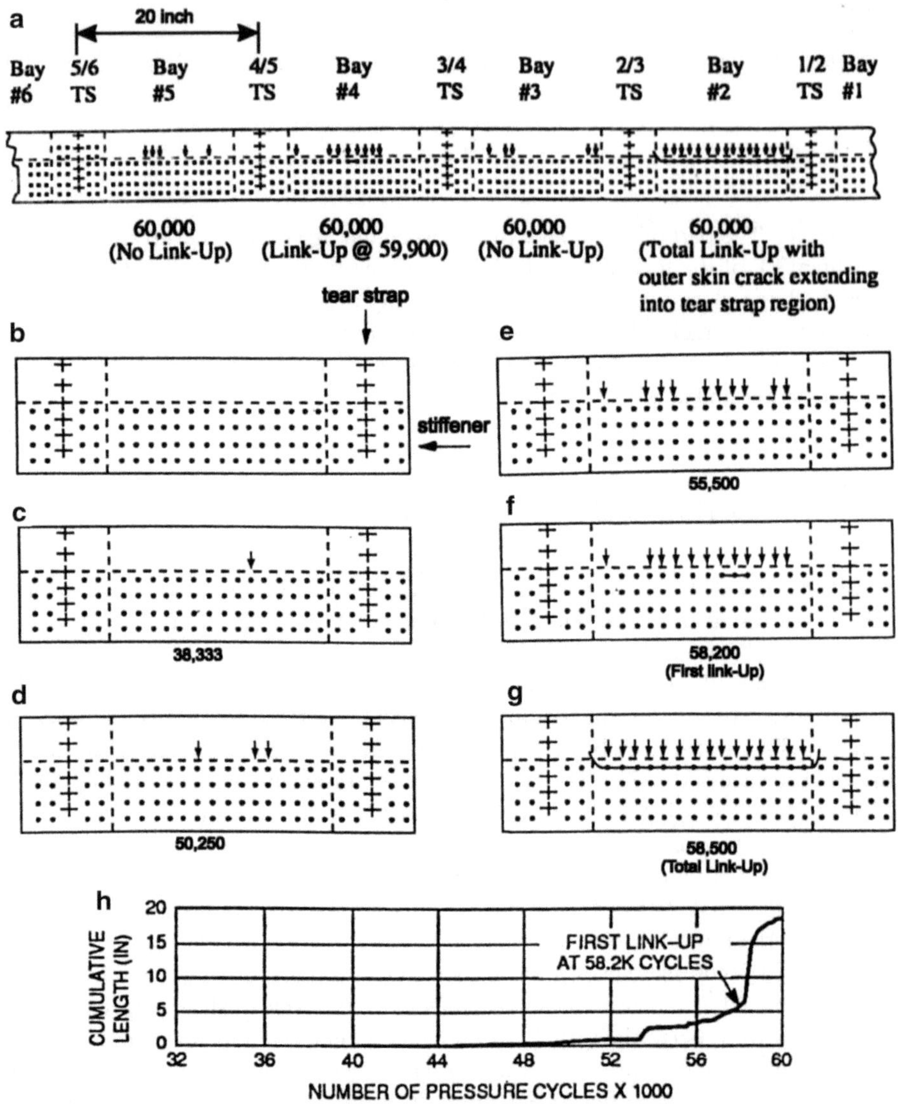

Fig. 8.15 Fatigue tests of a longitudinal riveted lap splice joint in the Boeing 747–400 fuselage panel: (**a**) schematic of the joint; (**b**) joint in bay 2 of the panel; (**c**–**g**) locations of outer skin fatigue cracks identified during visual inspections; (**h**) cumulative crack length (Piascik and Willard 1997)

Upon completing the test at 60,000 cycles, 33 of the totally 60 upper rivet row hole locations in Bays 2, 3, 4 and 5 exhibited visible cracks along the outer surface of the skin, Fig. 8.15a. No visible cracks were detected along other rivet rows.

A destructive examination has revealed that the cracks in the test panel were initiated underneath the rivet heads along the faying surface. Fretting damage could

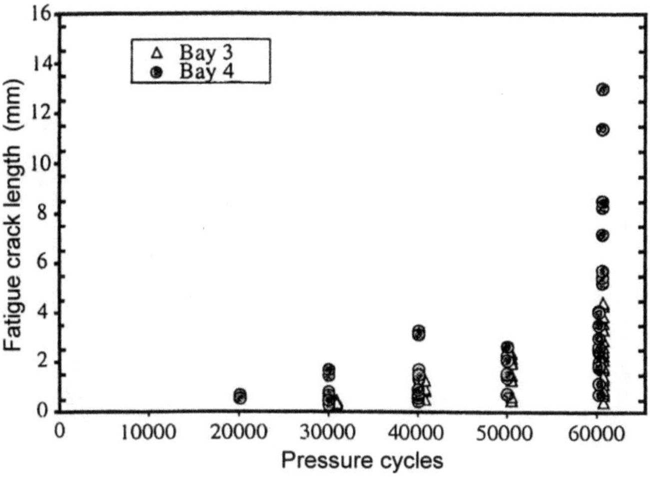

Fig. 8.16 Plot of fatigue crack length vs. pressure cycle number for bays 3 and 4 of the test panel (Piascik and Willard 1997)

be recognized as the primary mechanism for the crack initiation. Oxide debris was noted on the faying surface in the regions of crack initiation. Fatigue cracks of elliptical shape grew subsurface to a length equivalent to from one to three skin thicknesses before becoming through-the-thickness cracks. The fatigue cracks contained in the lower rows and tear strap regions detected during the destructive examination typically initiated at high stress concentration regions (corners of the holes, burrs, etc.). Out of altogether 281 rivet holes examined, 126 holes (i.e. 45%) were found to contain fatigue cracks.

On the basis of a detailed marker band analysis (marker bands introduced by performing a combination of the full and reduced pressure load), the crack length-load cycle history was determined for the upper rivet row cracks in Bays 3 and 4, including rivets of the Bay 3/4 tear strap, Fig. 8.16. The figure indicates the length of most cracks ranging from 0.5 to 5.0 mm at 60,000 pressure cycles. It was also observed that the fretting induced crack initiation was likely to have occurred between 10,000 and 20,000 cycles.

Samavedam et al. (1994) tested a panel (panel dimensions: length 120 in., width 68 in., radius 75 in.) designed to be representative of an older aircraft skin lap construction, Fig. 8.17a. A longitudinal joint consisted of three rows of countersunk rivets connecting 2024-T3 Alclad sheets 0.036 in. thick. The fatigue life of the panel loaded to a differential pressure of 8.5 psi at a frequency of 0.2 Hz was 75,263 cycles. Cracking in the upper row of rivets was first observed after 37,000 cycles. The first link-up of multiple cracks occurred after 69,400 cycles (92% of the total fatigue life) in the ligament between rivets 71 and 72, Fig. 8.17b. A similar link-up occurred in ligament 91–92 after 73,325 cycles and in ligament 46–47 after 74,493 cycles. The panel failure took place when several multiple cracks linked together to form a single crack, over 50 in. in length. After 50,000 cycles, 9% of the rivets in

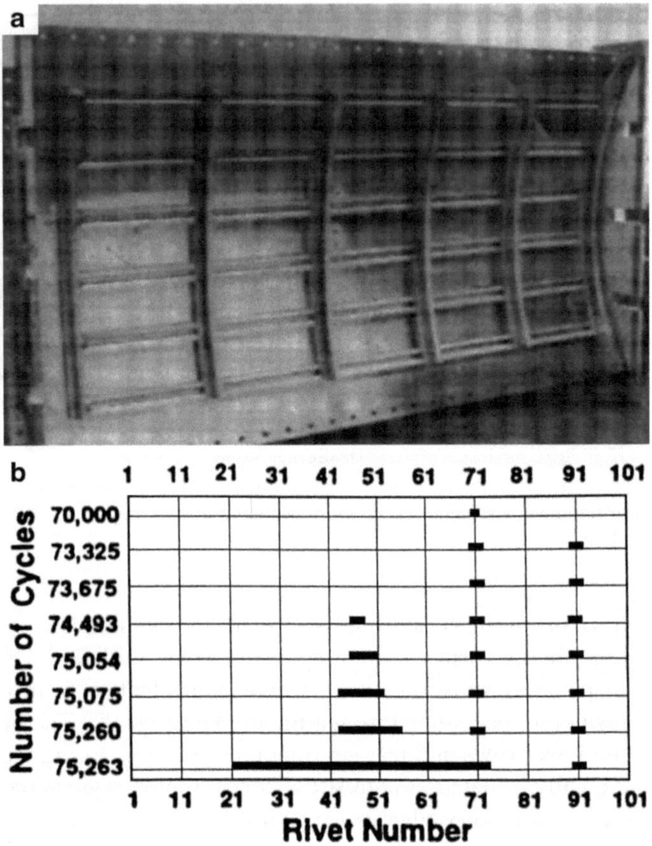

Fig. 8.17 Test panel (**a**) and the sequence of multiple crack link-up events (**b**) (Samavedam et al. 1994)

the upper row had cracks longer than 0.2 in. Further, after 70,000 cycles, when 93% of the panel fatigue life was consumed, 13% of the rivets in the upper row exhibited cracks longer than 0.4 in. (3% of these rivets had cracks longer than 0.8 in.).

Steadman et al. (1999) also studied MSD in a fuselage lap joint of a typical configuration (three rows of countersunk rivets) taken from an aircraft near the end of the design service goal (DSG). Based on a striation count using electron microscopy, they noted that the first outward MSD crack occurred when 83% of the DSG was consumed. The sequence of hole cracking within a bay was not random: the first crack commonly occurred near the centre of the bay, and a subsequent crack initiated at the adjacent hole.

The pristine panel 3.05 m long and 1.73 m wide with a radius of 1.68 m considered in experiments by Ahmed et al. (2003, 2005) and Johnson et al. (2007) is representative of a narrow body fuselage structure. The substructure of the panel included six frames with a 483 mm spacing and seven stringers with a spacing

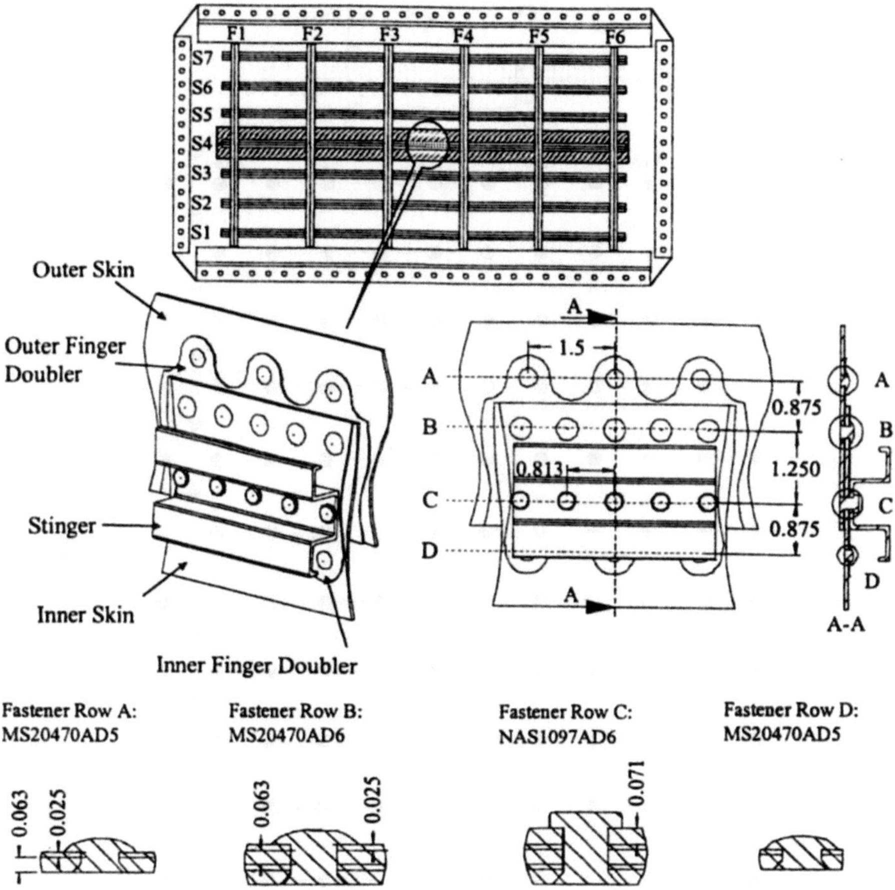

Fig. 8.18 Design of a fuselage panel tested by Ahmed et al. (2003)

of 191 mm. The skin (1.6 mm thick) was made of the 2024 T3 Al alloy and the substructure of the 7075 alloy. A longitudinal lap joint with two skin layers and two finger doubler layers (0.635 mm thick) was located along the middle stringer, Fig. 8.18. The joint contained four rivet rows labelled A, B, C and D. A test performed earlier had revealed that the upper rivet row A in the outer skin was the most susceptible to cracks. The fatigue test was conducted by applying cyclic loads including the internal pressure and reactive loads in the hoop and longitudinal directions. The maximum applied pressure was 110 kPa, resulting in a skin hoop stress of 96 MPa and a longitudinal stress of 58 MPa with a stress ratio of 0.1. Overload cycles were applied to generate marker bands on the fracture surface necessary to reconstruct the crack growth history.

The visually observed fatigue crack formation, crack propagation and crack link-up is presented in Fig. 8.19. The first damage in the critical upper rivet row A occurred after 80,550 load cycles on the right-hand side of rivet A23 located in

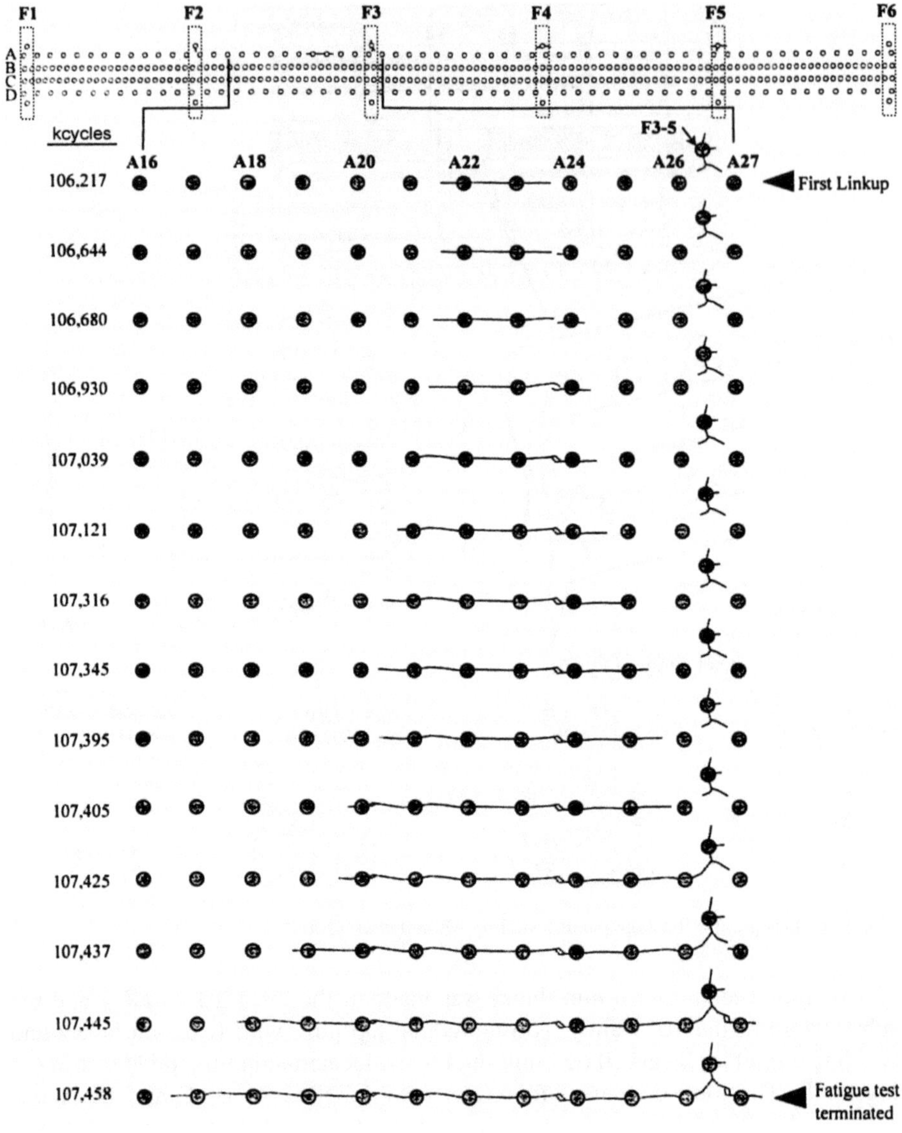

Fig. 8.19 Lead crack growth during fatigue test of the panel from Fig. 8.17 (Ahmed et al. 2003)

the second bay between frames F2 and F3. At a later stage, a through-the-thickness crack appeared on the left-hand side of this rivet hole. The cracks developed in a similar manner at neighbouring rivet hole A22. The first crack link-up occurred between rivet holes A22 and A23 at 106,217 cycles, forming a 72.9 mm long lead crack. After the first link-up, cracks emanating from rivet hole A24 (left-hand side after 106,641 cycles, right-hand side after 106,680 cycles) were detected. The

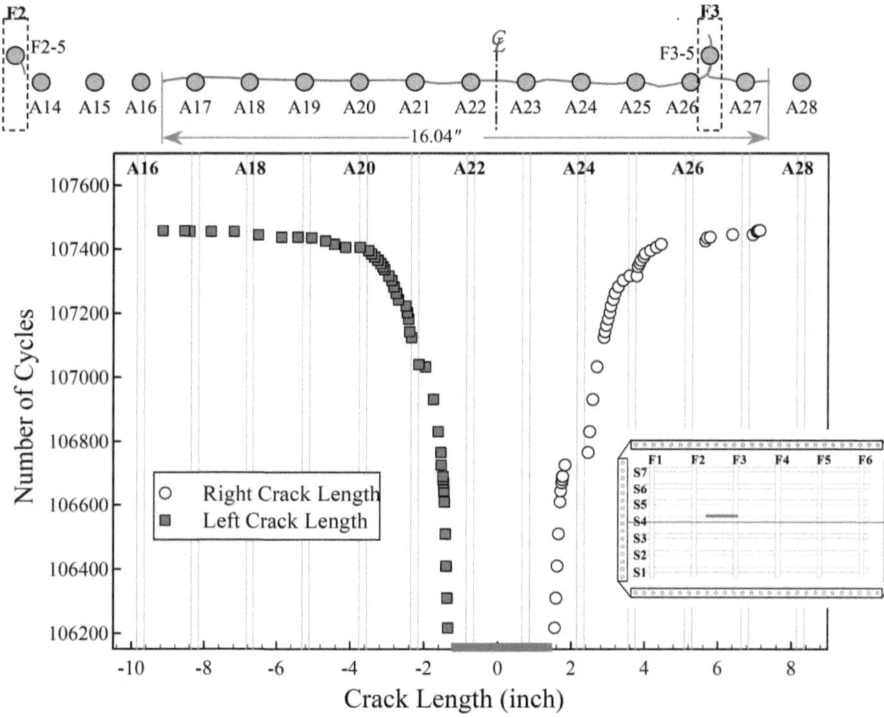

Fig. 8.20 Lead crack length as a function of fatigue cycles (Ahmed et al. 2005)

second link-up between rivet holes A22, A23 and A24 occurred 713 load cycles after the first link-up (106,930 cycles). After 109 additional fatigue cycles the lead crack grew into rivet hole A21 (107,039 cycles). At 107,425 cycles, the lead crack extended to the left (rivet hole A20) and to the right (rivet hole A26) and linked up with one of the existing cracks observed earlier at 79,915 load cycles near shear-clip rivet hole F3-5. Therefore, the lead crack crossed over into the next bay, forming a two-bay crack. Next, the lead crack grew into rivet hole A19 (107,437 cycles). It took only 8 additional fatigue cycles for the lead crack to link-up with the crack at the rivet hole A27 (at 107,445 cycles). At that time, the lead crack grew into rivet hole A18. At 107,458 load cycles, the lead crack spanned across rivet holes A17 to A27 with a total length of 407.4 mm. During the previous few cycles, the lead crack grew considerably with every load cycle, as it can be seen in Fig. 8.20. Therefore, the fatigue test was terminated and the residual strength test was performed. An initial extension of the 407.4 mm long crack occurred at 106.5 kPa. When the applied pressure reached 122.6 kPa, catastrophic failure occurred. A post-test inspection showed that frames F3, F4 and F5 all fractured under the lap joint. The crack extended explosively across bays 4 and 5 to the end of the lap joint at frame F6. The final length of the crack was 1.925 m.

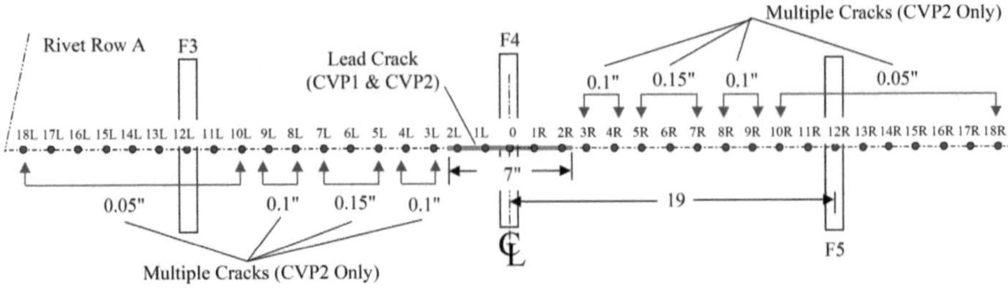

Fig. 8.21 Initial damage scenarios for MSD panel test program (Bakuckas et al. 2003)

It can be seen from the above investigation that crack growth up to the visible size consumed 75% of total fatigue life, while a period between the first link-up and the end of the test covered only 1% of the fatigue life. This observation is substantiated by the in-service experience of the Delta Air Lines operator (Steadman et al. 1999) indicating that first detectable cracks occur after 67–75% of DSG. These visible cracks were found in 90% of the fleet.

Panels shown in Fig. 8.18 were also tested by Bakuckas et al. (2003). Panel CVP1 contained a lead crack 7 in. long in the outer, critical rivet row of the lap joint. Panel CVP2 had the same lead crack as CVP1 with an addition of multiple, small cracks emanating from rivet holes ahead of the lead crack, Fig. 8.21. In both panels the central frame was severed. A unique Full-Scale Aircraft Structural Evaluation and Research (FASTER) facility was used for the experiments (Bakuckas 2002). This facility is designed to simulate the actual load an aircraft fuselage structure is subjected to during the whole flight, including differential pressure, longitudinal load, hoop load in the skin and frames and shear load. The applied loading simulated the cylindrical pressurization of 10.1 psi that a section of the fuselage along the neutral axis would experience (hoop load 554.6 lb/in.).

During CA loading (frequency 0.2 Hz, $R = 0.1$), lead crack propagation was observed. The data in Fig. 8.22 represent the lead crack position from the left and right sides as a function of the number of cycles. It is seen that the presence of MSD reduced the number of cycles to grow the lead crack to the final length by approximately 37%. Again the results demonstrate that the majority of fatigue life was spent in crack extension up to the first link-up. Once this link-up occurred, crack growth was very rapid.

After fatigue testing, residual strength of the panel was measured. For both panels, the initial damage consisted of a two-bay crack with a length of approximately 25 in. Results from the residual strength test are shown in Fig. 8.23. The data in this figure represent the crack extension for the crack tips as a function of applied pressure. During the test, pressurization was applied quasi-statically, and the crack extension measured up to panel failure. The test results for panel CVP1 indicated that there was a small amount of stable tearing at the beginning of testing, followed by rapid fracture. Next, the crack was arrested by frames. Final rupture occurred again under a slightly higher load of 11.14 psi. A similar scenario was observed

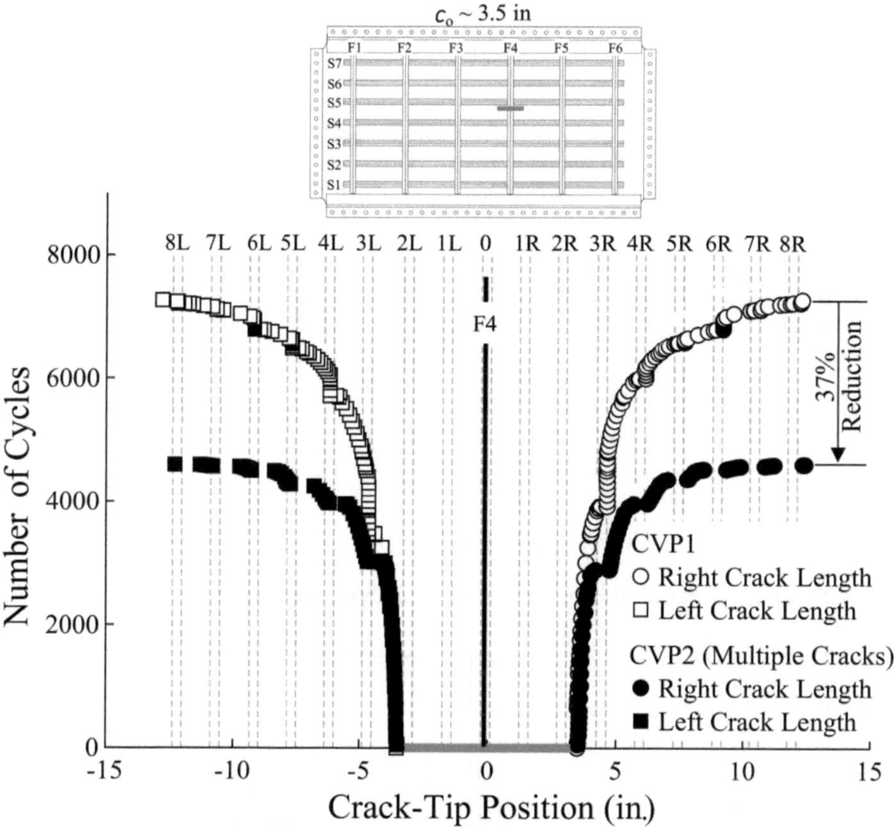

Fig. 8.22 Fatigue crack growth data for panels CVP1 and CVP2 (Bakuckas et al. 2003)

during the CVP2 test. The residual strength was 9.16 psi. As indicated in the figure, the presence of multiple cracks reduced the residual strength by approximately 20%.

A detrimental influence of MSD on the residual strength of fuselage panels was earlier reported by Maclin (1991). His tests were performed on flat wide (48 in.) panels with a three-row riveted lap splice. A baseline panel contained only a single 14 in. saw-cut in the skin, as shown in Fig. 8.24 (location 1). The second panel contained a 14 in. saw-cut along the row of rivets joining a stringer with the skin (Fig. 8.24, location 2) to test for the effect of holes without MSD. The long saw-cut together with small saw-cuts of approximately 0.02 in. was included in the top row of the lap joint in the third panel (Fig. 8.24, location 3). These small saw-cuts simulated the MSD situation. There were no differences in residual strength for the situations without MSD. When the MSD saw-cuts were added to the long crack, the amount of load tolerated decreased by about 17%.

In tests carried out under CA loading on wide flat panels with two and three countersunk rivet rows (supplied by Daimler-Benz Aerospace Airbus) Silva et al.

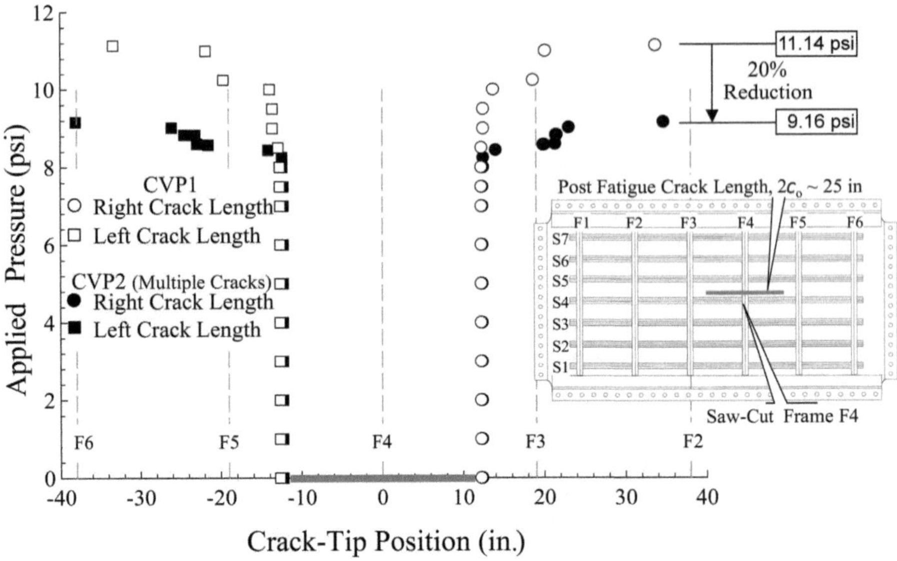

Fig. 8.23 Residual strength data for Panels CVP1 and CVP2 (Bakuckas et al. 2003)

Fig. 8.24 Flat panel for residual strength testing (Maclin 1991)

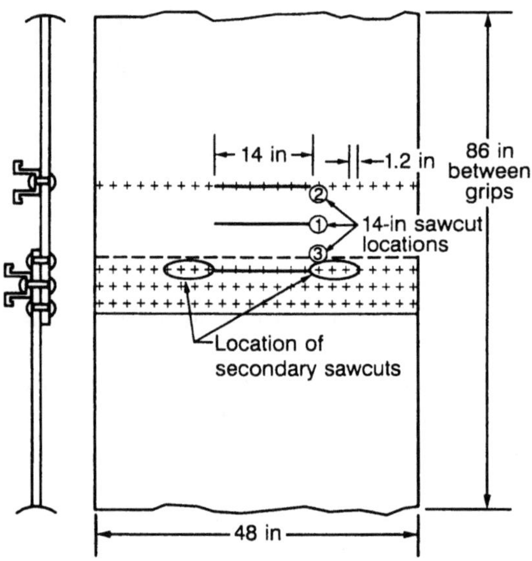

(2000) observed that the period between the first link-up and failure covered less then 2% of the total fatigue life. The percentage of life between the first visible crack and failure varied from 4% to 20%.

Conclusions from the fatigue tests described in this section are rather consistent. In all cases the upper rivet row of the longitudinal lap joints was critical and subjected to MSD. The first cracks visible at the outer surface of these joints were detected no sooner than after 50–75% of the panel total fatigue life or after about 70–80% of the DSG. The period between the first crack link-up and failure of the panel covered only 1–3% of the panel's total fatigue life. Fatigue tests on Boeing fuselage sections also showed a long damage detection period of 5–20 years (depending on the airplane type and assumed number of flights per year) between the time of the crack detection in longitudinal joints and the first link-up, provided that the cracks were noticed in their early stages (Maclin 1991).

8.2.5 *Effect of Fuselage Design on MSD*

As shown by experiments conducted at Boeing, the fuselage design can have an influence on final failure of a riveted longitudinal lap joint under an MSD condition (Miller et al. 1994). Two panels representing Boeing's typical narrow-body 427 and 437 fuselages (radius of 74 in.) were tested. The overall geometry of the panels was consistent with a typical fuselage design: 7075-T5 frames at a 20 in. pitch and 2024-T3 Clad stringers at a 9.25 in. pitch. The panels were divided into two halves by a single lap joint consisting of three rows of countersunk rivets. In the upper half of the panel the skin was 0.040 in. thick, while in the lower half of the panel it was 0.056 in. thick. The longitudinal lap joints were perpendicular to the rolling direction of the 2024-T3 Alclad skin. Internal bonded tear straps of 2024-T3 Clad were installed at each frame location. In the first panel (74-A) shown in Fig. 8.25a, all frames were "floating", namely connected to the stringers only by means of stringer clips (not connected directly to the skin, cf. Fig. 1.3). The second panel (74-B) differed from the first panel by the use of shear-ties to attach the frame directly to the skin and tear straps, as shown in Fig. 8.25b (for details see Fig. 1.2).

In the first panel each fastener hole had two small diametrically opposed longitudinal saw-cuts introduced during the original panel assembly, as seen in Fig. 8.25a. The dimensions of these saw-cuts varied as a function of the location along the lap joint and were smaller near the tear straps, as indicated by the fleet experience. A differential pressure of 90 psi (0.62 MPa) was applied during the test (test frequency one cycle per minute). After 7,000 pressure cycles a 5 in. saw-cut in the upper row of rivets was introduced, see Fig. 8.25a. Initially, during the further pressure cycling, crack growth from this saw-cut remained in the rivet row line. Subsequently, however, the crack began to grow significantly away from the rivet line as it was approaching the adjacent tear strap. Final failure was a controlled decompression ("flapping").

The initial condition of the second panel consisted of a 5 in. saw-cut and small diametrically opposed saw-cuts at each fastener hole, as described above. The final failure of the second panel is shown in Fig. 8.25b. It can be seen that both tear straps failed. One of these straps failed when the skin crack was around 23 in. in length. The other strap failed when the crack was 24.5 in. long. Only 85 pressure cycles were required to get both tear straps to fail, following the initial strap crack detection. Next, the crack continued to grow until it reached a length of 40.2 in. At

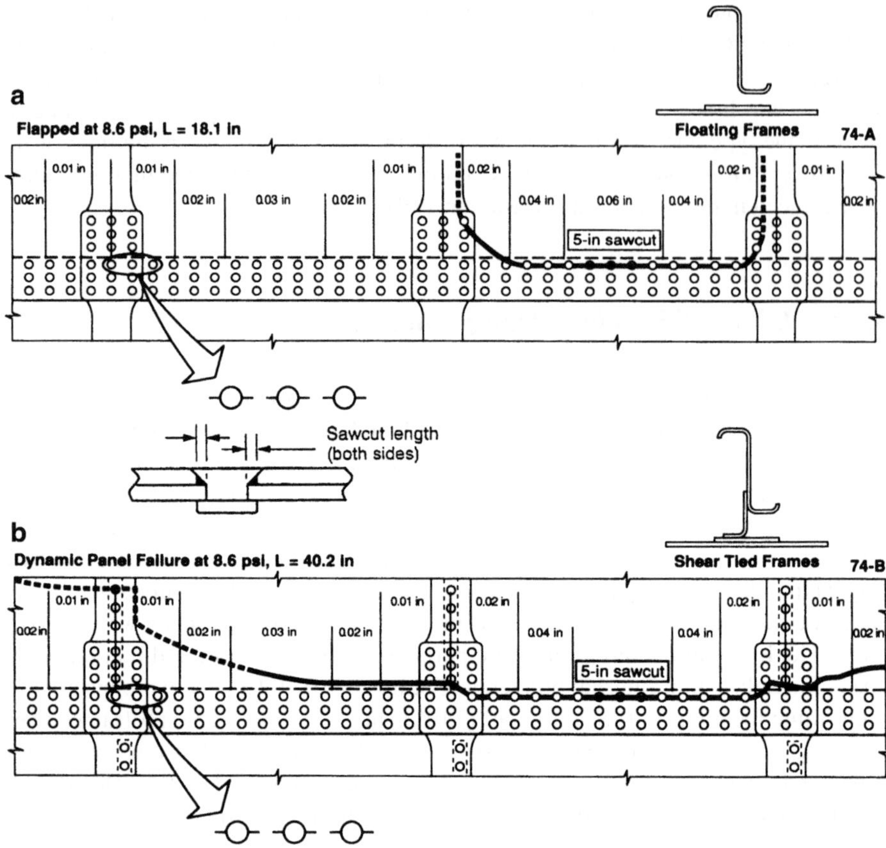

Fig. 8.25 Influence of fuselage design on fatigue behaviour of a riveted longitudinal lap joint: (**a**) floating frames; (**b**) shear-tied frames (Miller et al. 1994)

this time, an unstable crack extension resulted in final failure of the panel when the frames and stringers were broken.

Swift (1987) pointed out that in the case when a fuselage is noncircular, shear tied frame bending due to pressure may cause the skin stress to be higher locally near the frame than midway between the frames. When this situation exists, the crack is less likely to turn in a circumferential direction. This effect is enhanced by the presence of MSD ahead of the lead crack.

Based on comparisons of test results from various panels' geometries, Gruber et al. (1997) found that the panel geometry did not have a big influence on reduction of the residual strength due to MSD. Therefore, tests can be performed on simple panel geometries to get an acceptable representation of the effect of MSD on the residual strength of built-up panels.

8.2.6 Effect of Bending, Overloads and Underloads on MSD

As indicated by Terada (2001), bending loads caused by typical landing and gust loads affect the fatigue behaviour of riveted longitudinal lap joints. The corresponding test was carried out using a one-third scale model of the B 737 fuselage structure (5 bays, 6 frames, 6 tear straps and 16 stringers, sheet thickness 0.8 mm, 3 rows of countersunk rivets). The diameter and the length of the model were 1,250 and 2,500 mm respectively. The model was fixed at one end and a balancing weight was used at the free end to cancel the model weight. One specimen was loaded by pneumatic fatigue cycles only (maximum hoop stress 100 MPa, stress ratio 0.05, test frequency 1 cycle per minute). The final failure mode was the "flapping" at the centre bay of the top section after several link-ups of the cracks from the fastener holes. The total fatigue life was 118,930 cycles. The other specimen was tested under essentially the same conditions as the first one, but with additional bending cycles. The bending loads of 10.4 and 19.8 kN were applied at the minimum and maximum air pressure respectively. A first visible crack (crack size of 0.3–0.5 mm) occurred at 45,000 cycles, i.e. two times earlier compared to the first specimen. Unstable fracture occurred at 79,515 cycles, which was about two-thirds of the failure life for the first specimen. The magnitude of the maximum principal stress induced by the combined internal pressure and the bending loads was 106% of the hoop stress due to internal pressure only. Most cracks propagated at an angle of 10–20° to the longitudinal direction, namely perpendicularly to the maximum principal stress.

Terada (2001) studied also the effect of overloads and underloads on the fatigue life of three-row riveted lap joint specimens made of the 2024-T3 Al alloy (protruding head rivet 4 mm dia, sheet thickness 2 mm). The average fatigue life of the joints under CA loading (54 ± 42 MPa) was 322,000 cycles. In order to investigate the overload effect, two cycles of $S_{max} = 200$ MPa and one cycle of $S_{max} = 221$ MPa were applied at 40,000 and 66,400 cycles respectively. These maximum stresses corresponded to 80% and 90% of the tensile strength of the joint respectively. The specimens with overloads tolerated more than 440,000 cycles without any visible cracks. The author concludes that overloads do not necessarily cause a loosening of rivets, which he considers to be a primary reason for joint degradation (cf. Sect. 8.2.2). Instead, overloads can produce a favourable compressive residual stress field around the hole. The underload effect was examined by applying two cycles with a compressive stress $S_{min} = -25$ MPa corresponding to 90% of the specimen buckling load. In this case, the joint failed at 161,000 cycles, which was almost half of the life observed in the CA test. When the compressive load was applied, the specimen showed a considerable out-of-plane deformation. The associated enhanced secondary bending stresses could contribute to an observed decrease of fatigue endurance. Moreover, if the underload was applied in the presence of initiated fatigue cracks, the unfavourable load interaction effects (crack growth acceleration) were likely to occur during the post-underload smaller amplitude cycles (Skorupa 1998).

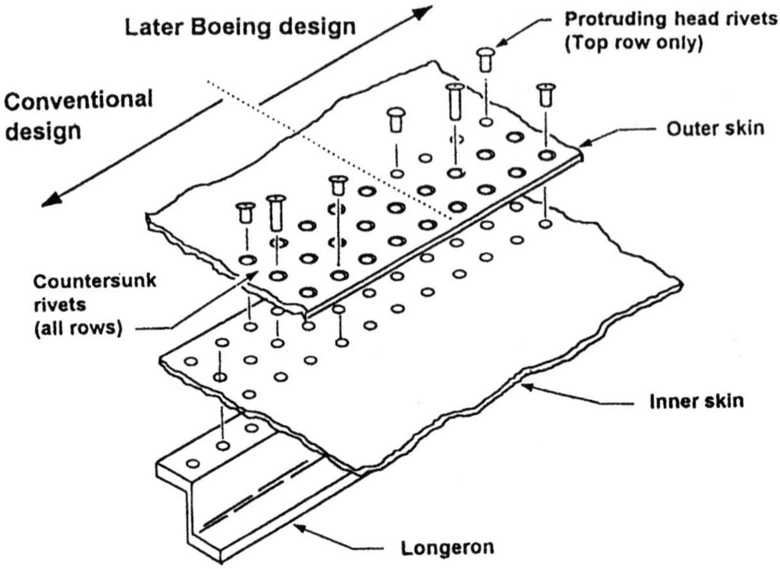

Fig. 8.26 Basic and modified for a terminating action three-row lap splice on Boeing aircraft (Müller and Hart-Smith 1997)

8.2.7 Fatigue Behaviour of Lap Joints Repaired by Riveting

There are a number of publications concerning riveted repairs to aluminium fuselage skin cracks. Some of these works have been reported by Fredell (1994). However, information about repairs to lap joints by means of riveting is extremely limited in the open literature. Consequently, results on the fatigue behaviour of repaired joints are scarce.

In 1987 Boeing introduced a remedial repair for fuselage lap splices known as the "terminating action" because its application terminates damage inspections (Samavedam et al. 1994). The basic idea behind this concept is that countersunk rivets in the upper row are replaced with universal rivets of a larger shank diameter, Fig. 8.26. The procedure was developed to eliminate the risk of widespread fatigue damage in the upper rivet row of a fuselage joint.

The usefulness of the terminating action was evaluated by comparisons between the fatigue test results for two panels shown earlier in Fig. 8.17a. The baseline test was described in Sect. 8.2.4. The terminating action was applied to the initially undamaged other test panel after 30,000 cycles. This implied that all countersunk rivets in the upper row were replaced with universal rivets. Although three small cracks were found at the time of the repair, they were completely drilled out when the holes were enlarged to accommodate the universal rivets. Also some

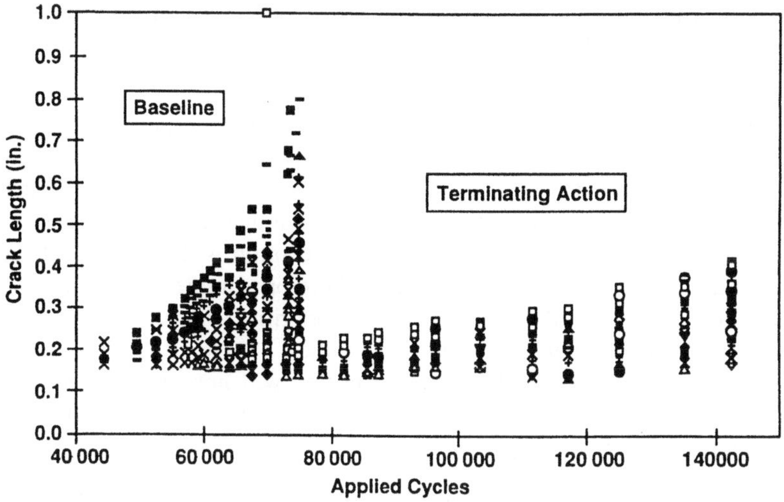

Fig. 8.27 Crack growth rates observed during a baseline test and a terminating action test (Samavedam et al. 1994)

of the debonded shear straps adjacent to the lap joint were re-attached using universal rivets. Cracking was observed in the middle row of rivets 37,800 cycles after the terminating action had been performed. Application of the terminating action appears to retard the CGR, Fig. 8.27. The CGR in the middle row after the terminating action was about 1/3 of the CGR in the upper row without the terminating action. After 70,000 cycles, 17% of rivets in the middle row exhibited cracks greater than 0.2 in. As pointed out earlier, for the baseline test 13% of the rivets in the upper row exhibited cracks greater than 0.4 in. after the same number of cycles. The panel subjected to the terminating action was fatigued for a total of 142,883 cycles. At that point, no link-up of the cracks occurred in the middle rivet row. The longest of these cracks was measured to be 0.41 in. It is worth noting that fatigue failure of the baseline panel occurred after 75,236 cycles. No cracks were detected in the upper rivet row in the panel subjected to the terminating action. The above results justify a conclusion that the terminating action repair effectively eliminates cracking in the upper rivet row and can lead to a significantly extended fatigue life of a joint.

8.2.8 Approach to MSD in Aging and New Aircraft

Statistics show that the number of old aircraft is large and will be increasing. However, there are no arguments to retire an aircraft from service if, by correct maintenance, the aircraft is kept in good condition. On the other hand, it must be realized that MSD cracks can occur in the old aircraft. According to Schijve (1995), several approaches to this problem are possible:

- the fatigue life is long enough to make the occurrence of MSD very unlikely. In this case results of full-scale testing, tear down inspections and service experience are essential to proof that an aircraft may have no MSD.
- MSD may occur, but it will be detected before critical failure takes place. The experiments described in Sect. 8.2.4 have shown that, in the presence of MSD, crack growth is relatively slow. The only exception is the period between the first crack link-up and the final rupture. Therefore, ample time exists for crack detection before the link-up leading to a large critical crack. It means that crack detection during regular inspections is possible.
- MSD may occur, but unstable crack extension will not be catastrophic, which can be a consequence of crack arrest (due to crack stopper elements) or of a safe decompression (by flapping).

With new fuselage structures, a design philosophy that allows for obtaining fatigue life sufficiently long to avoid MSD can be adopted. The extended fatigue life can be achieved in various ways. For example, a lower design stress level can be applied. This concept is associated with an unavoidable increase in fuselage weight, which leads to a conflict regarding economy.

Another option is offered by using more fatigue resistant materials. Nowadays, the typical Al alloys 2024-T3 and 7075-T6 can be replaced by new materials, e.g. fibre-metal laminates, like Glare. A Glare sheet consists of bonded thin aluminium alloy layers and glass-epoxy prepregs. Fibres of the prepreg layers can be oriented in different directions, depending on the dominant load on a structural component. For example, a cross ply is advantageous for the fuselage skin because the hoop stress is larger than the longitudinal stress (Vlot 2001; Schijve 2009a). Although the squeeze force to obtain the desired driven head dimensions of a rivet is comparable for a monolithic Al alloy and Glare, rivet hole expansion in Glare will be larger as a result of lower resistance of the fibre adhesive layers, inducing higher residual compressive tangential stresses around the rivet. As said in Sect. 3.3.5.1, the fatigue life of a riveted joint improves with increasing hole expansion. This explains excellent fatigue properties of Glare riveted lap joints proved experimentally, e.g. Fig. 8.28. Due to the high fatigue resistance of Glare, considerable weight savings can be obtained with reduced skin thickness. In addition, Glare shows very good properties with respect to corrosion durability and impact damage. It should be noticed that techniques used to produce aluminium alloy sheet materials are also applicable to Glare.

The extended fatigue life of a fuselage structure can also be obtained by an improved detail design and production. Additional costs are, however, required for testing of new concepts in order to avoid the risk of an inappropriate solution.

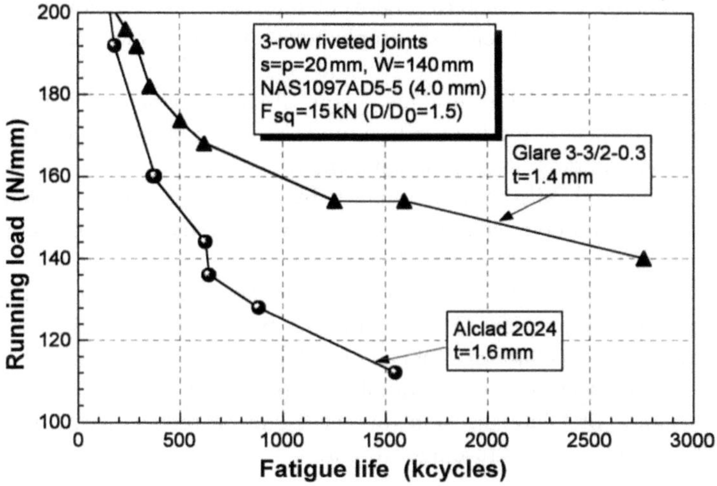

Fig. 8.28 Comparison between fatigue lives for Glare 3-3/2-03 and 2024-T3 three-row riveted lap joints (Müller 1995)

8.3 Summary of This Chapter

A situation when many cracks are initiated at about the same time in a critical rivet row, grow in a similar way and suddenly coalesce to form a single long crack, which can lead to catastrophic failure, is referred to as Multi-Site Damage (MSD). This particular type of failure is associated mainly with aging aircraft. On the contrary, Single-Site Damage (SSD) occurs when a crack grows from a rivet to a substantial size until failure and this growth is not influenced by cracks emanating from neighbouring rivets. Growth of medium-to-long MSD cracks is faster than that of an isolated crack of the same size. It means that the fatigue life for a joint with MSD is considerably shorter than for a joint under the SSD condition.

MSD in aging aircraft structures seems to stem from the clamping force deterioration (resulting from corrosion, fretting or excessive out-of-plane deformation), though some experiments suggest that poorly driven rivets do not necessarily contribute to MSD.

Flat specimens containing riveted lap joints with MSD cracks tested under biaxial loading show a longer fatigue life compared to similar coupons fatigued uniaxially. Bending loads and underloads have a detrimental influence on MSD in fuselage lap joints. Overloads do not necessarily cause a degradation of the riveted joint fatigue behaviour.

MSD tests on fuselage panels indicate that cracks visible on the outer surface are detected after 50–75% of the panel total fatigue life. The period between the first crack link-up and failure covers only a few percent of the total fatigue life.

However, a damage detection period between the time when a crack is visible and the first link-up is long. Therefore, ample time exist for crack detection before catastrophic failure.

Experiments show that the fuselage design (floating frame or shear tied frame) can have an influence on the final failure mode of longitudinal lap joints under the MSD condition (flapping or long unstable crack).

Riveted joints with MSD can be repaired by means of drilling out small cracks in a critical row with countersunk rivets and enlarging the holes to accommodate protruding head rivets.

An extended fatigue life of riveted joints in new fuselage structures can be achieved by replacing typical aluminium alloys by new materials like fibre-metal laminates (e.g. Glare). Glare exhibits an excellent fatigue behaviour and also good properties with respect to corrosion durability and impact damage. Due to the high fatigue resistance of Glare the skin thickness can be reduced, which leads to a considerable decrease in the fuselage weight.

Chapter 9
Predictions of Fatigue Crack Growth and Fatigue Life for Riveted Lap Joints

9.1 Introduction

Fatigue is a process in which a crack initiates at a particular location and grows slowly and progressively under service loading to a size at which the residual strength of a structure is insufficient to sustain the prevailing loads, and rapid failure occurs. Aircraft manufacturers were the first industrial branch to place reliance on damage tolerance designs. It was assumed that initial imperfections and flaws inherent in the material or introduced during manufacturing pre-exist in the structure and can become sources of cracking. Consequently, in recent years the main focus of the research related to fatigue of aircraft fuselage lap joints has been on modelling the crack growth stage rather than the crack initiation stage. Most crack growth prediction models are based on linear elastic fracture mechanics (LEFM) concepts and correlate crack growth using the stress intensity factor (SIF). It is well known that due to the anomalous behaviour of small cracks such predictions can be unreliable and lead to unconservative results in the early stage of crack growth from the initial defect. A concept of the equivalent initial flaw size (EIFS) developed by Rudd and Gray (1976, 1978) and Manning and Yang (1986) was conceived to cope with the lack of analytical means to deal with the crack initiation phase and small crack growth. The EIFS approach enables one to model the entire crack growth process using the LEFM methodology.

In the present chapter models and codes most commonly used to predict crack growth life and fatigue life of riveted lap joints, including the MSD crack problem, are presented. Available SIF solutions considered to be most appropriate for riveted lap joints are reviewed. A special attention is paid to the EIFS concept due to its significance for the prediction quality.

A. Skorupa and M. Skorupa, *Riveted Lap Joints in Aircraft Fuselage*, Solid Mechanics and Its Applications 189, DOI 10.1007/978-94-007-4282-6_9,

9.2 Crack Growth Prediction Models

Several crack growth prediction models available in software packages NASGRO (e.g. NASA 2000) and AFGROW (e.g. Harter 2006) can be utilized to predict crack growth in riveted lap joints.[1] Both interaction predictions, that account for load interaction phenomena occurring under VA loading, and non-interaction predictions, which neglect these effects, are possible. Models utilized in both packages are based on the LEFM methodology. Most often applied are the FASTRAN model (Newman 1992) available in somewhat differing versions both in NASGRO and AFGROW, and the Closure model in AFGROW.

FASTRAN employs the so-called strip-yield (SY) model, first proposed by Newman (1981). Models of this type are often considered to be a most versatile predictive tool for Mode I crack growth under arbitrary load histories. Extensively used by independent users are two SY models included in the NASGRO software, the one termed a constant constraint model being the modified FASTRAN model. A thorough evaluation of the NASGRO version 3.0 SY models' predictive capabilities for cracks in Al alloy sheets under a variety of CA and VA load histories has been provided by Skorupa et al. (2007).

The SY model is based on the Dugdale conception of crack tip plasticity, but it incorporates plasticity-induced crack closure (Elber's mechanism) by allowing a wedge of plastically deformed material to form on the surfaces of an advancing fatigue crack. It is assumed that the crack opening stress (S_{op}) in a current load cycle, that is the applied stress level at which the crack is fully open during uploading, depends on plastic deformations in the crack wake, which result from loads experienced previously. All plastic deformation is confined within an infinitely thin strip located along the fictitious[2] crack and embedded in perfectly elastic material. The strip stresses and deformations are solved using numerical methods by considering displacement compatibility along the fictitious crack surface, Fig. 9.1. To this end, the plastic strip is divided into a number of bar elements. The elements in the plastic zone can carry both tensile and compressive stresses, whilst the broken elements in the crack wake can only undergo compressive stresses referred to as the contact stresses. As illustrated in Fig. 9.1, the compatibility conditions imply $L_i = V_i$ ahead of the crack tip and in the contact zone behind the crack tip, and $L_i < V_i$ in the open crack region, where L_i is the length of element i and V_i is the fictitious crack opening displacement at location x_i. The S_{op} stress in FASTRAN can be computed either from the condition of equilibrium between the SIF due to an applied stress increment ($S_{op} - S_{min}$) and the (artificial) SIF due to the contact stresses, or alternatively from the so-called displacement method, that defines S_{op} as a level at which the last element in contact opens. In agreement with the crack

[1] The most recent versions are NASGRO Version 6.2 released in 2011 and AFGROW Version 5 released, as the first commercial version of AFGROW, in 2010.

[2] The fictitious crack length (c_{fict}) equals the physical crack length (c) plus the crack tip plastic zone size (r_p).

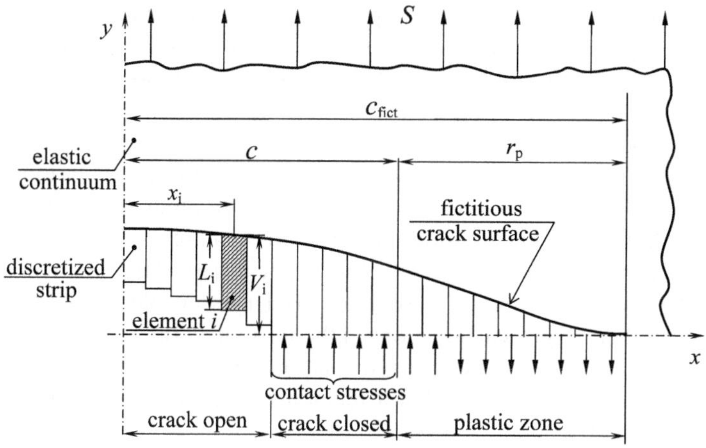

Fig. 9.1 Discretized plastic strip

closure concept, crack growth predictions are based on the assumption that the fatigue crack growth rate in a given load cycle is controlled by a current level of the effective SIF, ΔK_{eff}, namely

$$dc/dN = f(\Delta K_{eff}) \text{ with } \Delta K_{eff} = K_{max} - K_{op} \tag{9.1}$$

where the SIFs K_{max} and K_{op} correspond to S_{max} and S_{op} respectively.

Though the crack growth description with the AFGROW Closure model is in accordance with Eq. 9.1, the determination of S_{op} stems from premises very different than in the SY model applied in FASTRAN. As in FASTRAN, the S_{op}/S_{max}-value under CA loading is an empirical function of the stress ratio (R). However, with the Closure model the crack closure phenomenon is attributed to compressive residual stresses in the reversed plastic zone ahead of the crack tip rather than to the contact stresses behind the crack tip. The S_{op} stress is defined as a stress level at which these compressive stresses are overcome by applied tensile loading. Rules and constraints governing variations in S_{op} under VA loading stem mainly from the available experimental evidence on crack growth behaviour.

Both the AFGROW Closure model and the FASTRAN model require user adjustable parameters. Only some of these are material constants that can be determined empirically. To obtain appropriate values of the other parameters it is recommended that, whenever possible, the model should be tuned to fit a loading/geometry/material case "similar" to that considered in the actual predictions.

Besides the Closure and FASTRAN models, other interaction prediction approaches, namely the Generalized Willenborg, Wheeler and Hsu models, are available in the AFGROW software package. The first two models are widely discussed in textbooks. It is known that the Willenborg model will never yield a life prediction that is less than the non-interaction prediction and, hence, it leads to unconservative results when crack growth acceleration occurs. The Wheeler retardation model is one of the most empirical load interaction models and its results are strongly dependent on many parameters that must be determined from tests for a given material, spectrum, stress level, etc. Also the Hsu model is an empirically based approach that borrows and builds on the concepts offered by the Wheeler and Willenborg models, but also accounts for Elber's crack closure mechanism. The Hsu model is unable to predict crack growth under compression-compression cycles.

Though altogether five interaction crack growth prediction approaches are available in AFGROW, the Closure model is commonly referred to in the literature as the AFGROW model and this convention will be observed further on in this book.

The NASGRO and AFGROW software packages enable the user to choose between several crack growth equations in the form of d*c*/d*N* vs. ΔK relationships, where $\Delta K = K_{max} - K_{min}$. User-defined material crack growth data can also be input. Jones et al. (2008) claim that a relationship termed the generalized Frost-Dugdale law, which reads

$$\frac{dc}{dN} = C^* c^{(1-\gamma/2)} (\Delta K)^{\gamma}, \tag{9.2}$$

most adequately describes crack growth on the faying surface from the edge of a rivet hole observed in tests on simple laboratory coupons with riveted lap joints as well as in fleet data for fuselage lap joints with MSD. The material constants $C^* = 8.33 \times 10^{-12}$ and $\gamma = 3$ in Eq. 9.2 with c in mm and ΔKin MPa$\sqrt{}$mm were derived from laboratory test data, which is available in the literature, for Al 2024-T3 panels with centre and edge cracks.

Though in a model elaborated by Eijkhout (1994) only crack growth is described, it provides a way of estimating the crack initiation life. The approach is based on fractographic observations from different sources (riveted lap joints from F28-4000, F100, BAC 1-11, and B747-400 aircrafts investigated by NLR and NACA) described by Wanhill and Koolloos (2001) which indicate the following:

- MSD fatigue cracks tend to initiate on the faying surface near or at the rivet hole corners and grow in the thickness and longitudinal direction,
- crack growth rates (CGRs) in the thickness direction are nearly constant (cf. Fig. 7.19),
- CGRs in the thickness and longitudinal direction are initially similar,
- CGRs in either direction are above 10^{-8} m/cycle.

The trend line for the double logarithmic plot of the CGR versus faying surface crack length data (CGR between 10^{-8} and $3 \cdot 10^{-7}$ m/cycle, crack length from 0.1 to 5 mm) is linear, which implies the absence of the small crack effect.

Figure 9.2 shows a schematic of Eijkhout's model, where the *A* and *B* parameters must be empirically determined.[3] There are three main assumptions:

[3]Contrary to the conventional notation, *a* and *c* in Figs. 9.2 and 9.3 are the crack dimensions in the length and thickness direction respectively.

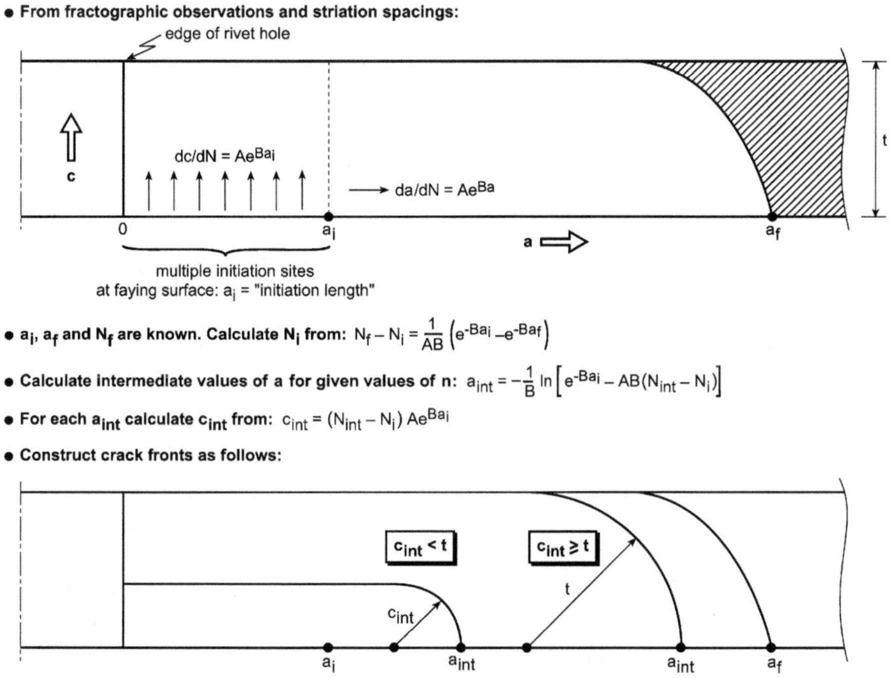

Fig. 9.2 Schematic of Eijkhout's model (Wanhill and Koolloos 2001)

- a constant CGR in the thickness direction, generally equal to the initial CGR in the longitudinal direction $dc/dN = Ae^{Ba_i}$,
- crack depth $c = 0$ at the initial length a_i,
- quarter circular crack front after the crack break-through.

The model can be also used for a variable d*c*/d*N* and for crack initiation at rivet hole corners ($a_i = 0$). An example of the model application is shown in Fig. 9.3. It demonstrates that this approach enables one to estimate fatigue lives at which MSD fatigue cracks become through-thickness, which is useful information for in-service inspection.

Like other prediction concepts considered in this chapter, Eijkhout's model is strongly empirical, which implies that its transferability is limited. Specifically, it should not be used for crack growth predictions in other structural areas with differing lap joint geometries, faying surface conditions and local stress levels.

Special codes developed to predict MSD crack growth in mechanically fastened joints are addressed in Sect. 9.5.

Predictions of the lap joint fatigue behaviour accounting for the contribution of fretting remain in an academic interest stage. Theoretical and experimental investigations on fretting damage indicate that the source of crack nucleation caused by fretting is a sharp peak in the tangential stress due to fretting at the edge of the contact zone (Szolwinski et al. 1997; Szolwinski 1998). The effect of

outer sheet rivet hole 17981P nDOF				
a_i (mm)	a_f (mm)	da/dN (μm/cycle)	N_f-N_i (cycles)	N_i (cycles)
0	5.2	$0.05311e^{0.1593a}$	67,000	59,250

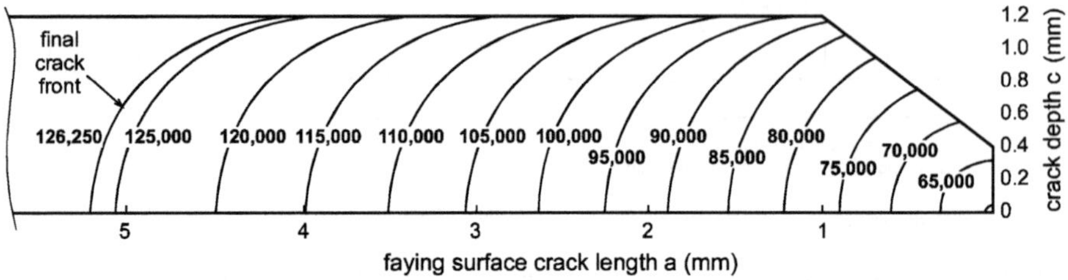

Fig. 9.3 Example of Eijkhout's model prediction on fatigue crack front development in an F100 full scale lap splice (Eijkhout 1994)

tensile load applied in the direction of fretting movements can be approximated by superimposing the local tangential stress caused by that loading with the tangential stress due to fretting. As under cyclic loading the stress due to fretting changes proportionally to the ratio of the applied load to the clamping force, the cyclic stresses and strains are nonproportional in nature, leading to conditions of multiaxial fatigue. Similarities between multiaxial fatigue crack nucleation mechanisms and observations of fretting cracks led to the application of multiaxial fatigue concepts to the fretting problem. Nucleation is assumed to occur at the edge of contact on the plane oriented 90° from the surface at a location where a parameter assumed to govern the crack nucleation adopts a maximum value. Usually the so-called Smith-Watson-Topper parameter $\sigma_{\max}\varepsilon_a$, ($\sigma_{\max}$ – local maximum cyclic stress, ε_a – local strain amplitude) is considered as most appropriate to correlate the crack nucleation under multiaxial fatigue conditions. The crack nucleation period is computed from the following equation resulting from multiplication of the Manson-Coffin equation by the Basquin equation:

$$\sigma_{\max}\varepsilon_a = \frac{(\sigma'_f)^2}{E}(2N_f)^{2b} + \sigma'_f\varepsilon'_f(2N_f)^{b+c} \tag{9.3}$$

where σ'_f, ε'_f, b and c are the material parameters.

Based on available theoretical and experimental investigations, Szolwinski (1998) proposes that, within the N_f period predicted by the above equation, a semi-elliptical, about 1 mm deep crack is nucleated. Subsequent crack growth is affected by the applied load rather than by the contact phenomena and can be predicted by an appropriate crack growth model.

As shown in Fig. 5.26, the applied tensile load on a lap joint leads to a nonsymmetric stick zone, the effect being enhanced with the increasing squeeze force (Brown and Straznicky 2009). Their analyses indicate that the peak maximum principal stress occurs at the loaded state edge of contact and that with increasing the squeeze force that stress location moves away from both the net section and the hole edge, which is consistent with Fig. 7.7. It remains as the next step to evaluate the effectiveness of Eq. 9.3 at predicting the fretting fatigue life and the location of crack nucleation.

9.3 Stress Intensity Factor Solutions

In order to account for the effect of applied loading on crack tip stress conditions for a crack at a rivet hole in a lap joint, the total SIF is usually derived by superposition as

$$K = K_{\mathrm{BP}} + K_{\mathrm{BR}} + K_{\mathrm{bending}} \tag{9.4}$$

where the indices BP, BR and bending correspond to contributions from the bypass load, bearing load and secondary bending respectively, Fig. 9.4. The superposed SIFs are usually expressed in the form

$$K_{\mathrm{BP}} = S\,(1 - R_{\mathrm{TR}})\, f_{\mathrm{BP}}\sqrt{\pi c}, \tag{9.5a}$$

$$K_{\mathrm{BR}} = S\frac{w}{d}\,R_{\mathrm{TR}}\, f_{\mathrm{BR}}\sqrt{\pi c}, \tag{9.5b}$$

$$K_{\mathrm{bending}} = S\,K_{\mathrm{b}} f_{\mathrm{b}}\sqrt{\pi c} \tag{9.5c}$$

where S is the applied stress, $R_{\mathrm{TR}} = T_{\mathrm{TR}}/P$ is the load transfer ratio, K_{b} denotes the bending factor (cf. Eq. 6.1), c is the crack length on the faying surface (cf. Fig. 7.17), $f_{\mathrm{BP}}, f_{\mathrm{BR}}$, and f_{b} are correction coefficients, and the meaning of other symbols follows from Fig. 9.4. It is assumed in Eq. 9.5b that the bearing load is identified with the transfer load, which implies that contribution of friction to load transmission is neglected (cf. Fig. 5.1).

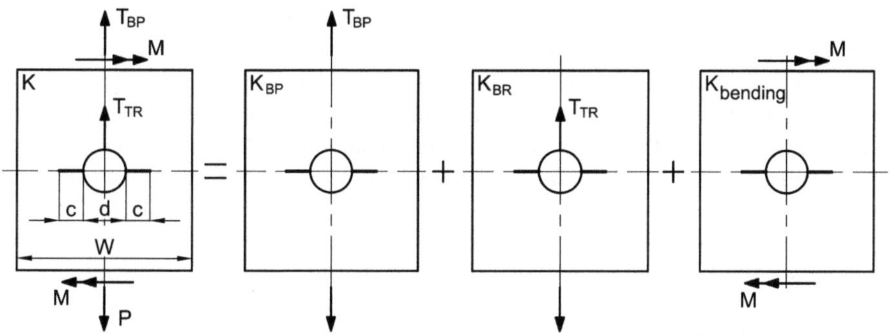

Fig. 9.4 Decomposition of loads to determine SIF for a crack at rivet hole in a lap joint

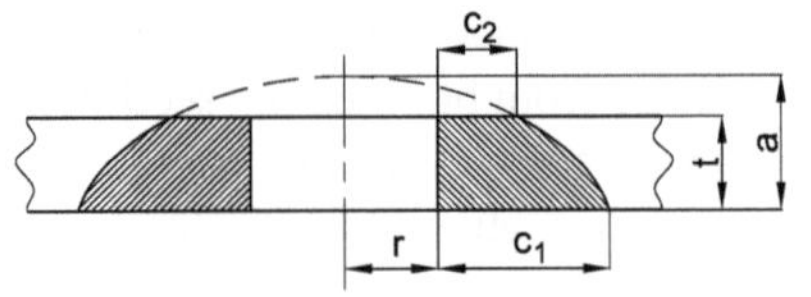

a/t	1.05, 1.09, 1.13, 1.17, 1.21, 2, 5, 10
a/c_1	0.2, 0.3, 0.4, 0.6, 1.0, 2.0
a/r	0.5, 1.0, 2.0

Fig. 9.5 Part-elliptical through crack and crack front geometries considered by Fawaz (1997)

As reviewed in Chap. 7, an extensive fatigue test evidence indicates that cracks in riveted joints commonly nucleate as corner cracks at the intersection of the faying surface and the rivet hole bore, cf. Fig. 7.7a. Because a closed form solution for the corresponding SIF is not known, engineering estimates or numerical methods are used. A K-solution most widely applied for two symmetric corner cracks at a hole was developed by Newman and Raju (1986). It covers a wide range in crack size (a/t) and crack shape (a/c) for two ratios of the hole radius to thickness ($r/t = 0.5$ and 1.0). To convert a K-solution obtained for two symmetric (corner or through-the-thickness, straight front) cracks on either side of a hole to a solution for a single crack (and vice versa) a correction coefficient developed by Shah (1976) is usually applied.

After through cracks are obtained at rivet holes in lap splice joints, they continue to grow with oblique fronts due to combined tension and bending (cf. Fig. 7.15). A FEM solution for such cracks under combined loading conditions including remote tension (uniaxial and biaxial, several biaxiality ratios), pin loading (point load, cosine, and cosine squared traction distribution on a fastener hole), and out-of-plane bending was developed by Fawaz (1997) and introduced to the AFGROW model (Harter 2006). Based on fractographic investigations the crack front was approximated as partly elliptical shaped, $a > t$, with the two axes along the sheet surface and the edge of the hole, Fig. 9.5. The SIF correction coefficients from Eq. 9.5 were derived for crack shape geometries given in the table in Fig. 9.5.

In response to fractographic information indicating that predictions of two unsymmetric corner cracks at opposite sides of a fastener hole are necessary (see Sect. 7.3), Fawaz and Andersson (2004) applied the *hp*-version FEM to generate accurate SIFs for diametrically opposed unsymmetric corner cracks at a centrally located hole in a finite width sheet subject to tension, bending, and bearing, as shown in Fig. 9.6. Altogether 5.6 million K-solutions were developed with numerous combinations of a/c, a/t and r/t ratios. Finite dimension (width and height) effects were eliminated using a large plate width and height relative to the hole diameter. The authors claimed that all structurally significant crack shapes were considered, specifically a/c of 0.1–10.0, a/t of 0.10–0.99, and r/t of 0.1–10.0. The Fawaz and Andersson solution has been incorporated into the AFGROW code (Harter 2006).

Empirically determined SIF correction coefficients for asymmetric through-thickness, straight front cracks at opposite sides of a fastener hole have been provided by Stefanescu et al. (2003) based on crack growth tests on Al 7050-T76 specimens with precracked open holes under remote tension, which implies that the solution can be utilized only for the K_{BP} component of the SIF. The correction

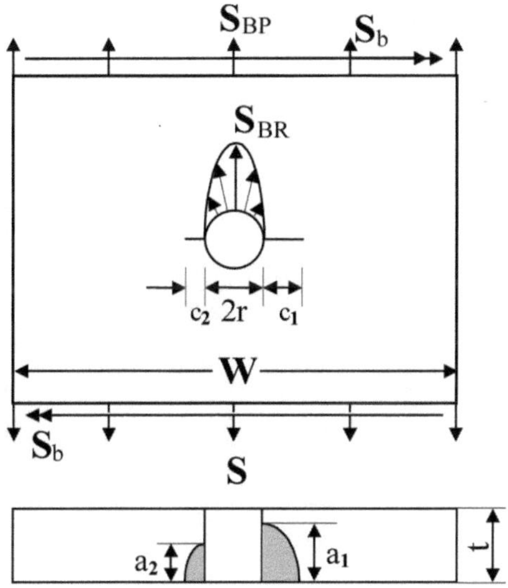

Fig. 9.6 Two unequal corner cracks at centrally located hole in a finite width sheet subject to bypass (S_{BP}), bearing (S_{BR}, $\cos^2$ distribution) and bending (S_b) applied stresses considered by Fawaz and Andersson (2004)

coefficients for a smaller crack of length c_s and the larger crack of length c_p were presented in terms of the normalized eccentricity defined as $(c_p - c_s)/2[r + (c_p + c_s)/2]$, where r is the hole radius.

Bakuckas (1999) compared the Newman and Raju (1986) solutions on f_{BP}, f_{BR}, and f_b to other, available in the literature, solutions for corner cracked holes obtained by various approaches and found agreement within ±3%. However, Fawaz and Andersson (2004) reported the literature solutions to be in error as much as 50% compared to their accurate *hp*-version FEM results for some specific cases including bending loading and bearing loading, deep cracks ($a/t > 0.8$), crack depth location, and intersection of crack front and free surface. The influence of differences between the Newman/Raju solution and the new K's according to Fawaz and Andersson (2004) on fatigue life predictions using the AFGROW code was assessed by Fawaz (2007). Combined remote tension and out-of-plane bending was applied, and CA loading and several VA loading histories, in either case for two different reference stress levels, were considered. The crack geometries included symmetric corner cracks at both sides of the hole and a corner crack at one side of the hole. Exemplary comparisons for a shallow initial crack ($a_i/t = 0.2$) and a deep initial crack ($a_i/t = 0.8$) are presented in Fig. 9.7a, b respectively. Here, the error e_0 is defined as $(N_{f,N/R} - N_{f,F})/N_{f,F}$, where $N_{f,N/R}$ and $N_{f,F}$ are the predicted fatigue lives (until the crack penetrates 99% of the sheet's thickness) according to the Newman and Raju (1986) and Fawaz and Andersson (2004) solutions respectively. The positive e_0-values indicate that the Newman and Raju solution yields fatigue lives always longer than the Fawaz and Andersson solution, the divergences being much more pronounced for the deep crack compared to the shallow crack. For the double-crack case this can be attributed to differences in the K's from either solution. Figure 9.7 reveals that the single-crack results show larger than the double-crack

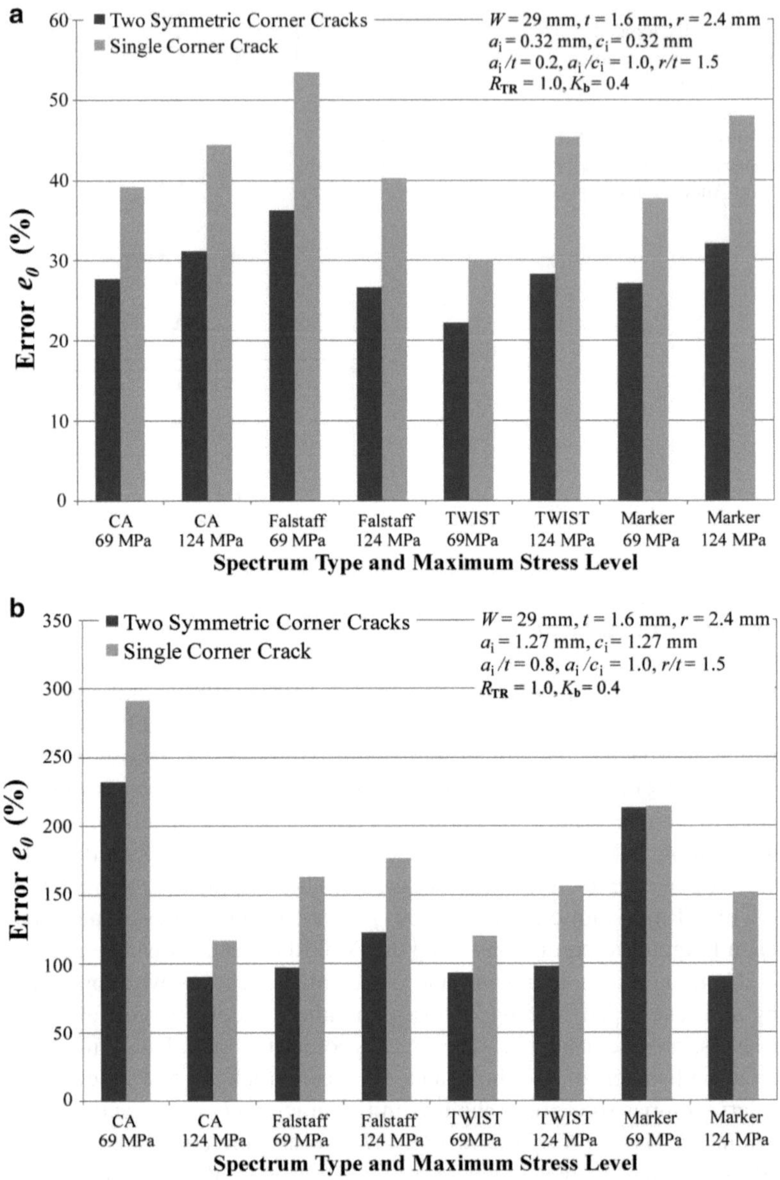

Fig. 9.7 Divergences in fatigue life estimates based on Newman and Raju (1986) and Fawaz and Andersson (2004) SIF solutions for double symmetric and single corner crack at a hole: (**a**) small shallow crack; (**b**) large deep crack (Fawaz 2007)

results differences between the two sets of solutions, which obviously stems from an inadequacy of the Shah correction coefficient used to convert the Newman/Raju SIF derived for two symmetric cracks to an SIF for a single crack.

Development of cracks alters the load transfer mechanism, causing shifting some of the load from cracked to uncracked rows. An analytical approach to handle this effect has been proposed by Wang (1998) based on the assumption that the compliance of a structure increases with crack growth, which leads to an additional displacement due to crack opening, δ_{crack}. Consequently the compatibility condition given by Eq. 5.1 should be rewritten as

$$l_{k,i} + \Delta_{k,i} + \delta_{k,i+1} = l_{k+1,i} + \Delta_{k+1,i} + \delta_{k,i} + \delta_{crack} \tag{9.6}$$

with

$$\delta_{crack} = P \cdot C \tag{9.7}$$

where P is the applied load and C is the cracked sheet compliance.

Utilizing the LEFM relationship between the energy release rate G and the SIF

$$G = \frac{K^2}{E} = \frac{P^2}{2t}\frac{\partial C}{\partial c} \tag{9.8}$$

the following expression for C is obtained if only the K_{BP} and K_{BR} terms are accounted for in the solution for K (cf. Eqs. 9.4 and 9.5):

$$C = \frac{2\pi}{Et}\int_0^c c\left(\frac{1-R_{TR}}{w} f_{BP} + \frac{R_{TR}}{d} f_{BR}\right)^2 dc. \tag{9.9}$$

Substituting Eq. 9.7 with C given by Eq. 9.9 to Eq. 9.6 enables one to solve for R_{TR} at a given crack length, and thus to account for the load transfer redistribution with crack growth. Computed R_{TR} variations for a lap joint with three rivet rows utilizing the NASGRO (NASA 2000) TC05 solution for K (through crack in a row of holes) are shown in Fig. 9.8, where it is assumed that cracking develops in the top row. It is seen that R_{TR} in that row drops from 37% to 10% when the crack approaches the middle line of the rivet pitch. Such a crack extension would, however, imply final failure of the sheet due to link-up of all cracks as the above approach implicitly assumes that cracks of the same size are present in all rivets in the critical row. Results shown in Fig. 9.8 are in quantitative agreement with FE results produced by Beuth and Hutchinson (1994).

In the case of MSD in a fastener row, a redistribution of load transfer occurs also within that row (Kuo et al. 1986; Ingram et al. 2002). This issue is further addressed in Sect. 9.5.

As amply documented in this book, the manufacturing process, that to a large extent manifests in the magnitude of hole expansion (*he*), has a profound effect on the fatigue behaviour of riveted joints. Based on an FE analysis of the effect of *he* on tangential stresses at a rivet hole, Newman et al. (1997) postulated that stress concentration due to applied loading becomes effective only above the lift-off

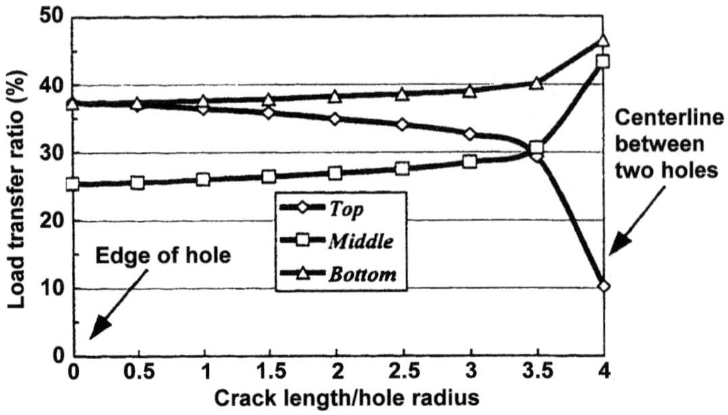

Fig. 9.8 Load transfer ratio against crack length for a three-row lap joint (Wang 1998)

stress (S_{LO}) defined as the applied stress level at which first separation of the rivet shank and hole surface occurs during uploading. They concluded that it would be reasonable to neglect the cyclic stress range below S_{LO}, similar to the rationale used in the crack closure concept. Accordingly, the effective SIF range in the SY model for crack growth predictions should be redefined as (cf. Eq. 9.1)

$$\Delta K_{\text{eff}} = K_{\max} - K_{LO} \quad \text{if} \quad S_{LO} > S_{OP},$$
$$\Delta K_{\text{eff}} = K_{\max} - K_{OP} \quad \text{if} \quad S_{LO} \leq S_{OP} \tag{9.10}$$

where K_{LO} is the SIF at the S_{LO} stress.

Elastic FE analyses by Newman et al. (1997) showed the S_{LO} stress to be independent of the crack length and linearly related to *he*. As *he* remains generally unknown, they suggested determining S_{LO} by trial and error to make the SY model (FASTRAN) predictions of crack growth fit the test data. According to this concept, once S_{LO} is derived for a specific lap joint, it should be valid for a similar joint in another structure. Applicability of this approach to real predictions is conditioned by formulation of a criterion for the joint similarity, which is not a straightforward task.

Wang (1998) applied the Prandtl-Reuss plastic flow model to estimate analytically *he* if the squeeze force and clearance between the rivet and hole prior the rivet installation are given. Then residual pressure on the prospective crack path (tangential stress distribution) was determined utilizing a solution by Hsu and Forman (1975), and the corresponding residual SIF (K_R) was estimated by the weight function method. The total SIF was computed as the sum of K given by Eq. 9.4 and K_R. The (fictitious) negative K_R-value associated with compressive residual pressure on the crack faces contributed to a reduction in the SIF. The results of Wang for a 1.5% interference (countersunk rivet diameter 0.192 in., sheet thickness 0.09 in., rivet pitch in row 1 in.) indicate that its beneficial effect on the SIF vanishes after the crack has grown less than 0.1 in. from the hole edge.

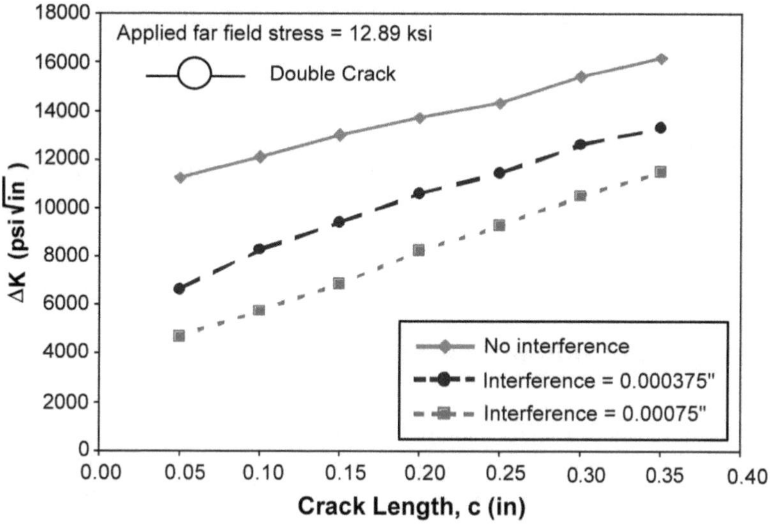

Fig. 9.9 Computed effect of interference on the SIF (Trego and Cope 2001)

This is in difference to (elastic) FE results provided by Trego and Cope (2001) for a joint geometry similar to that of Wang except that the sheet thickness was 0.04 in.. For both interference levels they considered, namely 0.4 and 0.8%, the difference between K-solutions without and with interference was more pronounced than reported by Wang for 1.5% and persisted still for the crack length of 0.35 in., Fig. 9.9. Accounting for the presence of residual stresses around the rivet hole (input by the user in the crack plane) is possible when the AFGROW software package is utilized (Harter 2006). The K_R-value is determined either by the Gaussian integration technique or the weight function method.

As illustrated in this section, the SIF for a lap joint is typically derived by the compounded solution method, which relies on superpositioning solutions for individual effects. To account for various indeterminate contributions to the SIF, such as friction between the mating surfaces, contact stresses on the hole, or a countersink hole geometry, a semiempirical approach is sometimes used, as first proposed by Hsu and Yu (1998) and detailed in the next section. An alternative empirical approach was used by Nesterenko et al. (2007) who assumed the SIF for a longitudinal lap joint in the form $K = S\beta_1\beta_2\beta_3\beta_4\beta_5\sqrt{\pi c}$, where the crack length c is measured from the hole centre and the correction coefficients account for: β_1 – finite width, β_2 – open hole, β_3 – filled hole, β_4 – load transfer without bending, β_5 – secondary bending. A reference fatigue test series on centre crack specimens subject to remote tension was performed to generate dc/dN-ΔK data. Specimens with symmetric sawcut precracks at an open hole and at a filled hole were used to obtain β_2 and β_3 respectively. β_4 was obtained from a non-bending double strap specimen with two protruding head rivet rows, while a lap joint with two rows of the same type and diameter rivets was used in the tests to determine β_5. The β_1-coefficient according to the literature solution was assumed to be accounted for in the SIF for any test series. A β-value for a given influence was derived by extracting the mth root from the ratio of dc/dN observed in the test corresponding to that influence to dc/dN from the reference tests.

The SIF solutions considered above are applicable to cracks initiated at rivet holes. For cracks nucleating on the faying surface somewhat away from the hole, as occurs for tight rivets, Newman and Ramakrishnan (2011) suggest applying a K-solution for a surface crack in a thin sheet without a hole subject to remote tension and bending.

9.4 Equivalent Initial Flaw Size

An important reason for a considerable research effort put into the EIFS concept is that the EIFS distribution is among basic stochastic input data required in aircraft structural risk and reliability analyses that have gained increasing interest in recent times (cf. Sect. 10.7). EIFS is a hypothetical crack size that is assumed to exist in a structural part prior to service. For a specimen with a single fatigue crack, crack size (c) versus number of cycles (N) data can be obtained by in situ and/or fractographic measurements. The EIFS can be derived by extrapolating an c *vs.* N curve fitted to the experimental data back to the beginning of the test, i.e. to N of zero (Manning and Yang 1986). The procedure is illustrated in Fig. 9.10. An alternative way is to back extrapolate the dc/dN *vs.* N curve. The back extrapolation is usually carried out using a crack growth prediction code. Fractographic examinations indicate sometimes an unsymmetric crack geometry at rivet holes, which can result from different times of the crack nucleation on either side of the hole (Fawaz 2000). Although the K-solution for two unsymmetric corner cracks by Fawaz and Andersson (2004) addressed in Sect. 9.3 would be more appropriate in such cases, two symmetric quarter circular cracks are often assumed in EIFS predictions.

The time to crack initiation (TTCI) distribution is sometimes used instead of the EIFS distribution. The TTCI is defined as the number of cycles or flight hours it takes for a non detectable, pre-existing crack to grow to a reference crack size. The latter is usually selected on the basis of a crack detectable by the NDI technique.

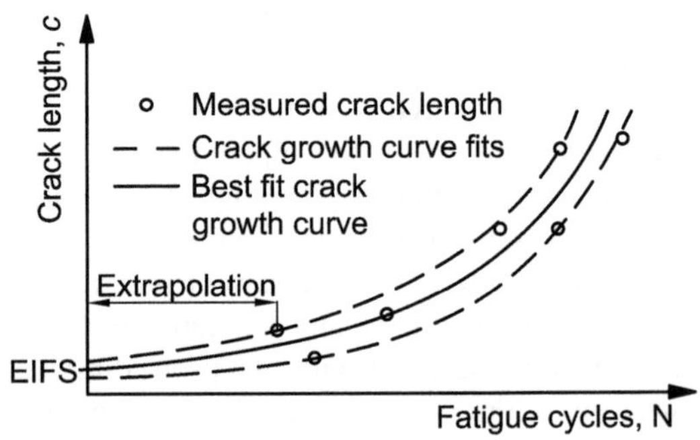

Fig. 9.10 Method of EIFS determination

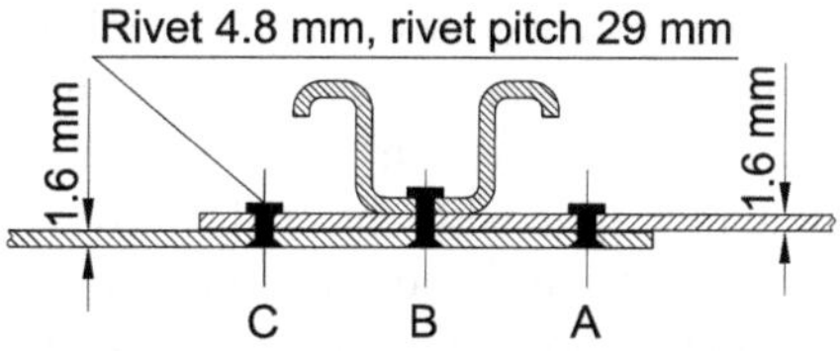

Fig. 9.11 Lap joint configuration in a flat panel considered by Fawaz (2003)

Therefore, the TTCI can be easily obtained by experiments. The TTCI distribution function can be transformed to a compatible EIFS distribution using a power crack growth law (Tong 2001).

For a component with multiple cracks, the EIFS distribution can be obtained from crack growth data for the individual cracks. Investigations addressed below indicate that EIFS-values for riveted joints can be affected by the manufacturing quality, sheet material, fastener hole type, fastener type, fastener/hole interference, design solution and loading conditions, and that different crack growth prediction methodologies yield different EIFS values. In practice, the EIFS distribution is used to predict a distribution of crack growth lives or a distribution of MSD after a fraction of fatigue life for a structure and loading conditions similar to those for which the EIFS was derived. Manning and Yang (1986) recommend that the minimum crack size determined from fractography should be consistent with the minimum crack size still amenable to LEFM, typically on the order of 0.125 mm. Ideally, the EIFS distribution should be fitted so that the best fit to the data is at the upper tail portion of the population because it is the largest EIFS that will govern durability of the structure (White 2006).

It is well known that crack growth analysis results depend on the methodology used. Hence, it is appropriate to apply the same code for both the EIFS derivation and the "real-work" predictions. Fawaz (2003) used two crack growth prediction codes, AFGROW and FASTRAN characterized in Sect. 9.2 to compare their influence on the EIFS estimate for wide, flat panels ($559 \times 1422 \times 1.6$ mm) from 2024-T3 Alclad with a typical three-row riveted lap joint, Fig. 9.11. Row C was always critical. Because single overloads applied between CA baseline cycles did not adequately mark the fracture surface, another spectrum containing underloads was applied to aid a post-crack growth history reconstruction. Blocks of CA baseline cycles were separated by blocks with underload markers, the latter blocks consisting of clusters of underloads inserted between clusters of the baseline cycles. The R-ratio was 0.02 for all cycles. Using a SEM, a crack length as small as 9 μm could be detected. The distribution of crack histories for various rivets appeared quite uniform indicating that the cracks were growing in a similar manner. The details of the experimental investigation are given in an earlier work of this author (Fawaz 2000). A consistency between $(c + r)$ *vs.* N curves, where r is the hole radius, for cracks in both simple riveted coupons and fuselage panel lap joints was also reported by Jones et al. (2008) from fractographic examinations. Based on the fractographic evidence, Fawaz (2003) assumed, as many other researchers, two symmetric quarter circular cracks at a hole for the EIFS predictions. As the fatigue growth codes were incapable of accounting for multiple part-through cracks growing in one structural detail, the cracks were modelled as two diametrically opposed part-through cracks emanating symmetrically from a single, straight shank hole in a finite width plate. The SIF according to Eqs. 9.4 and 9.5 was utilized. The bypass, bearing[4] and (maximum) bending stresses normalized by the applied stress equalled 0.66, 2.09 and 0.36 respectively. In order to improve correlation between the predicted and actual fatigue lives, an additional, empirically based SIF correction coefficient β_u derived by Hsu and Yu (1998) was introduced into both codes to account for influences not covered in the theoretical K-solution. The derivation of β_u is explained later on in this section. The EIFS determined using AFGROW ranged from 7.29 to 62 μm, whilst with FASTRAN the minimum and maximum EIFS equalled 6.35 and 39.9 μm respectively. An obvious source of the differences in the results was different definitions of ΔK_{eff} in either code (see Sect. 9.2). An additional cause was distinct assumptions about the crack shape development. For the part-through stage of crack growth, the Newman and Raju (1986) K-solution for symmetrical corner cracks growing from both sides of a straight shank hole was used in both codes. AFGROW, however, lets the crack shape develop naturally (a/c varying with crack growth), while with FASTRAN the aspect ratio is held constant until the breakthrough ($a/t = 1$), somewhat inconsistently with in-service observations and test results (see Sect. 7.3). For the through crack portion of the fatigue life, the AFGROW code again permits the crack shape to develop naturally (a/c_1 variable, cf. Fig. 9.5) utilizing the part-elliptical crack solution of Fawaz (cf. Fig. 9.5). In FASTRAN, a straight front, through crack is assumed once the crack has grown through the thickness, again in disagreement with experimental results (cf. Fig. 7.15). Because of the differences in crack growth modelling in either of the codes, the β_u correction coefficient determined by Hsu and Yu (1998) using FASTRAN may not be fully adequate for AFGROW.

The lower EIFS-values obtained by Fawaz are close to the 6-μm flaw assumed as the initial defect size needed to predict the fatigue life of a pristine, circular hole specimen made of 2024-T3 Al alloy (Newman 1999). This semi-circular flaw had an area equal to the average inclusion-particle sizes observed by Piascik and Willard (1999) to initiate MSD in riveted lap joints. Also Hoeppner (2011) points out that fatigue cracks in aircraft structural materials most often form at initial discontinuities including pores and particles. It is obvious, however, that there is no direct correlation between the microstructural defect size and the EIFS value because, as said in the beginning of this section, EIFS is a fictitious quantity dependent on multiple influences.

A dependency of the EFIS value on the crack initiation site, namely at the rivet hole edge or on the faying surface at a distance from the hole (cf. Fig. 7.7a versus Fig. 7.7b), is documented in several investigations. Harris et al. (1999) attempted to predict the fatigue life of a Type-1 flat panel used in tests by Furuta et al. (1997). 1 mm thick, 2024-T3 Alclad sheets were lap jointed using two rows of countersunk

[4]The bearing stress was defined as $S \cdot R_{TR} \cdot w/d$, cf. Eq. 9.5b.

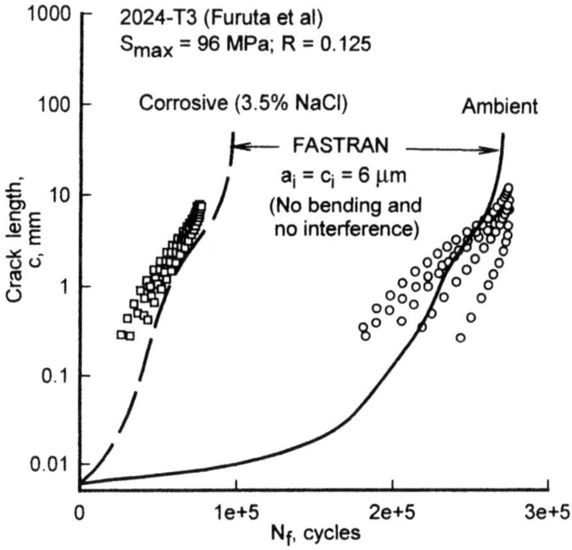

Fig. 9.12 Measured and calculated fatigue crack growth curves for Type 1 panel lap joint under ambient condition and salt water solution (Harris et al. 1999)

MS20470AD5 (4 mm dia) rivets. The loading conditions were described in Sect. 2.2. The FASTRAN model assuming a 6-μm EIFS (two symmetric quarter circular crack at a straight shank hole) was applied. For this two-row lap joint the bypass and transfer load ratios equalled 0.5. The rivet/hole interference was assumed to be small and its effect was neglected. Because out-of-plane deformation of the splice area was prevented during the fatigue test, secondary bending was not accounted for in the predictions. Figure 9.12 reveals that the FASTRAN results for salt water (dashed line) and laboratory air (solid line) are in a reasonable agreement with the experimental results. Satisfactory predictions were also obtained for other panel types considered by Furuta et al. and shown in Fig. 2.6.

In the same work Harris et al. (1999) applied the EIFS concept to predict by the FASTRAN code crack growth at rivets of lap joints in a full-scale fuselage test article. The majority of cracks were initiated by fretting damage in the faying surface clad. The 6-μm EIFS yielded a substantial underprediction of crack lengths observed at a given cycle number. An EIFS of 50 μm equal to the nominal thickness of the clad was found to be adequate in the case of the cracks initiated later in life. For a joint exhibiting much greater fretting damage, even an EIFS of 100 μm was needed to reasonably match the predictions with the experiment, Fig. 9.13.

An EIFS of 50 μm had been already used earlier by Piascik and Willard (1997) in fatigue crack growth predictions for the riveted lap splice joint described in Sect. 8.2.4, Fig. 8.15. The experimental data were obtained from a full scale fuselage test article that was subjected to 60,000 pressurization cycles. As in the Harris et al. (1999) work referred to above, a fractographic examination at rivet hole locations in the critical row revealed that cracks initiated in the clad layer underneath the rivet head at a distance from the hole edge and exhibited a semi-elliptical crack

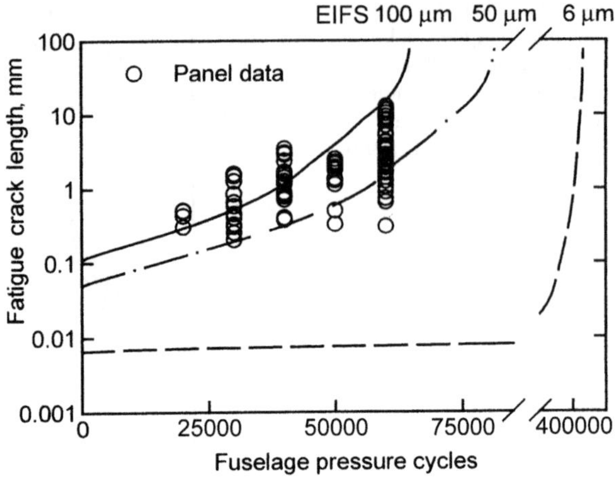

Fig. 9.13 Measured and calculated fatigue crack growth at rivets in aircraft fuselage lap joint (Harris et al. 1999)

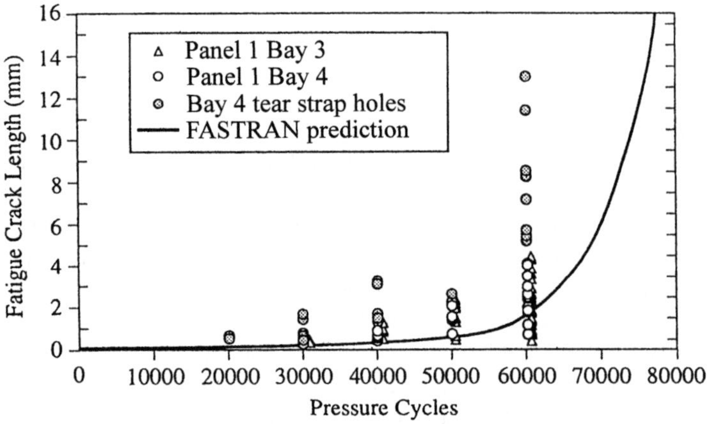

Fig. 9.14 Comparison between measured and predicted fatigue crack lengths in riveted fuselage lap joints (Piascik and Willard 1997)

front shape. A fretting debris was noted on the faying surface in the region of crack initiation. Crack growth histories for bays 3 and 4 were determined from marker bands found on the outer skin, Fig. 9.14. The fatigue crack growth prediction for the maximum applied stress was performed using the FASTRAN code. The transfer load ratio of 0.29 for the critical rivet row based on an FE analysis of a four-row lap joint and the bending factor of 0.5 based on the model of Schijve (1972) were adopted in the computations. A non-interference rivet/hole fit and, before the crack had penetrated the outer skin surface, two symmetric quarter elliptical cracks ($a/c = 0.8$) at a straight shank hole were assumed. The predicted crack growth curve for S_{max} of 90 MPa is plotted in Fig. 9.14. A fair correlation with the bay 3 and 4 measured data is observed in the long crack region, but the computed results underpredict the length of short cracks. As in the predictions by Harris et al. (1999)

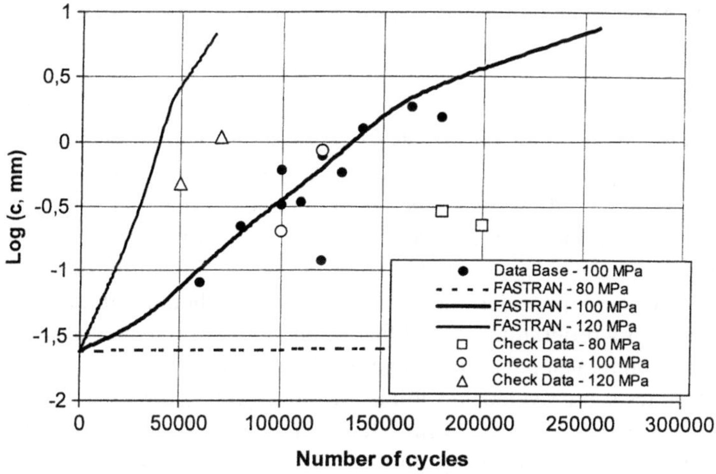

Fig. 9.15 Comparison of test results and predictions (Cavallini et al. 2001)

addressed above, a large EIFS of 50 μm partly compensates the evident inadequacy of applying the *K*-solution for a corner crack at a hole to model a surface crack initiated at a distance from the hole, especially in the early stage of crack growth. Neglecting the countersink hole configuration is another source of the prediction error for the short crack region. Figure 9.14 further demonstrates an underprediction of crack growth for cracks emanating from rivets in a tear strap, but the origin of these cracks is severe stress concentration rather than fretting damage.

Investigations by Cavallini et al. (2001) and Moreira (2004) indicate a dependency of EIFS on loading conditions. Cavallini et al. used the EIFS concept and the FASTRAN code to predict crack growth for a simple three-row riveted lap joint. Al 2024-T351, 1.6 mm thick sheets were connected using 19 columns of MS2026D5-6 (4 mm dia) rivets with a rivet pitch of 20 mm. Two fatigue test series were performed. The one devoted to the determination of the EIFS distribution involved CA loading with $S_{max} = 100$ MPa at $R = 0.1$. The corresponding test data are shown as full circles in Fig. 9.15 (notation data base). Another test series was intended to generate results used for comparisons with the predictions (check data, Fig. 9.15). Again CA loading with three different S_{max}-values of 80, 100 and 120 MPa was applied (two specimens for each S_{max} stress). Specimens from the first test series were used for a reconstruction of crack growth histories. When a desired percentage of the fatigue life was reached, the test was stopped, the critical rivet row was broken, and crack dimensions at the rivet holes were measured. Despite a large rivet head diameter to rivet shank diameter ratio of 1.6, the fatigue cracks initiated at rivet holes in the critical row of the outer sheet faying surface and no fretting damage was observed. The EIFS distribution was obtained by a back extrapolation of the measured results using the FASTRAN code and the SIF defined according to Eqs. 9.4 and 9.5. The rivet/hole interference was accounted for by assuming in

FASTRAN ΔK_{eff} given by Eq. 9.10. A quarter elliptical crack shape with a varying aspect ratio obtained from the destructive examination was incorporated into the FASTRAN code and an EIFS mean *c*-value of 24 μm was assumed in crack growth predictions for the check test series. The actual and predicted results are compared in Fig. 9.15. Each data point represents a single test result obtained by averaging the crack length measurements in the critical row. It is seen that the predictions agree with the experiment only for the check data obtained at the load level of 100 MPa, the same as used in the base test series for the EIFS estimation. For the other two stress levels, especially $S_{max} = 80$ MPa, the predictions are fully inconsistent with the measurements. A poor correlation between the predictions and test results was also obtained for a programme, simplified flight simulation sequence (large $R = 0.1$ cycle interrupted by 9 small amplitude, high *R* ratio cycles).

From CA tests on riveted lap joints Moreira (2004) described a dependence between EIFS and S_{max} in the form EIFS $= A\cdot S_{max} + B$, where *A* and *B* are the fitted constants. This relationship implies that the EIFS increases with the stress level. For example, EIFS mean values were 34 and 49.3 μm for S_{max} of 90 and 160 MPa respectively. The EIFS calculations were based on the back extrapolation and Paris' crack growth equation. Crack growth histories were obtained from tests on small, single rivet column, three-row lap joints (2024-T3 Alclad sheet 1.2 mm thick, NAS1097AD4 rivet, row pitch 20 mm). Therefore, the EIFS values obtained may not be fully representative for wider three-row lap joints.

Yang et al. (1985) proposed a back extrapolation methodology termed stochastic crack growth approach to calculate the EIFS distribution. Fractographic data were acquired from a lower wing skin structure of two retired B707-100 and -321 airplanes. The Al 2024-T3 skin was connected to Al 7075 stringers using 0.25 in. diameter aluminium and steel fasteners. The wing structure was disassembled and all occurrences of cracking and corrosion damage at the fastener holes were inspected. The SIF for each cracking site was calculated and an equivalent double, quarter circular, symmetric crack size was determined. From the fractographic data for cracks at aluminium fastener holes grown under spectrum loading, the fatigue crack growth equation was obtained in the form

$$\frac{dc(t)}{dt} = XQc^{b} \tag{9.11}$$

where *t* denotes time, *Q* and *b* are material related parameters, and *X* is a random variable accounting for contributions to the crack growth rate made by the statistical variability of material properties, crack geometry, crack modelling, manufacturing factors, loading, etc.

Equation 9.11 was used to back extrapolate the distribution of measured cracks $c(t)$ at fastener holes in the retired aircraft to obtain the $c(0)$ data, i.e. the EIFS distribution. The latter could be fit by a three-parameter Weibull distribution. A maximum EIFS-value of 239 μm obtained by Yang et al. (1985) was more than two times that needed in the aforementioned work by Harris et al. (1999) to match the predicted and observed crack growth for a joint exhibiting a severe fretting

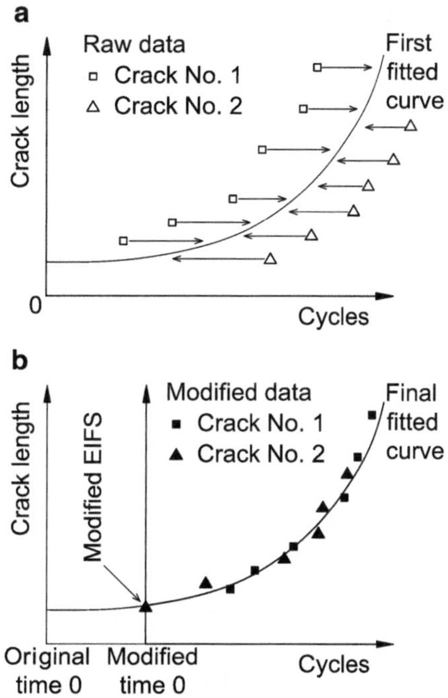

Fig. 9.16 Synchronization of crack growth history: (**a**) shifting measured data to form a single crack and fitting appropriate polynomial equation; (**b**) modification of the fitted curve (Anderson et al. 2004)

damage. A median EIFS of 76 μm in the Yang et al. investigation was close to the maximum EIFS value found by Fawaz (2003) using the AFGROW model (62 μm), see earlier in this section. The above results again substantiate a conclusion that EIFS values for riveted joints are heavily dependent on manufacturing and loading conditions and structure design.

One of the objectives of analyses by Anderson et al. (2004) performed using the FASTRAN code was to determine characteristics of EIFS, in terms of sizes and distributions, that could be useful for crack growth/fatigue life predictions for similar aircraft structures. To this end experimental results by Fawaz (2000) on flat panels containing riveted lap joints, referred to earlier in this section, were utilized. As already said, most of the crack growth curves measured at various rivet holes had the same overall shape, but were separated from each other. To make the EIFS analysis of these dispersed data possible, the curves were shifted sideways along the cycle axis to form a single crack curve. Next, a semi-logarithmic polynomial equation for the collapsed crack growth data points was fitted, as shown by the line labelled First fitted curve in Fig. 9.16a. In the following step, the fitted curve was modified so that the smallest number of cycles was set to zero, which shifted right the origin of the cycle axis, Fig. 9.16b. This curve labelled Final fitted curve was assumed to represent every crack in the same specimen. The SIF correction coefficient β_u was estimated, as proposed by Hsu and Yu (1988), by comparing growth rates based on the final fitted curve equation representing the measured data to FASTRAN predictions assuming the modified EIFS value shown in Fig. 9.16b.

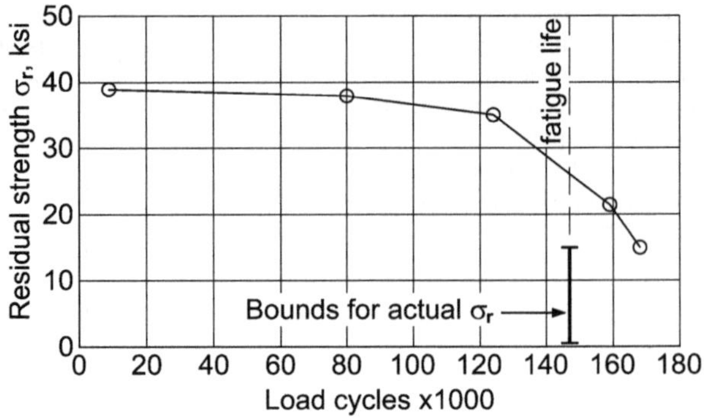

Fig. 9.17 Residual strength prediction for riveted wide flat panel using the EIFS concept (Anderson et al. 2004)

This concept yields $\beta_u = [(dc/dN)_{test}/(dc/dN)_{predicted}]^{1/m}$, where m is the exponent in Paris' equation, $m = 4.08$ being used for Al 2024-T3. The β_u-coefficient was intended to account for the influence of various factors neglected in the FASTRAN predictions based on the K-solution of Newman and Raju for a double corner crack at a hole and Eqs. 9.4 and 9.5.

On the basis of the corrected K-solutions the EIFS-values for each individual rivet in the critical row were calculated. These values were found to range between 0.00014 and 0.00089 in. (3.6 and 20 μm). Anderson et al. (2004) point out that due to such a variation in EIFS, it would not be unusual to see variations of several lifetimes even when using the same as for the EIFS determination crack growth analysis tools for fatigue live predictions. The EIFS-values derived for the individual rivets were utilized to reconstruct the MSD distribution and sizes along the critical row after a particular, chosen number of cycles. From this analysis the residual strength as a function of time was estimated using the STAGS code developed by Rankin et al. (1998).[5] Figure 9.17 shows the plot of the computed residual strength (σ_r) versus number of cycles for one of the panels with a joint as in Fig. 9.11. According to the STAGS simulations, the panel tended to fail at the first link-up. The expected trend of gradual degradation of the structure in time due to the development of MSD cracks is evident in Fig. 9.17. However, even though no residual strength test data were available to substantiate the analyses, the results in Fig. 9.17 suggest a significant overprediction of σ_r. The panel failure occurred after 147,410 cycles. This implies that the residual strength at this cycle number could not exceed 15 ksi (baseline stress maximum level), which is much below the computed trend line.

Newman and Ramakrishnan (2011) applied the FASTRAN and AFGROW models to calculate crack growth at rivet holes for a longitudinal three-row riveted joint in a curved panel of a retired Boeing 737 aircraft fuselage during its 60,000 pressure cycle history. The bypass, bending and bearing stresses normalized by the remote applied tensile stress of 0.63, 0.85 (according to Schijve's model) and 0.37 respectively were assumed. The SIF was determined according to Eqs. 9.4 and 9.5

[5]Residual strength analysis methodology is detailed in Chap. 10.

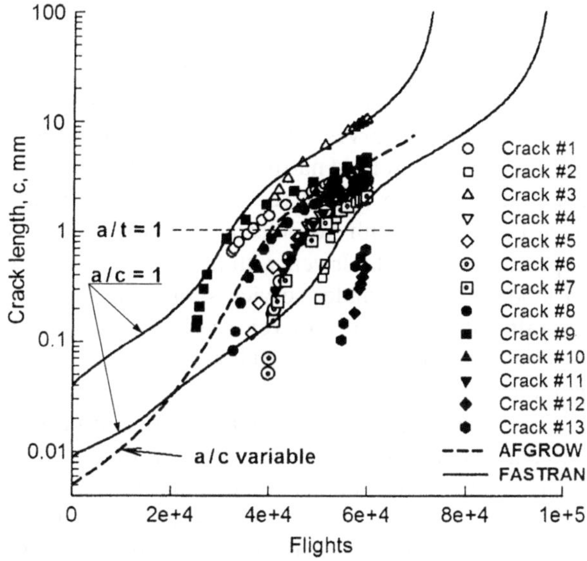

Fig. 9.18 Measured and calculated crack growth for riveted joints in a narrow-body aircraft fuselage panel using the EIFS concept (Newman and Ramakrishnan 2011)

except that a correction coefficient for biaxial loading (longitudinal stress of 0.5 hoop stress) developed by these authors was included in K_{bypass} in the FASTRAN computations. The rivet/hole interference was neglected as the fractographic examination indicated loose rivets. As said earlier, the FASTRAN code assumes a constant a/c ratio of unity until the breakthrough and a straight front through crack subsequently. Though in AFGROW an initially quarter circular crack is allowed to grow independently in the a- and c-directions, it was predicted to grow at a/c $\sim$ 1. The computation results are compared with the fractographic inspection data in Fig. 9.18, where the dashed line gives results from AFGROW obtained for an EIFS of 5-μm, double, quarter circular crack. This value was chosen to best fit the mean of the data taken from the fractographic examinations. The solid curves show calculations made with FASTRAN for EIFS-values ranging from 9 to 40 μm. It is seen that in the small crack regime AFGROW predicts much faster crack growth than FASTRAN. As a result, for *c* above 30 μm the AFGROW curve, though based on the lowest EIFS-value, lies between the FASTRAN curves. It is not clear to what extent the differences between the results from both codes spring from neglecting in AFGROW the influence of biaxial loading.

It is evident that the largest errors in SIF solutions due to inadequate assumptions about geometric effects and contact stresses on the hole occur at small crack sizes. As such, they can heavily affect the computed EIFS-values since the EIFS remains within the small crack range. In addition, in the presence of MSD a scatter in EIFS values can be very much enhanced by various phenomena difficult to be rationalized and, hence, to be covered in analyses. For example, Piascik and Willard (1997) reported that cracks in adjacent bays or adjacent rivet holes in retired aircraft fuselage panel lap joints initiated at significantly different times (between 10,000 and 20,000 cycles) within the life of the lap splice joint. At the same time, it has been revealed that the same initiation mechanism occurred at all crack locations and all fatigue cracks grew at similar d*c*/d*N*, as deduced from the crack growth curves "shifted", as shown in Fig. 9.16a. Based on a destructive examination of a retired narrow-body aircraft fuselage after 60,000 flights Newman and Ramakrishnan (2011) reported that the left side of the fuselage appeared to have tight rivets with no detectable cracks, whereas on the other side the rivets appeared to be underdriven with a large number of cracks initiated at rivet holes. In laboratory tests under service simulating loading, panels from the side of the aircraft with the tightly driven rivets were able to withstand a large number of pressurization cycles (e.g. 120,000 for one of the panels) without cracking in the lap joints.

It can be concluded that while the tools used to analyze single-site and multi-site damage are essentially the same, predictions of MSD are incomparably more complex, due to the number of locations that may be cracked and interaction of these cracks, randomness of crack initiation and a variability of the crack initiation/growth mechanism at rivets which can occur within the same fuselage structure and even within the same rivet row.

Estimates of the EIFS considered so far are based on a back extrapolation methodology. With a novel EIFS concept proposed by Liu and Mahadevan (2009) and Xiang et al. (2009) back extrapolation is avoided. This approach utilizes a model of El Haddad (1979) according to which the fatigue limit $\Delta\sigma_f$ is related to the fatigue threshold ΔK_{th} through the equation

$$\Delta K_{th} = \Delta\sigma_f Y \sqrt{\pi c_{th}} \tag{9.12}$$

where Y is a geometry correction coefficient and the crack size c_{th} is assumed to be an intrinsic material property.

Equation 9.12 can be rewritten as

$$c_{th} = \frac{1}{\pi}\left(\frac{\Delta K_{th}}{\Delta\sigma_f Y}\right)^2. \tag{9.13}$$

Liu and Mahadevan (2009) and Xiang et al. (2009) postulate that c_{th} represents the EIFS. If a specimen under the stress range of $\Delta\sigma_f$ has an initial crack length of EIFS, the calculated fatigue life according to fracture mechanics based approach is infinite (in practice it is on the order of 10^7–10^8 cycles). Because commonly used experimental methods to determine ΔK_{th} lead to divergent results, it is proposed to obtain ΔK_{th} by linear extrapolation of the long crack growth curve (Paris regime) to a crack growth rate of 10^{-10} m/cycle. Up to now, the validation of this new concept is limited to simple, small laboratory notched specimens.

9.5 Predictions of MSD Crack Growth

As said in Sect. 9.3, MSD crack growth, which is a typical damage scenario for riveted joints of the aircraft structure, may cause a redistribution of load transfer in a fastener row with MSD, in addition to load shedding from the cracked row to uncracked rows (cf. Fig. 9.8). For a single strap joint (geometry as for joint D in Fig. 6.25) with seven rivets in row, FE results by Kuo et al. (1986) revealed that when a crack was growing on one side of the middle (fourth) fastener hole towards fastener 3 in one of the end rows, the load distribution in this row, and also in the neighbouring row, became highly non-uniform. For example, when two ligaments (between fastener 4 and 2) became broken, the load transferred by fastener 3 dropped to 0.55 the value for the intact joint, while the load transferred by fastener 1 increased to 1.4 that value. Also FE analyses by Ingram et al. (2002) for a double-row lap joint with a lead crack and MSD cracks in one row indicated that the load shedding in the lead crack path was accompanied by shifting some of the load to fasteners ahead of the lead crack, in addition to the uncracked row. However, Beuth and Hutchinson (1994) found the effect of load shedding insubstantial when each rivet hole in the end row of a three-row lap joint was allocated a longer crack on one side and a slightly shorter crack on the other side. Evidently, a significant load redistribution along a rivet row occurs only in the case of larger disproportions between sizes of cracks in that row.

In addition to the load redistribution with crack growth, interactions between growing cracks and between the cracks and still uncracked holes can occur, which should be reflected in SIF solutions. Compared to analyses assuming that each crack is growing independently, accounting for the interaction effects provides higher SIFs and crack growth rates, resulting in shorter fatigue lives (Liao et al. 2009; Bombardier et al. 2011).

It is obvious that MSD cracking scenarios require appropriate SIF solutions and prediction tools capable of propagating multiple cracks simultaneously. When the AFGROW code (Harter 2006) is used for crack growth predictions, user-introduced correction coefficients can be applied on top of K solutions available in the software. Alternatively, it is possible to use an external K-solver that communicates with AFGROW. According to the AFGROW manual, the software contains a basic infrastructure to handle any number of cracks. The life prediction process must be applied to each crack tip treated as a separate entity. The computation procedure is, however, far from being straightforward and is subject to various constraints.

Essentially, it is possible to account for all important characteristics of the MSD problem in lap joints with mechanical fasteners using FE modelling. However, such an approach can be very time-consuming and, hence, impractical in the case of structural risk analysis utilizing the Monte Carlo simulation technique (cf. Sect. 10.7). Within this methodology, which requires performing crack growth predictions for a large number of initial MSD crack scenarios, computation time constraints are very important and a balance between accuracy and computation speed needs to be achieved. For that reason, approaches involving direct numerical computations of

SIFs are typically rejected and closed-form or tabular SIF solutions, usually relying on the compounding method, are preferred in order to ensure fast execution of the simulations (Sampath and Broek 1991; Trey and Schmidt 2001; Liao et al. 2009; Renaud et al. 2011; Cavallini and Lazzeri 2012). The need for using a K-solution of this type, as well as various approximate assumptions required in the Monte Carlo simulations to limit the calculation time, practically preclude applying this method to lap joints in curved fuselage panels.

A compounded K-solution for analysis of MSD crack growth at rivet holes has been developed at the National Research Council Canada (NRC). SIF correction coefficients available in the literature for the baseline problem (e.g. crack at a hole in a finite width plate, crack approaching a hole or a free edge, two collinear cracks, etc.) and correction coefficients (β_f) which only are related to "special features" of a given structure are included (Liao et al. 2009; Renaud et al. 2011). A β_f-coefficient is obtained for different crack lengths as a ratio of the SIF from a FE model including these features and the SIF from a model without the features. The total SIF correction coefficient is determined as a product of β_f and the baseline coefficient from available solutions. The above methodology was applied to determine SIFs for interacting, quarter circular MSD cracks of various sizes propagating from several holes in a plate under remote tension and pin loading on the holes (Bombardier et al. 2011), and to compute SIFs for fatigue-critical locations of an aircraft wing. In the latter case, the β_f-coefficient was necessary to cover the effect of adjacent structural components on cracks growing in a wing skin panel (Renaud et al. 2011).

An algorithm for compounding SIF correction coefficients (β) according to the method outlined above was incorporated in a software for crack growth predictions, which was developed by NRC to analyze the MSD problem in the CC-130 wing panel by executing Monte Carlo simulations (Liao et al. 2009). The table look-up method uses baseline β-coefficients stored in a table by spline interpolation. In addition to this pre-defined β-library, user-defined correction coefficients can be used. A special analysis tool linked with a p-version FE package was developed to allow calculations of the β_f-coefficients (Renaud et al. 2011). The NCR crack growth prediction model is capable of simultaneously growing multiple cracks. Analysis examples reported by Liao et al. (2009) and Renaud et al. (2011) present applications of the NRC methodology to a flat wing panel with MSD cracks in a single row of fastener holes and open (unloaded) holes located between the fastener holes. No information on accounting for load redistribution due to crack growth is given. Results showing application of the NRC methodology to MSD cracks in lap joints have not been reported in the open literature.

A methodology to predict MSD crack growth at fastener holes in lap joints by means of Monte Carlo simulations has been included in an analysis aiming at the MSD/WFD (Widespread Fatigue Damage) development for the pressurized fuselage of Airbus large transport category airplanes (Trey and Schmidt 2001) and into a PISA (Probablisitic Investigation for Safe Aircraft) code oriented toward the risk assessment of simple aircraft components, including riveted panels with lap joints subject to CA loading (Cavallini and Lazzeri 2012). With the Airbus methodology, the initial damage scenario is defined by assigning to each potential crack location a TTCI obtained from a log-normal distribution of fatigue lives for simple coupons. SIFs are computed using the compounding method. The effect of stress redistribution due to crack growth is accounted for by increase of stresses at still uncracked holes, which yields earlier crack initiation at that locations, but the paper of Trey and Schmidt lacks information on how the above is effected and how load transfer through the joint is determined. Also, it is not reported if the influence of riveting process (rivet/hole interference) is allowed for in crack growth predictions. An example produced in the paper demonstrates a good correlation between computed and observed growth of MSD cracks at holes in a flat panel. The authors claim that also MSD cracking scenarios at riveted joints could be successfully predicted.

The stochastic input to the Monte Carlo simulations incorporated in the PISA code are distributions of the EIFS, the C-constant in the power crack growth law used for crack growth predictions, the fracture toughness (K_C) and the inspection reliability (probability of crack detection). In order to account for the influence of rivet installation $\Delta K_{eff} = K_{max} - K_{LO}$ (cf. Eq. 9.10) is assumed as the crack driving force parameter, the S_{LO} stress being evaluated according to Wang (1988), based on rivet hole expansion measurements. A compounding SIF solution for straight front through cracks is used, which superimposes correction coefficients from ancillary K-solutions applicable to joints with countersunk and protruding head rivets collected by Kuo et al. (1986). The SIFs due to the bypass load, transfer load and secondary bending are determined, as proposed by Sampath and Broek (1991). Load transmission is obtained using FE modelling for the uncracked joint and remains invariable during crack growth. The EIFS distribution is derived by back extrapolation of crack size versus number of cycles data observed in fatigue tests on representative, simple riveted specimens utilizing the same crack growth model as for the real predictions. Two collinear cracks are assumed to link up according to the Swift criterion (cf. Sect. 10.2.1). Final fracture occurs either due to crack growth instability ($K_{max} \geq K_C$) or net section yielding.

A code developed by Sampath and Broek (1991) to assess the probability of detection of MSD in fuselage lap-splices in ageing airplanes incorporates a probabilistic model for MSD crack growth predictions. This approach is distinct from others in that the initial MSD scenario is defined by the distribution of fatigue damage $\sum \frac{n}{N}$. Here, n is the number of cycles to grow at a certain applied stress level a crack of detectable size c_o, which is the smallest crack visible on the upper sheet, $c_o = (D - D_o)/2$, and N is the fatigue life at that same stress. The distribution function for $\sum \frac{n}{N}$ was obtained from fatigue tests on riveted lap joint specimens under CA and VA loading. In order to allow for the effect of frequency and environment, a factor of 3 in lives was imposed on S-N data from these laboratory tests. Another stochastic input to the prediction model is a distribution of the C-constant from the Paris equation. The probability of distribution data for both $\sum \frac{n}{N}$ and C are in the form of curve-fitting (Weibull) equations which do not make any statements about the statistics involved. The probability of exceedance functions for $\sum \frac{n}{N}$ and C are divided between 0 and 1 into m intervals of $1/m$, where m is the number of rivet holes in a critical row. The deterministic input data for each hole, namely a hoop stress, transfer load and bending stress are calculated using a FE modelling of the intact structure. Sampath and Broek considered 100 fastener holes in the critical row of a three-row lap joint in a curved panel (5 frame spacings and 10 strap spacings) subjected to internal pressure. Random numbers, each representing a hole, are picked between 1 and m and the number (hole) picked is assigned values of $\sum \frac{n}{N}$ and C from the corresponding intervals of the probability distributions, the damage assignment being made for both the left side and the right side of the hole. However, the lowest $\sum \frac{n}{N}$ (of 0.5 in the Sampath and Broek study) is assigned to a hole picked at random until one is found with a hoop stress higher than average. The analysis is triggered by allocating a crack of size c_0 to one side of that particular crack. Its occurrence is at an (immaterial) number of cycles n corresponding to $\sum \frac{n}{N} = 0.5$, which implies that all holes initially have experienced n cycles. With knowledge of n for every hole, the damage $\sum \frac{n}{N}$ for all other holes can be found, given the hoop stress and fatigue life (N) at a particular hole. Should the damage be larger than 0.5, the hole is cracked, otherwise it is not. With an example produced in the considered work, that begin-situation involved from 1 to 5 cracks. During subsequent cycling, the increased damage at each hole is assessed every 100 cycles and new cracks of size c_0 are introduced with some accounting for stress redistribution due to cracks at adjacent holes. The latter is effected by defining the hoop stress as the net stress averaged over three fastener spacings. Growth of all cracks is also accumulated every 100 cycles, and a compound K-solution, accounting for the bypass and transfer load, secondary bending and the interaction of co-linear cracks is utilized. The link-up criterion is taken as the exceedance of net section yield allowing for stress redistribution, as said above. Final failure is defined as the occurrence of a total of 10 links. Because of a relatively large size of the reference crack, the effect of riveting on crack growth should be insignificant and is not addressed in the prediction model.

Due to random assignments of $\sum \frac{n}{N}$ and C, the results of each analysis run will be different even when the statistical parameters and assumptions are the same. Exemplary results for a series of runs with identical parameters produced in the paper of Sampath and Broek show a considerable variability in terms of the total crack growth life, number of cracks versus number of flights, crack size versus number of flights and crack size at a particular location. Significance of the predictions for evaluating the probability of MSD crack detection is considered in Sect. 10.7.

9.6 Fatigue Life Predictions

As with fatigue crack propagation predictions considered in the preceding sections, a common feature of fatigue life predictions for the lap joints is their strongly empirical character. Available approaches, e.g. Jarfall (1986), Homan and Jongebreur (1993), Das et al. (2001), are based on the assumption of similarity between the actual design, for which predictions are made, and a reference design, for which

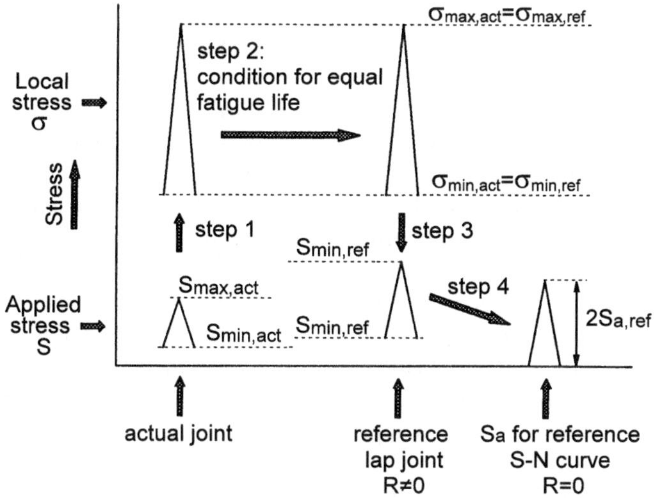

Fig. 9.19 Schematic of fatigue life prediction method according to Fokker (Homan and Jongebreur 1993)

the response to fatigue loading is known from tests. A success of the predictions relies on adequacy of the similarity criterion and on the availability of appropriate reference experimental data. According to the Fokker (van Dalen et al. 1992; Homan and Jongebreur 1993) and Boeing (Das et al. 2001) concepts, the fatigue life of a joint is governed by local cyclic stresses in a critical location. This implies that if the local maximum and minimum peak stress will be equal for the actual and reference design, the fatigue life will be identical for both cases. That fatigue life can be obtained from the available *S-N* data for the reference design.

Like the SIF (cf. Fig. 9.4), the local stress (σ) is determined by superposing the effects of the bypass and transfer load and of secondary bending as:

$$\sigma = S\left[\alpha_{\mathrm{BP}}(1 - R_{\mathrm{TR}})K_{\mathrm{t,\,BP}} + \frac{w}{d}\alpha_{\mathrm{BR}}\, R_{\mathrm{TR}} K_{\mathrm{t,}\,BR} + \alpha_{\mathrm{bending}}\, K_{\mathrm{b}} K_{\mathrm{t,}\,bending}\right] \quad (9.14)$$

where α_{BP}, α_{BR}, and $\alpha_{\mathrm{bending}}$ are empirically determined coefficients, $K_{\mathrm{t,BP}}$, $K_{\mathrm{t,BR}}$, and $K_{\mathrm{t,bending}}$ are the stress concentration factors for a finite width plate with hole under load cases shown in Fig. 9.4, and the meaning of other symbols is as in Eqs. 9.5.

The prediction procedure according to the Fokker code is illustrated in Fig. 9.19. In step 1, the local cyclic peak stresses ($\sigma_{\mathrm{max,act}}$ and $\sigma_{\mathrm{min,act}}$) are calculated from Eq. 9.14 for the actual joint under the $S_{\mathrm{max,act}}$ and $S_{\mathrm{min,act}}$ nominal stresses. In step 2, the calculated peak stresses are applied on the reference joint, which implies $\sigma_{\mathrm{max,ref}} = \sigma_{\mathrm{max,act}}$ and $\sigma_{\mathrm{min,ref}} = \sigma_{\mathrm{min,act}}$. Remote nominal stresses $S_{\mathrm{max,ref}}$ and $S_{\mathrm{min,ref}}$ for the reference joint are computed for $\sigma_{\mathrm{max,ref}}$ and $\sigma_{\mathrm{min,ref}}$ in step 3 utilizing Eq. 9.14. An iteration procedure is required as the bending factor K_{b} is *S*-value dependent (cf. Fig. 6.5). The reference *S-N* curve is available for R of 0. If the R-ratio for the stress cycle computed in step 3 ($S_{min,ref}/S_{max,ref}$) is different from zero, the Goodman correction is applied in step 4 to compute the reference stress amplitude $S_{a,ref}$ corresponding to $R = 0$.[6] Finally, the reference *S-N* curve is entered to derive the fatigue life of the reference joint for $S_{a,ref}$. By assumption, this fatigue life is identical to that of the actual joint under the actual loading.

With the Fokker procedure, the α_{BP}, α_{BR}, and $\alpha_{bending}$ coefficients of unity are used as it is required that the riveting process, which affects the amount of hole fill (hole expansion) and the clamping force (friction), should be the same for the reference specimens as for the actual joint. Also the rivet type (protruding or countersunk) should be the same for both cases. The main potential of the method manifests if the joint geometry is the only varying parameter (van Dalen et al. 1992; Homan and Jongebreur 1993).

When the Boeing approach is applied, $\alpha_{BR} \neq 1$ allows for the pressure distribution on the hole, and α_{BP} and $\alpha_{bending}$ of 0.6 are assumed based on experimental data to account for the fact that the holes are filled, not open. When the fatigue life is derived for the actual design, the next step is to determine its fatigue quality. From fatigue test results on representative laboratory specimens, detail fatigue rating (DFR) *S-N* curves have been empirically developed for various members of the structural population. They identify for any level of operational stress a minimum fatigue life that can be comfortably exceeded by all but the 5% weakest members (Goranson 1997).

Like the crack growth prediction methodology considered earlier, the fatigue life prediction concepts are oriented towards crack initiation at rivet holes. If for an actual design riveting process yields crack nucleation at a distance from the rivet hole, satisfactory predictions based on Eq. 9.14 cannot be expected even if an appropriate reference *S-N* curve is available.

9.7 Summary of This Chapter

Approaches developed to predict the fatigue behaviour of riveted lap joints are of a strongly empirical character. Their transferability is limited to cases when the actual design, for which predictions are made, is similar to that for which the prediction model has been tuned. A precise formulation of a criterion for the joint similarity is not possible due to many loading/geometry/material/surface condition/riveting process dependent factors, sometimes interrelated, known to affect the joint fatigue response.

Prediction concepts for riveted joints mainly concentrate on crack growth. Most models are based on the linear elastic fracture mechanics methodology and correlate crack growth using the SIF.

---

[6] With the Boeing procedure, the reference curve for the actual stress ratio must be available.

The SIF solution for a riveted lap joint is typically derived by superpositioning of solutions for the bypass load, transfer load and secondary bending obtained for a crack in a finite width plate with a single hole. SIF correction coefficients needed to be applied on top of that solution to account for the effect of riveting (contact stresses on the hole, residual stresses in the vicinity of hole, load transmission by friction) and of the countersink geometry can only be obtained experimentally.

In the case of riveted joints with MSD, SIF solutions for individual cracks should additionally take account of the effect of other cracks and holes as well as of load shedding from a cracked rivet row to uncracked rows and load redistribution within the cracked row.

A concept of the equivalent initial flaw size (EIFS), i.e. the size of a hypothetical crack assumed to be present in a structural part prior to service, was conceived to cope with the lack of analytical means to deal with the crack initiation phase and the anomalous behaviour of small cracks. Somewhat inconsistently with the fractographic evidence, the EIFS for riveted lap joints is usually represented by two symmetrical, quarter circular cracks at either side of a rivet hole.

Investigations indicate that EIFS-values for riveted joints are affected by manufacturing related factors, design solution, loading conditions, and a crack growth prediction methodology used to compute the EIFS. In the case of MSD cracks, a possible (even within the same rivet row) variability of the crack initiation life and of the crack initiation mechanism causes a significant scatter in EIFS values derived for the individual rivets.

While tools to analyze single-site and multi-site damage are essentially the same, predictions of MSD crack growth are incomparably more difficult due to the number of crack locations and complex characteristics of the MSD problem. The Monte Carlo simulation technique is a tool most often applied to analyze the fatigue damage development due to MSD. This methodology requires various simplifying assumptions in order to achieve a balance between accuracy and computation speed.

Fatigue life prediction models are based on the assumption that fatigue life of a riveted lap joint is governed by local cyclic stresses in a critical location. This implies that if the local peak stresses are equal for the actual design and for a similar, reference design, for which the *S-N* data are available, the fatigue life will be identical for both cases.

Both the crack growth prediction methodology and fatigue life prediction methodology are oriented towards the crack initiation at rivet holes. Predictions of the lap joint fatigue behaviour in the case of cracks nucleated at a distance from the hole, as well as accounting for the contribution of fretting remain in the academic interest stage.

Chapter 10
Residual Strength Predictions for Riveted Lap Joints in Fuselage Structures

10.1 Introduction

In general, aircraft fuselages experience fatigue crack growth. At the upward part of the very last cycle, however, crack propagation involves a considerable amount of stable tearing followed by fast final failure. The term residual strength corresponds to the applied load level at which unstable crack growth occurs. As already said in Chap. 8, the increasing usage of aging aircraft triggered the attention to the problem of MSD. A host of experimental works on simple specimens and complicated fuselage structure panels with riveted joints have demonstrated that, compared to the case of a single lead crack only, the crack arrest capability and residual strength is degraded when the same lead crack is accompanied by smaller collinear MSD cracks (cf. Sect. 8.2.4). In parallel, an extensive research effort has been dedicated to quantifying the reduction of residual strength due to MSD. Analytical predictions of residual strength require the application of appropriate criteria to determine crack link-up and final failure. A number of various methodologies for predicting residual strength in the presence of MSD, some of them employing very sophisticated analysis tools, have evolved. In this chapter, however, only approaches most thoroughly confronted with experimental results and claimed to be most suitable for aircraft fuselage panels are considered. Next, selected from the literature comparisons between predicted and observed results on residual strength of flat and curved panels with riveted lap joints containing either a single crack or a lead crack and MSD cracks are presented.

Though the mainstream research has been concentrated on self-similar (straight-line) crack growth, curvilinear crack paths have been also observed in some fuselage structures. Crack flapping at tear straps is desired as it can prevent explosive decompression of the fuselage. Hence, it is of a practical importance to recognize conditions that result in crack kinking and eventually crack arrest due to flapping at a tear strap. In the present chapter crack directional criteria considered to be most suitable for stiffened shell structures and approaches to evaluate the crack arresting potential of tear straps are reviewed.

A. Skorupa and M. Skorupa, *Riveted Lap Joints in Aircraft Fuselage*, Solid Mechanics and Its Applications 189, DOI 10.1007/978-94-007-4282-6_10,

10.2 Crack Link-Up and Failure Criteria

Experimental investigation of residual strength involves pulling to failure with a monotonically increasing load or displacement a test coupon containing a single precrack or multiple precracks, while loads and crack propagation are recorded. The specimen should be large enough to allow significant crack growth prior to reaching residual strength. Buckling guides to prevent out-of-plane displacements are required in these experiments, if conducted on flat panels, as elastic-plastic FE analyses (e.g. Wang et al. 1997c) indicate that significant compressive stresses along the crack growth direction develop above the entire main (lead) crack and above the MSD crack ligaments. Often, the precracks are saw-cuts with possibly sharp ends to simulate the fatigue crack tip geometry. Sometimes the saw-cuts of a small length are grown by fatigue loading to the desired precrack size, which is a more realistic option in the case of the aircraft fuselage structure.

Already experiments on simple flat panels with multiple collinear cracks subjected to uniaxial tensile loading can illustrate a dependence of the crack link-up behaviour and residual strength on the crack configuration. Figure 10.1 shows initial crack patterns and residual strength (σ_r) for four of altogether twelve 0.04 in. thick Al 2024-T3 sheet panels with a width of 20 in. comprising various multiple crack configurations considered in tests by Broek et al. (1994). Due to the symmetry of the crack configuration only half of the cracked cross section is shown. The four panels have the same MSD crack length of 0.25 in. The lead crack lengths for Panels P5, P6 and P7 are similar (3.6, 3.8 and 3.7 in. respectively). The residual strength of Panel 6 with two MSD cracks is considerably lower compared to Panel 5 with only a single MSD crack, but adding a third MSD crack in Panel 7 does not cause a further reduction of σ_r. However, for Panel 6 the final failure occurred only after the absorption of the second MSD crack while for Panel 7 already first link-up immediately caused final failure. For complex fuselage panels the crack coalescence and residual strength become further affected by the presence of stiffeners, crack stoppers (tear straps) and riveted connections.

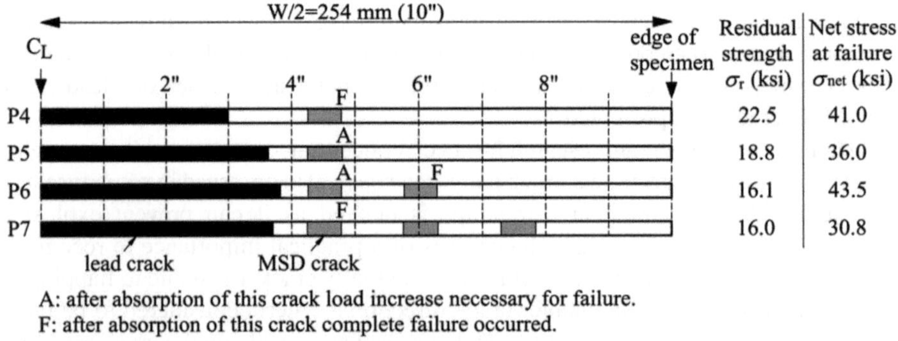

Fig. 10.1 Crack configurations and residual strength of flat panels considered by Broek et al. (1994)

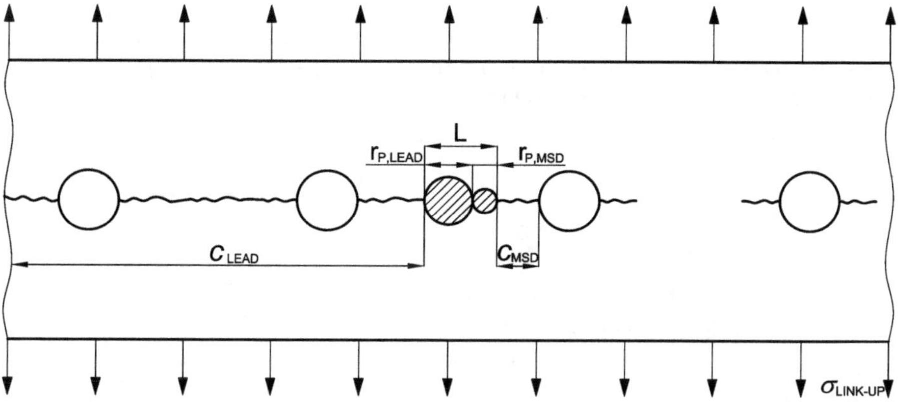

Fig. 10.2 Illustration of Swift (1993) criterion

10.2.1 Plastic Zone Link-Up (PZL) Criterion

Crack configuration dependent values of the net stress at failure (σ_{net}) given in Fig. 10.1 suggest that the net section failure criterion is not representative for the considered data. For Panels 4 and 6, σ_{net} is close to the material yield stress of 43.65 ksi, but the other two panels failed at σ_{net} considerably below the yield stress. Evidently, fracture of panels with MSD cracks is not necessarily governed by the net section yielding, as also observed by others (e.g. Anderson et al. 2004). One of the earliest concepts of predicting the coalescence of a lead crack with a neighbouring, smaller MSD crack in a fuselage panel and the effect of this coalescence on residual strength was offered by Swift (1993). According to his crack link-up criterion, further referred to as the plastic zone link-up (PZL) criterion, a lead crack would absorb an MSD crack when the entire ligament between the two crack tips has yielded, i.e. when the two plastic zones touch each other, as indicated in Fig. 10.2, where L is the ligament length and the indices LEAD and MSD refer to the lead and MSD crack respectively. Assuming the SIF equation for either crack in the form

$$K_I = S\beta\sqrt{\pi c} \tag{10.1}$$

and the Dugdale equation for the plastic zone size

$$r_p = \frac{\pi}{8}\left(\frac{K_I}{S_Y}\right)^2. \tag{10.2}$$

where S_Y is the material yield stress, the applied stress at which crack link-up occurs can be derived as

$$\sigma_{LINK-UP} = \frac{S_Y}{\pi}\sqrt{\frac{8L}{\beta_{LEAD}^2 c_{LEAD} + \beta_{MSD}^2 c_{MSD}}}. \tag{10.3}$$

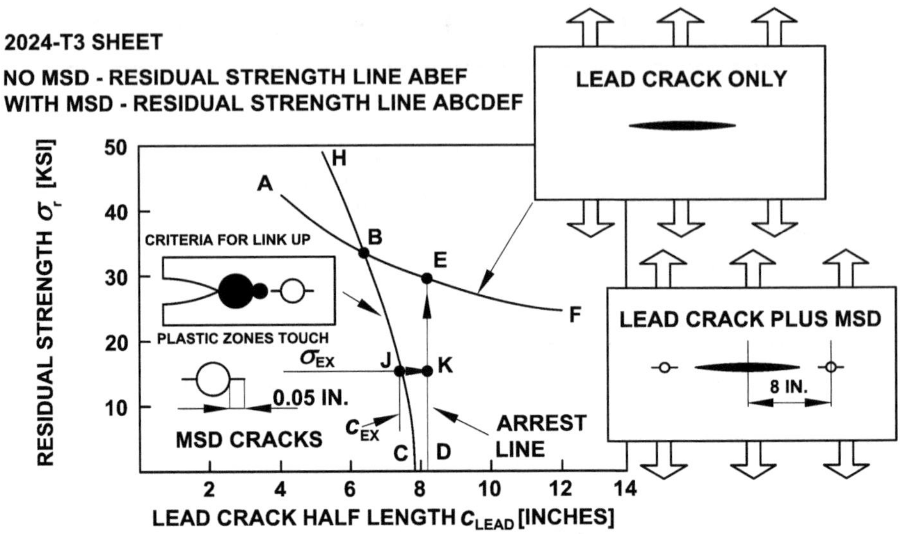

Fig. 10.3 Effect of MSD on residual strength computed according to Swift's criterion (Swift 1993)

The β-factors must include correction for crack interaction. Figure 10.3 illustrates the Swift criterion application for a simple case of an infinitely wide Al 2024-T3 unstiffened panel with MSD cracks. The curve ABEF represents the residual strength for the lead crack alone given by $K_C/\sqrt{(\pi c)}$ with the plane stress fracture toughness[1] K_C assumed to be 150 ksi$\sqrt{\text{in}}$. Line HBC is a plot of the $\sigma_{\text{LINK-UP}}$ stress given by Eq. 10.3 at which the plastic zones at the tips of the pair of two 0.05 in. long MSD cracks propagating from 0.19 in. diameter holes touch the lead crack plastic zone. If for a lead crack length of c_{EX} the remote stress is increased to σ_{EX} corresponding to point J, then link-up will occur and the lead crack length will suddenly increase to point K. The residual strength for lead cracks beyond point B will therefore drop to point E. If there are more MSD cracks, an individual $\sigma_{\text{LINK-UP}}$ vs. c_{LEAD} line analogous to HBC in Fig. 10.3 should be plotted for each consecutive MSD crack under the assumption that the preceding MSD crack has been absorbed and the link-up stress can be determined as shown in Fig. 10.3. If the current link-up stress is lower than the preceding one, it is assumed that the linkage of the preceding and current MSD cracks has occurred simultaneously. When all MSD cracks have been absorbed by the lead crack, it behaves like a large single

[1]The fracture event should be characterized by a fracture toughness corresponding to the final crack length at σ_r. Customarily however, the residual stress diagram employs the initial crack size which for thin sheets can be significantly smaller than the final crack size. To compensate this inconsistency a toughness value lower than the true plane stress toughness K_C should be used to construct the residual stress diagram.

crack and the failure is predicted according to line ABEF. The highest link-up stress in a multiple link-up event is referred to as the predicted residual strength.

Calculation results produced by Swift (1993) employing this simple "intuitive" concept (term used by this author) reflect a considerable loss in residual strength even with extremely small MSD crack sizes. For a fuselage panel with a longitudinal crack along the row of fastener holes, the gross applied stress at link-up of a lead crack with an MSD crack appears to be very much influenced by the structural configuration (rivet spacing in row, the presence of crack stoppers), while for a given configuration it is relatively insensitive to the size of MSD.

Broek at al. (1994) applied Swift's criterion to predict $\sigma_{\text{LINK-UP}}$-values for multiple cracks in Al 2024-T3 flat panels considered in the tests addressed above (see Fig. 10.1). On average, the predictions overestimated the observed values by 4.7%, the prediction error increasing with the ligament length. The predictions averaged within 2% of the experimental data when the Swift criterion was modified to include the effect of stable tearing. The latter approach leads to a reduced distance between cracks at link-up and, consequently, to increased K-factor values due to interaction. The resulting predictions of the link-up stress and residual strength with stable tearing will be, therefore, lower than those without this effect. The amount of stable tearing (Δc) was calculated using a resistance curve (R-curve) derived from a regression analysis of the panel test data. With the R-curve approach the condition for stable crack growth can be expressed as

$$G = R \quad \text{and} \quad \frac{dG}{dc} \leq \frac{dR}{dc} \tag{10.4}$$

whilst unstable crack growth occurs if

$$\frac{dG}{dc} > \frac{dR}{dc} \tag{10.5}$$

where G is the energy release rate that under LEFM conditions is related to K by

$$G = K^2/E\,. \tag{10.6}$$

Equation 10.4 implies that the stress level required for stable tearing at a given crack length corresponds to the tangency point of the $G(c)$ line and the R-curve. With the Swift criterion modified to include stable tearing, an iterative procedure must be used to solve Eq. 10.3 because the distance L depends on Δc that is a function of $\sigma_{\text{LINK-UP}}$.

Similar analyses were also performed by Broek et al. (1994) to account for the effect of stiffening for curved panels (1 mm thick, 1,728 mm along the circumference, 3,048 mm along the width) loaded by pressure. The stiffening was due to tear straps riveted by two rows of rivets and adhesive bonded to the sheet. SIFs in the cracked stiffened panels were obtained employing a displacement compatibility method proposed by Swift (1974) to account for rivet flexibility (in circumferential direction), biaxial stress, and broken or intact stiffeners. The effect of crack face bulging caused by the pressurisation and the damping of bulging at stiffener locations was also accounted for in the SIF utilizing available literature solutions. For the unstiffened curved panels, the modified Swift criterion, allowing for stable tearing prior to link-up, yielded again improved predictions compared to the original concept. Only the modified approach was applied to the stiffened curved panels. The predictions on link-up appeared to be affected by the distance between the cracks (L) and were within 10% and 20% of the test results for L below and above 0.5 in. respectively. In terms of the overall panel failure (residual strength) the prediction accuracy was much better for "light" tear straps than for "heavy" tear straps. The light straps were located at a smaller spacing and had a lower width and thickness than the heavy straps. The predictions on residual strength for panels with the light tear straps were within 25% of the test data, while for the heavy straps an overestimation by as much as 77% was noted. According to the authors, a main reason for the latter discrepancy could be uncertainties associated with the effect of bulging on the SIF in the case of heavy straps.

De Wit et al. (1996) applied the Swift criterion to predict $\sigma_{\text{LINK-UP}}$ and σ_r for six extra large (3,988 mm high, 2,286 mm wide, 1.016 mm thick) Al 2024-T3 panels with multiple through cracks saw-cut horizontally at mid-height and located symmetrically on either half of the cross section. Each MSD crack had a circular 5.6 mm-diameter hole in its centre to simulate a rivet hole. The panels were pulled to failure under displacement control. Antibuckling guides were used to restrain out-of-plane displacements. The prediction method was as described above in the context of Fig. 10.3. The analysis provided slightly underestimated $\sigma_{\text{LINK-UP}}$ and σ_r-values, the average prediction errors equalling −10% and −13% respectively. Additionally, predictions were also made employing the R-curve, as illustrated in Fig. 10.4 for a panel containing three MSD cracks. The R-curve is assumed to originate from each MSD crack (full line plots 2, 3 and 4) as well as from the main central crack (plot 1). Each curve is given by the same equation derived from tests on three panels of the same dimensions containing single cracks. Then G-lines (according to Eq. 10.6) tangent to each R-curve are drawn to determine link-ups and final instability in agreement with conditions given by Eqs. 10.4 and 10.5. The G-plot is a straight line only for crack no. 4, which is a single crack. At the preceding crack tip positions the G-lines show a significant curvature just before the instability point because the β-factor in Eq. 10.1 must account for interaction with the MSD cracks. Stresses corresponding to the consecutive tangency points computed combining Eqs. 10.1 and 10.6 are given in Fig. 10.4. Also provided in this figure are initial and final crack sizes (c_0 and c_c) corresponding to stable tearing prior to each linkage as well as critical G values (G_c) at the instability points. It is seen that $\sigma_{\text{LINK-UP}}$ for the third MSD crack is lower than for the second. This indicates that after link-up of the second MSD crack, the panel is supercritical and will run straight through after absorbing the second crack. The same behaviour was predicted for this panel from the Swift criterion. In general, predictions from both methods were reported to be in fair agreement indicating that both the linkage and fracture behaviour exhibited in these experiments were amenable to the LEFM conditions.

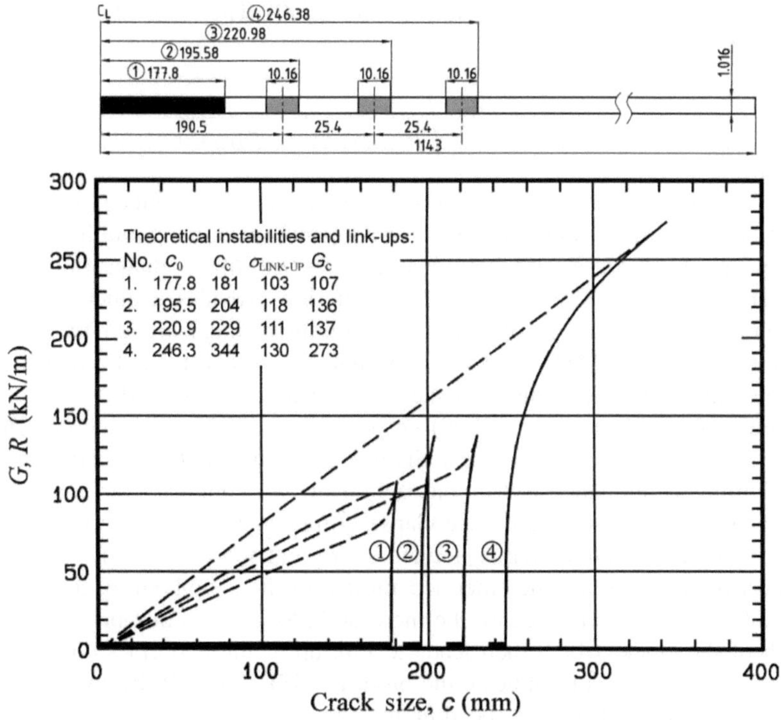

Fig. 10.4 Evaluation of link-up stress and residual strength from the *R*-curve for a panel with MSD cracks (de Wit et al. 1996)

At the critical load, the plastic zone size ahead of the lead crack tip in an aircraft fuselage, as estimated from the Irwin or Dugdale equation, is on the order of the typical rivet spacing (25 mm). Therefore, it is not negligible when compared to the size of ligament ahead of the lead crack and the MSD cracks, which implies that the validity conditions of LEFM are not satisfied. Actually, the case of touching plastic zones of the neighbouring cracks, as required by the PZL criterion, is beyond the LEFM validity limits, as pointed out already by Schijve (1995). Elastic-plastic FE analyses by Wang et al. (1997c) for 90 in. wide and 0.04 in. thick Al 2024-T3 panels with multiple collinear cracks indicate that when the main crack is half way through the ligament between its tip and the first MSD crack, this ligament is already fully plastic. Moreover, the ligament between the first and second MSD crack is also fully plastic. The PZL criterion would predict that all cracks have linked up at this point, while the experiment shows that stable crack growth occurs prior to each link-up. An engineering approach to allow for this observation is the PZL criterion modified by Broek et al. (1994), as presented earlier in this section. Limitations of the original Swift approach have been confirmed experimentally by Dawicke and Newman (1997), Ingram et al. (1999), Smith et al. (2000), and Tan et al. (2003), and concepts to improve predictions based on the PZL criterion have been proposed, e.g. by Ingram et al. (1999) and Smith et al. (2000). The corrected link-up stress was presented as (cf. Eq. 10.3)

$$\sigma_{LINK-UP} = CS_Y \sqrt{\frac{L}{\beta_{LEAD}^2 c_{LEAD} + \beta_{MSD}^2 c_{MSD}}} \tag{10.7}$$

where C is an empirical factor (a function of L or $c_{\mathrm{LEAD}}/\mathrm{L}$) obtained by correlating the predictions with tearing test results for flat panels containing multiple cracks.

Compared to elastic-plastic fracture mechanics (EPFM) residual strength criteria considered in the next section, the PZL criterion provides a quick and simple solution to assess the effects of MSD cracks on the residual strength of a structure with a lead crack and MSD cracks and for that reason it is still utilized even in the case of complex fuselage structure panels, as considered in Sect. 10.5.

More recently Wang et al. (2009) proposed an LEFM based link-up criterion similar to the concept of Swift in that the link-up event is not identified with actual crack impingement. The onset of interaction of two collinear cracks at adjacent rivet holes is assumed to cause the collapse of ligament between these cracks. The beginning of crack interaction is defined as a point where two SIF vs. c curves, one for a single crack and the other accounting for the neighbouring crack, start to diverge from each other. Because the increase in SIF results in rapid and unstable growth of the crack, which then leads to link-up of the two cracks, the onset of interaction can be used to indicate the onset of MSD. The SIF behaviour must be tracked by FE simulations. To speed up the computation, a semi-empirical method to determine the crack length increment at which the SIF update is needed during the crack growth simulation was proposed. The authors claim that the new criterion is more conservative than Swift's PZL concept, but not overly conservative, as shown by representative fuselage lap joint tests by Jones et al. (2008).

10.2.2 Elastic-Plastic Fracture Mechanics Failure Criteria

According to a pretty common opinion, essential characteristics of the very last stage of crack growth in pressurized aircraft fuselage structures, which include large scale yielding, a substantial amount of crack extension, and out-of-plane deformation (bulging at the crack edges), should be accounted for in residual strength predictions. The large scale yielding conditions eliminate the SIF as a crack tip parameter and EPFM parameters must be considered. The J-integral has been used as a fracture criterion for large scale yielding with small amounts of crack growth. However, the J-integral resistance curve shows a configuration dependence for large crack extensions observed in thin ductile materials (Anderson 2005). In the context of the residual strength analysis of aircraft skin metallic structures, the largest computational effort over the last decades has focused on the applicability of the crack tip opening angle (CTOA) as a failure criterion first proposed by de Koning (1975). Another EPFM parameter also considered to be suitable for that purpose is the T^*-integral first proposed by Brust et al. (1985).

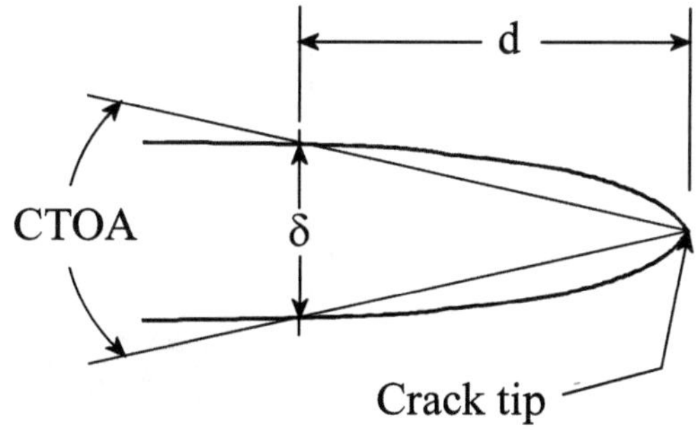

Fig. 10.5 Illustration of CTOA definition

10.2.2.1 CTOA Failure Criterion

The CTOA fracture criterion evolved from the well known crack tip opening displacement (δ) concept proposed for large scale plasticity and large amounts of stable tearing. It is defined as

$$\text{CTOA} = 2\tan^{-1}\frac{\delta}{2d} \tag{10.8}$$

where δ is measured at a specific distance (d) behind the crack tip, Fig. 10.5. The CTOA criterion was found a mild function of the d-distance. Usually d of 1 mm is most commonly adopted.

The CTOA criterion assumes that crack extension occurs when the CTOA reaches a critical value ψ_c. The critical CTOA can be obtained experimentally if the CTOA on the specimen surface and crack extension are monitored during a fracture test, exemplary results for two specimen types of various widths being shown in Fig. 10.6. For a given material ψ_c decreases when the sheet thickness is increasing (Newman et al. 2003). For example, Starnes et al. (2003) reported ψ_c-values (determined at $d = 1$ mm behind the crack tip) of 5.6°, 5.4° and 5.25° for C(T) specimens from the 2024-T3 Al alloy 1, 1.6 and 2.3 mm thick respectively. As seen in Fig. 10.6, surface measurements produce the ψ_c-value with a considerable scatter. According to Dawicke and Newman (1998), a variation of $\pm1\%$ they observed in measured ψ_c-values, if implemented in fracture analysis, would result in an unacceptably large variation of 20% in the residual stress value. Hence, they suggested that a more accurate method of deriving ψ_c is to simulate the fracture behaviour of a laboratory specimen with an elastic-plastic FE analysis and to determine a CTOA-value that best correlates the experimental load-crack extension results. Obviously, an agreement of ψ_c obtained independently from experimental measurements and numerical analyses at the same specific distance d greatly increases the confidence in the chosen critical CTOA value. 3D analyses are required as neither a plane stress nor a plane strain analysis can predict the specimen size effect with a constant CTOA value (Dawicke and Newman 1998). An alternative illustrated in Fig. 10.7 can be a 2D analysis employing a "core" of plain strain elements along the crack plane while the rest of the specimen is under

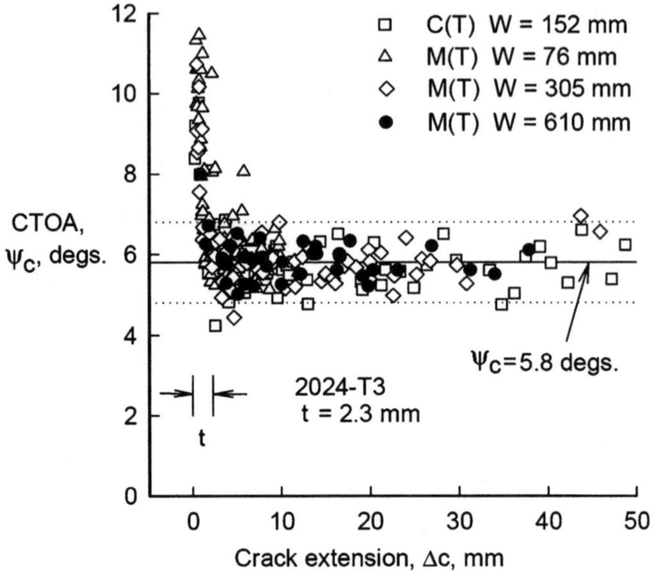

Fig. 10.6 CTOA measurement results for several Al 2024-T3 specimen types (Newman 1999)

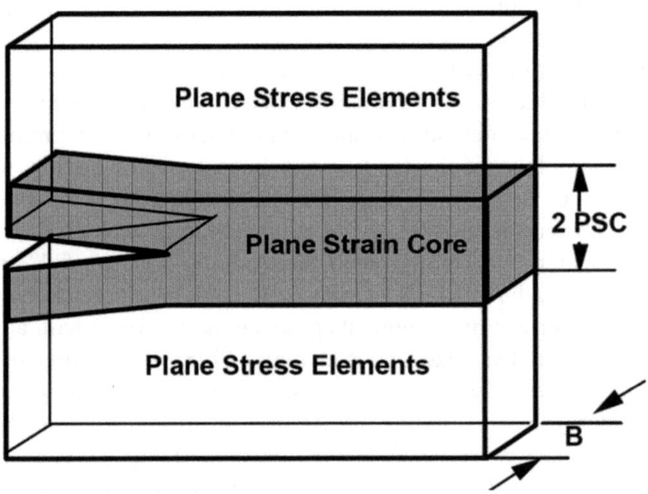

Fig. 10.7 Illustration of the plane strain core concept

plane stress (Dawicke 1997). The plane strain core is intended to approximate the 3D constraint at the crack tip. Its height (PSC, Fig. 10.7) can be obtained through comparison with 3D analyses. PSC is close to the specimen thickness and generally increases with thickness (Dawicke 1997).

3D FE simulations of stable tearing indicate that the transitional CTOA versus crack extension behaviour observed experimentally during initial crack extension

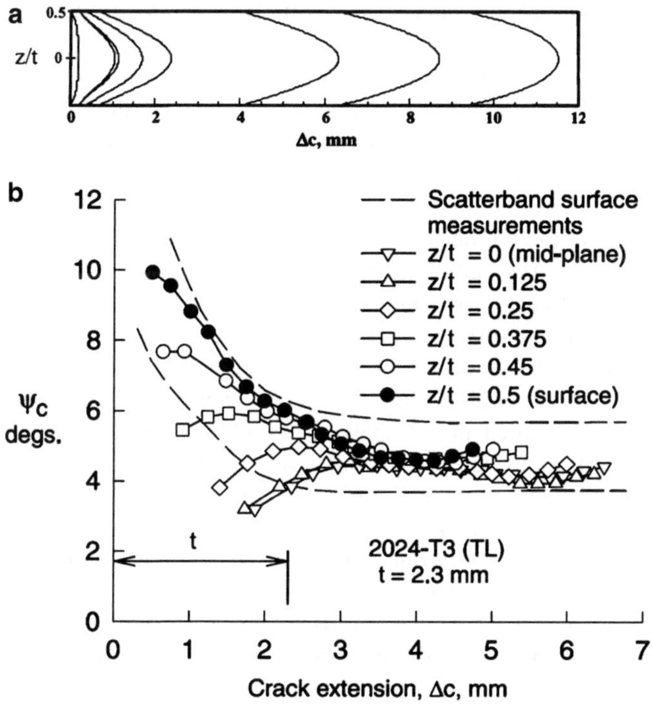

Fig. 10.8 Crack tunnelling effect on critical CTOA: (**a**) observed crack front shapes (not to scale through thickness); (**b**) computed critical CTOA values at various through-the-thickness locations (Dawicke et al. 1995a)

(cf. Fig. 10.6) is associated with the crack tunnelling and flat to slant crack growth transition (Dawicke and Newman 1998; Newman et al. 2003; James and Newman 2003). Fatigue loading on thin Al alloy sheet material produces crack fronts that are only slightly tunnelled. However, in the early stage of stable tearing a severe tunnelling is observed, Fig. 10.8a, presumably due to low (plane stress) constraint at the free surface and a higher constraint in the interior. More crack tunnelling is exhibited for the flat fracture surface than for the slant fracture surface (James and Newman 2003). Dawicke et al. (1995a) used a 3D, elastic-plastic FE analysis code to simulate tunnelling observed during the stable tearing of 2024-T3 specimens, Fig. 10.8a, and to calculate the ψ_c-values at various through-the-thickness locations along the crack front, Fig. 10.8b. Multiple specimens were fractured to obtain the crack front shape and extension at a given load level, one crack front contour per specimen. As seen in Fig. 10.8b, the calculated ψ_c decreased through-the-thickness in the mid-plane direction. On the surface, the calculated ψ_c-values agreed with the measured values. After crack extension at the free surface of 3 mm, which was close to the specimen thickness, all through-the-thickness calculations of CTOA approached an average value of about 4.5°, equal to the average CTOA measured on the surface.

Measured ψ_c-values oriented in the rolling direction of the sheet (LT) are somewhat higher than for cracks oriented transverse to that direction (TL), e.g. Seshadri and Newman (2000). It is not clear if this behaviour is linked with the amount of crack tunnelling and fracture surface morphology. Dawicke and Sutton (1994) found that for a 2.3 mm thick Al 2024-T3 sheet, the TL orientation consistently displayed flat fracture and the LT orientation consistently displayed slant fracture. At the same time, James and Newman (2003) reported considerably more tunnelling for the flat than for the slant fracture surface.

Typically, load versus crack extension results obtained from a 3D analysis assuming a straight crack front underestimate the load for tearing initiation and, therefore, overestimate surface crack growth observed in early stages of stable tearing tests on Al-alloy sheet material, e.g. Dawicke and Newman (1998), Chen et al. (1999), Newman et al. (2003). James and Newman (2003) hypothesize that the straight front assumed in 3D analyses represents the "average" crack length at any given load. However, their results showed that if an area-averaged crack extension correction was accounted for in the experimental data presentation, based on crack front measurements on the fracture surfaces, an improved agreement with predictions (3D, straight crack front) only occurred in the case of flat cracking, while with slant cracking a little difference was seen.

It can be concluded that the technique used to measure the critical CTOA remains in a stage of development, and the measured data are still subject to interpretation by the researchers.

If a fracture test is conducted on specimens with fatigue precracks rather than with saw-cut precracks, the aforementioned overestimation of crack growth in the initial phase of stable tearing can additionally stem from the fact that load history effects on the crack opening profile are not accounted for in the simulation. Because fatigue loading generates a wake of plastically extended material behind the crack tip, a higher applied load is required to open the fatigue crack to the extent corresponding to the critical CTOA-value than in the case of an "ideal" (saw-cut) crack with sharp ends (Chen et al. 1999). Consistently with the above effect, Dawicke et al. (1995b) showed that a higher applied stress during fatigue precracking increased the resistance of stable crack growth initiation, which could result from more plastically elongated material in the crack wake. As demonstrated experimentally by these authors, the initial crack tip profile (saw-cut with blunt tips or fatigue crack) and the initial load history (fatigue precracking level) have a stronger effect on the initial portion of stable crack growth and link-up than on residual strength. They observed more differences in residual strength values due to the initial crack history for specimens with multiple cracks than for single crack specimens.

10.2.2.2 *T**-Integral Failure Criterion

The T^*-integral is a measure of the total energy release rate inside the crack tip region enclosed by a contour Γ_ε, Fig. 10.9, and is defined as

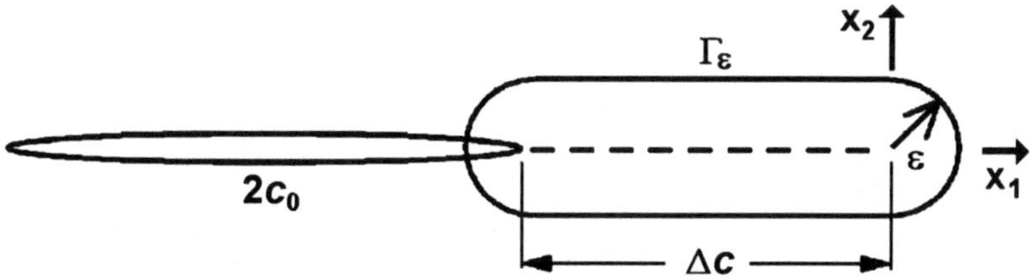

Fig. 10.9 Contour path for determination of T^*-integral

$$T^* = \int_{\Gamma_\varepsilon} (Wn_1 - \sigma_{ij} u_{i,1} n_j) d\Gamma \tag{10.9}$$

where W is the strain energy density evaluated along Γ_ε ($W = \int_0^{\varepsilon_{ij}} \sigma_{ij} \varepsilon_{ij}$, σ_{ij} and ε_{ij} – stress and strain tensor), n_1 is the x_1 component of the unit vector normal to Γ_ε, x_1 and x_2 define the local coordinate system centred at the crack tip, and the subscript ε indicates the contour size. Different ε-values lead to different values of the T^*-integral. For numerical reasons it is convenient to adopt an elongated contour path, as shown in Fig. 10.9 (Wang et al. 1997b). For a stationary crack, T^*-integral is identical to the J-integral and becomes independent of the choice of the contour path.

Stable tearing and residual strength can be predicted based on the T^*-integral resistance curve concept. It is assumed that once the computed T^*-integral reaches a critical value according to the T^*-resistance curve, the crack will extend by a predetermined increment. The critical T^*-integral value increases with the amount of crack extension. The T^*-integral resistance curve is assumed to be an intrinsic material property. The stable tearing behaviour for other loadings and geometries can then be predicted by forcing the fracture process to follow the same resistance curve provided that the same contour size ε is used in the analyses.

Since the T^*-integral cannot be measured experimentally, the T^*-resistance curve must be evaluated in a generation mode of the analysis through the numerical simulation of a stable tearing test on a laboratory specimen (Wang et al. 1997b). In the simulation, the test specimen model is first loaded to the level at which stable tearing initiates in the experiment and the initial T^*-integral value is computed. Then the crack is extended a small amount, while the load is changed simultaneously according to the load versus crack extension experimental data, and the T^*-integral is computed at each step of the crack growth.

10.3 Crack Growth Directional Criteria

The failure criteria presented so far deal with the so-called self-similar elastic-plastic crack growth when the crack trajectory is known *a priori*. However, a crack in a shell structure is likely to propagate in a non-self-similar fashion. As discussed by

Swift (1987), the curvature of a cylindrical fuselage panel containing a longitudinal crack results in shell bulging. This term refers to large out-of-plane displacements occurring in the vicinity of the crack in response to the loss of the hoop stress component along the free edge of the crack face. Bulging deformation causes significant tensile membrane stress in the longitudinal direction (parallel to the crack). Because the bulging effect is enhanced as the crack propagates, this stress in the longitudinal direction increases with crack growth while the hoop stress remains unchanged. Consequently the crack tends to turn in the circumferential direction and this turning causes a portion of the shell wall to flap open. Because the bulging effect is weaker in thicker sheets than in thinner sheets, thicker sheets are less prone to flapping (Maclin 1991). A tear strap locally reduces the hoop stress (cf. Fig. 2.2), thus causing the longitudinal normal stress to become greater than the hoop stress for a crack length that is less than the length for flapping of the unstiffened cylinder.

Available experimental evidence suggests that the occurrence and mode of flapping depends on the panel design. Fatigue testing of a retired Boeing 737 aft fuselage was concluded with a two bay crack in a longitudinal lap splice riveted joint and safe decompression by flapping (Maclin 1991). In this test, MSD was present in adjacent holes with a non-uniform distribution of crack sizes typical of fatigue scatter. However, reported in the same work fatigue test results on Boeing 737 panels with small saw-cuts added in the critical rivet row to simulate more severe MSD demonstrated that the worst case MSD due to normal fatigue scatter may prevent flapping. Such an effect of MSD on the crack path can be expected because, compared to the case of a single crack, the mode I SIF is elevated by interaction with a row of MSD cracks. The crack will then continue to propagate in the longitudinal direction despite the presence of bulging and the resultant mode II SIF, as exemplified by the famous Aloha Airlines aircraft accident (Aircraft Accident Report 1989). Addressed in Sect. 8.2.5 tests by Miller et al. (1994) on a curved fuselage panel with floating frames indicated that flapping of a crack propagating in a riveted lap joint with MSD cracks occurred (cf. Fig. 8.25a). Such a crack turning was not observed for a fuselage panel with shear tied frames (cf. Fig. 8.25b). Also Steadman et al. (1999) and Bakuckas et al. (2003) did not witness flapping at shear tied frames for cracks propagating in fuselage panel lap joints with MSD. Flapping at floating frames of a fuselage also occurred in the case of a longitudinal crack located in the skin adjacent to a stringer (Miller et al. 1994), but it was not observed at frames riveted on skin when the crack was midway between machined stringers integral to the skin (Munroe et al. 2000). Experimental evidences (e.g. Aircraft Accident Report 1989) show that for a crack in a longitudinal lap joint fracture will typically occur only in the upper skin with the lower skin and the remaining portion of the upper skin still attached to the fasteners. Bulging and the associated tensile longitudinal membrane stress along the crack edge develop mainly on the upper crack side due to the lower stiffness of the crack edge on this side compared to that of the opposing crack edge (Kosai et al. 1992). Similarly, in the case of cracking close to a stringer bulging will be restrained on the crack edge nearest to the stringer (Miller et al. 1994; Potyondy et al. 1995). However, for a longitudinal crack located symmetrically relative to stringers and tear straps, bulging deformation is symmetrical. Consequently, if flapping at tear straps occurs, it develops symmetrically on both sides of the crack due to bifurcated crack paths turning to the circumferential direction parallel to the straps (Kosai et al. 1999). Available literature evidence (Swift 1987; Maclin 1991) suggests that flapping cannot be relied upon on skin gauges thicker than 2.5 mm (0.063 in.). Cracks in skin gauges of 1.6 mm (0.04 in.) or less, reinforced with tear straps or shear ties, show a strong tendency to form flaps, except when the crack appears in a row of fasteners containing a large amount of MSD (Maclin 1991).

Flapping is a failure mode that limits fracture to the affected bays. Because the associated controlled depressurization of the fuselage can be considered as the last defence of an axially rupturing fuselage, analyses of crack turning are undertaken in order to asses the crack curving efficacy of tear straps. There is no consensus on the most satisfactory theory to predict crack growth direction. In the case of cracks in fuselage panels, usually the maximum tangential stress criterion derived by Erdogan and Sih (1963) is a starting point to evaluate the crack growth direction. According to Erdogan and Sih, the crack extension starts in the direction along which the tangential stress $\sigma_{\theta\theta}$ is maximum, which yields

$$K_{\mathrm{I}} \sin\theta_{\mathrm{c}} + K_{\mathrm{II}}(3\cos\theta_{\mathrm{c}} - 1) = 0 \tag{10.10}$$

where K_{I} and K_{II} are the mode I and II SIFs respectively and θ_{c} is the angle of crack kinking.

Equation 10.10 fails to predict the crack path instability experimentally observed in pressurized cylinders in the absence of K_{II}. There is substantial evidence (e.g. Ramulu and Kobayashi 1983; Knops 1994) that to more accurately model the crack turning behaviour in fuselage structures, a second order term or the T-stress, that is neglected in the Erdogan and Sih theory, should be included in the asymptotic stress field in the vicinity of the crack tip. For two-dimensional mixed-mode problems, the tangential stress $\sigma_{\theta\theta}$ up to the order of the T-stress is

$$\sigma_{\theta\theta} = \frac{1}{\sqrt{2\pi r}} \cos\frac{\theta}{2}\left(K_I \cos^2\frac{\theta}{2} - \frac{3}{2}K_{II}\sin\theta\right) + T\sin^2\theta. \tag{10.11}$$

The crack kinking angle can be obtained by maximizing the tangential stress which results in

$$\frac{K_{II}}{K_I} = \frac{-\sin\frac{\theta_c}{2}}{3\cos\theta_c - 1}\left(2\cos\frac{\theta_c}{2} - \frac{16\sqrt{2\pi}}{3}A\cos\theta_c\right); \quad A = \sqrt{r_c}\frac{T}{K_I} \tag{10.12}$$

Ramulu and Kobayashi (1983) have shown that r_{c} is a material dependent parameter which must be determined experimentally. An alternative way is to choose r_{c} that yields a best correlation between the FE simulations and the observed crack path for a double cantilever beam specimen (Chen et al. 1999). Physically, r_{c} represents the maximum radial distance where the maximum tangential stress criterion is applicable. A parametric study by Chen et al. (1999) indicates that r_{c}

strongly affects analysis results. However, the physical meaning of this parameter is not well understood and the appropriate experimental method to measure r_c is still not fully determined, as evidenced by widely differing measured r_c-values reported in the literature (Pettit and Ingraffea 1999).

From Eq. 10.12, θ_c increases with T-stress, and negative T-stress tends to stabilize the crack path. Note also that using Eq. 10.12 crack path instability in pure Mode I environment is predicted if

$$r_c > \frac{9}{128\pi}\left(\frac{K_I}{T}\right)^2. \tag{10.13}$$

The directional theory was further extended by Chen et al. (1999, 2002) to include the effect of fracture toughness orthotropy. Crack propagation is assumed to occur in the direction of maximum normalized tangential stress, such that at $\theta = \theta_{c,}$

$$\text{Maximum}\left[\frac{\sigma_{\theta\theta}(K_I, K_{II}, T, r_c, \theta)}{K_c(\alpha)}\right] = \left(\frac{\sigma_{\theta\theta}}{K_c}\right)_{critical} \tag{10.14}$$

where the angle α characterizes the material grain orientation and $K_c(\alpha)$ is the anisotropic fracture toughness.

The above concepts are strictly valid for LEFM conditions and flat panel geometry. Determination of SIFs and T-stress in the case of elastic deformations of arbitrary magnitude, which occur in the case of curved shell structures, involves various simplifying assumptions (Potyondy et al. 1995; Chen et al. 1999). A rigorous elastic-plastic directional criterion for non-self-similar stable crack growth simulation is still lacking. In order to evaluate the adequacy of the directional criterion according to Eq. 10.12 for fuselage structures, Kosai et al. (1999) drove an elastodynamic FE code with geometric nonlinearity in its generation mode by prescribing the measured crack extension together with the applied loading. The measurements were made during a residual strength test on a cylindrical laboratory specimen (360 mm dia, 914 mm long, 0.3 mm thick) under internal pressure. The specimen contained the rudiments of a fuselage construction with riveted and bonded stringers and tear straps. K_I, K_{II} and T extracted in the generation phase were then used to obtain the crack kinking angles and the resultant crack path, which in turn was compared with the actual crack path. Despite the obvious large scale yielding in crack growth of the laboratory specimen, the actual and predicted crack paths almost coincided. Moreover, only a maximum of 2% difference in the crack tip radial deformations was found between results from an FE model with both geometric and material nonlinearity, and from a model with only geometric nonlinearity. Since the crack tip bulging, as represented by these deformations, governs the crack tip parameters involved in the crack directional criterion, the authors claim their results to justify the use of elastic FE codes (with geometric nonlinearity) in crack direction analyses for fuselage structures. Another validation of the maximum tangential stress criterion provided by Keesecker et al. (2003) is considered in Sect. 10.4.

10.4 Computational Issues

Various computational approaches to evaluate residual strength of thin cracked sheets have been proposed in the literature. Addressed below will be, however, only those demonstrated in the literature to be applicable to lap joints with mechanical fasteners and verified experimentally, i.e. enabling a transfer of methodology developed from basic research to real-work application. These analysis procedures, developed with the support of the Federal Aviation Administration (FAA) and the National Aeronautics and Space Administration (NASA), use concepts considered in Sect. 10.2, namely the PZL, CTOA and T^*-integral failure criteria to predict stable tearing and fast fracture in the presence of MSD in aircraft structures. The CTOA criterion is implemented into the STAGS (STructural Analysis of a General Shell) program developed by Lockheed Missiles and Space Co. (Rankin et al. 1998). The prediction employing the T^*-integral resistance curve is performed using a TKALT code, which is an analysis module of the 2D-EPFEAM (Elastic-Plastic Finite Element Alternating Method) program (Wang et al. 1997a, b, c).

The STAGS FE code can be applied for a general purpose analysis of shell structures of arbitrary shape and complexity subjected to arbitrary loads including internal pressure. It is particularly well suited for analyzing shells that have structural features such as frames, stringers, tear straps, stringer clips, fastener and bonded connections, etc. STAGS includes both geometric nonlinear analysis necessary to account for out-of-plane bulging experienced by a long crack in a pressurized fuselage structure and material nonlinear analysis. Shell, plate and beam elements are used. The shell elements in STAGS are formulated for all six degrees of freedom (three translations and three rotations). For lap splice connections of the plates, each layer in the overlap area is modelled with its own separate shell elements. When a crack is present in a fuselage structure, the plane stress assumption inherent in the shell theory is still valid for the greater part of the structure, but it is violated in the immediate neighbourhood of the crack tips. With STAGS, it is possible to apply a 3D formulation in a selected region by using volume (brick) elements. STAGS also supports plane strain elements for modelling a plane strain core surrounding a crack (cf. Fig. 10.7). The code has been developed especially for nonlinear stable and unstable crack growth analysis with the CTOA as the failure criterion. Using the critical CTOA, equilibrium states for stable tearing along a user defined crack path are computed through a nodal release technique. Along the crack path, each node is actually two nodes. At locations where the crack is closed, the degrees of freedom for both nodes are the same. The CTOA is calculated for every increase in pressure. Once the calculated CTOA becomes within 5% of ψ_c, STAGS attempts to release the pair of crack nodes at the crack tip and equilibrium is established for the model with the longer crack length. Provided the CTOA for the new crack tip is below $0.95\psi_c$, the process of increasing the pressure resumes. In addition, STAGS allows the calculation of equilibrium points during a region of unstable crack propagation. The unstable tearing algorithm is triggered

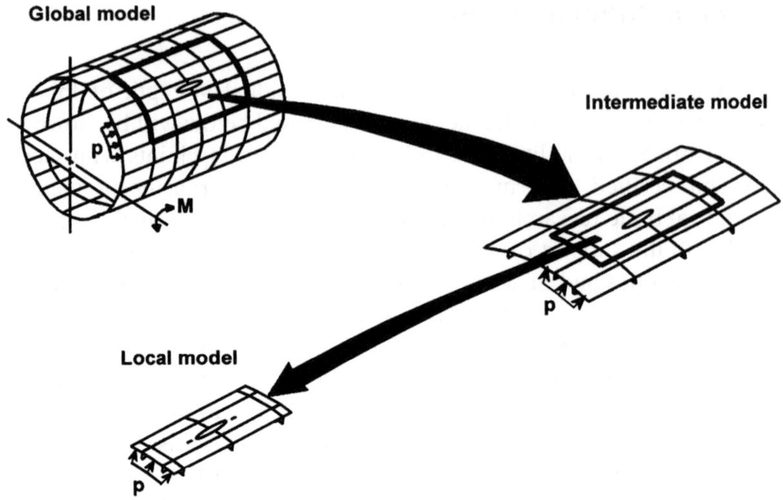

Fig. 10.10 Illustration of hierarchical modelling

if the CTOA calculated immediately following the crack extension is equal to or larger than ψ_c. A nonlinear solution procedure is implemented to find after each crack increment a decreased pressure such that CTOA equals ψ_c.

A topology based program FRANC3D (FRacture ANalysis Code for 3D solids and shells) developed by the Cornell University Fracture Group (Wawrzynek 1991; FRANC3D 2003) is often applied as a pre- and postprocessor for STAGS, both codes being integrated in the FRANC3D/STAGS software (Potyondy 1993; FRANC3D & STAGS 2003). FRANC3D generates a mesh consisting of general shell elements for the STAGS analysis. In a multilayer or stiffened structure, structural components are connected through fasteners or adhesive bonds.

An efficient strategy applied sometimes in residual strength analyses of aircraft fuselages in the presence of cracks is the global-local hierarchical modelling illustrated schematically in Fig. 10.10. The hierarchical modelling levels range from a relatively coarse global shell model to a highly refined local shell model for performing the crack growth analysis. The lead crack (cracks) is introduced already to the global model, while MSD is accounted for in subsequent modelling levels or only in the local model. At any modelling level, kinematic boundary conditions stem from the solution for the preceding level model.

The 2D-EPFEAM code combines FE analyses with a closed form solution employing the superposition principle. The problem is first solved using the linear elastic analysis. The solution to the original problem, which is a finite cracked body with prescribed boundary conditions, is obtained as the sum of solutions to two simpler problems: (a) a finite uncracked body with the prescribed original boundary conditions, and (b) a finite cracked body with tractions acting on the crack surfaces. The crack surface tractions are determined as the negatives of the tractions acting at the prospective crack locations in problem (a). Problem (a) is solved exactly by the conventional FEM. Problem (b) is considered as the sum of: (c) an infinite cracked body with tractions as in (b) acting on the crack surfaces, and (d) a finite cracked body with negative displacements and tractions prescribed on its boundary, where these displacements and tractions are determined from problem (c) at the prospective finite boundary location. Problem (c) is solved analytically, as detailed by Wang et al. (1997a), while problem (d) is solved using the same procedure as for the original problem. The procedure is repeated until a convergent solution for the original problem is obtained. The extension of the linear elastic solution to the elastic plastic solution is accomplished using the initial stress method known in the theory of plasticity. Thus, the EPFEAM involves two analysis loops. The first loop is associated with using the superposition principle in the linear elastic solution referred to as the FEAM, while the second loop is associated with the iterative scheme of the initial stress method. The computation time can, therefore, be very extensive. However, according to Anderson et al. (2004), since the FE analysis part of the EPFEAM involves an uncracked body, the modelling effort is significantly less than for a conventional elastic-plastic FEM. The 2D-EPFEAM code can only cope with a 2D elastic-plastic problem. Therefore, this code must be used together with other FE codes to predict the residual strength of complex structures. The hierarchical modelling is applied. For each particular lead crack length the global model, that contains the full panel with all significant features (stringers, frames, tear straps, splice joints), is solved using a conventional linear elastic FEM to compute tractions and displacements at the boundaries of the 2D local model, including the fastener transfer loads. Next, the 2D-EPFEAM analysis is applied to the local model, as described above, and stable tearing is simulated using the TKALT code. The local model represents a small region containing a portion of the lead crack including the crack tip, and the adjacent fastener hole.

Riveted connections between structural components are modelled using elastic-plastic 1D, two-node beam or spring elements. Components such as skin, doubler or stiffener are modelled using plate or shell elements at the respective mid-thickness planes. Hence, the simulated fastener length is set equal to the average thickness of the joined layers. Fasteners that connect more than two components are divided into smaller segments, each segment connecting only two adjacent layers. The fastener elements are modelled with six degrees of freedom (three translations and three rotations). The stiffness of each degree of freedom is defined by prescribing a force-deflection curve assuming that the fastener behaves like an elastic beam of the diameter equal to that of the fastener. However, in the codes referred to in this section, the shear stiffness of the rivet is computed using a semi-empirical formula for fastener flexibility according to Swift (1971). Typically, the geometry of rivets and their interference with the sheets is not faithfully represented in residual strength analyses. Instead, a more simple modelling of the fastener-sheet connection is applied. It can be a single point connection, Fig. 10.11a, or a distributed connection, Fig. 10.11b, of the fastener element nodes to the adjacent shell elements. In the latter case the connection is accomplished by defining rigid links (spring elements an order of magnitude stiffer than the rivet spring) between the rivet element node and the surrounding shell element nodes. The area of the transfer load distribution should

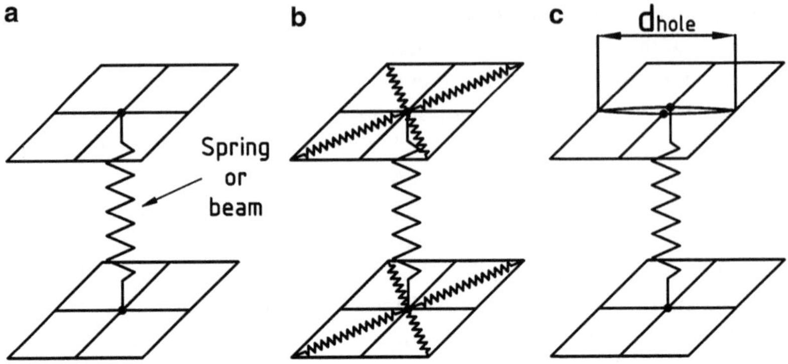

Fig. 10.11 Modelling of rivets in residual strength FE analysis: (**a**) tight rivet with single point connection to the sheet; (**b**) tight rivet with distributed connection to the sheet; (**c**) loosened rivet

be close to the rivet hole area. As a crack propagates through the rivet hole in one of the joined sheets, all coincident nodes on the crack path between the crack tips are disconnected in that sheet. In that case it is usually assumed that the rivet stays intact and the rivet element is connected to a single nodal point in the centre of the rivet hole on the bearing side of the cracked sheet. Therefore, with the idealization according to Fig. 10.11b, only shell elements on this side of the crack are used to model the distributed connection.

Seshadri and Newman (2000) postulate that modelling according to Fig. 10.11a, b is representative of more tight rivets, while for loosened fasteners an approach applied by Anderson et al. (2004) and shown in Fig. 10.11c is more appropriate. With the latter approach, fastener holes without cracks are modelled as single slits (or cracks) with a length equal to the physical length of the hole diameter (d_{hole}), as illustrated in Fig. 10.11c, whilst in the case of fastener holes with MSD cracks the crack length equals the physical length of the two cracks plus the diameter of the hole. In either case the rivet element is connected to the node on the bearing side. According to Seshadri and Newman (2000), the idealizations shown in Fig. 10.11a or b on one side and the modelling according to Fig. 10.11c on the other side bound the possible behaviour of actual riveted joints.

Interestingly, Anderson et al. (2004) report that under a uniform remotely applied tensile stress SIFs for symmetrical cracks emanating from a hole and SIFs for a slit of the same size as the hole diameter and MSD cracks combined differ by only about 3%.

If a fracture analysis based on the CTOA criterion is performed using the STAGS code, all types of modelling the rivets considered Fig. 10.11 are feasible. With the T^*-integral criterion, the TKALT code imposes limitations as only a single layer of the overlap can be analysed. Rivet holes are not modelled and the fastener action on the plate is represented by the bearing load determined from the global model, as said above, and applied as a point force. For holes with MSD cracks the bearing force is applied at a distance equal to the radius of the hole away from the crack surfaces on the bearing side of the sheet. For holes without MSD cracks it is applied in the centre of the fastener. For each length of the lead crack the bearing loads are calculated from the global model based on the shears in fastener elements idealized according to Fig. 10.11a.

The effect of some choices involved in the numerical modelling performed using the FRANC3D/STAGS system was investigated by Chen et al. (1999). The analyses were validated through comparisons with experimental results obtained for two curved fuselage panels 12-stringer-bay (pitch of 9.25 in.) wide and 5-frame-bay (pitch of 20 in.) long with a radius of curvature of 127 inches designed to simulate typical wide-body fuselage crown structures. The floating frames were connected to hat section stringers with stringer clips. Tear straps were hot-bonded to the skins at each frame station. The skins and tear straps were made from 0.063 in. thick 2024-T3 Alclad, and the stringers, frames and stringer clips were made from 7075-T3 Alclad. The skins were lap jointed by three rows of 3/16 in. diameter, 100° countersunk rivets.

In either panel a 5 in. long initial saw-cut was inserted along the upper rivet row in the outer skin. For the panel with MSD cracks, small saw-cuts were additionally inserted in the outer skin after the rivet holes had been drilled, but prior to the application of the fay sealant and rivet installation. The panels were subjected to pressure cycling until the length of the lead crack reached about two frame bays, namely 38.2 in. and 41.7 in. for the panel without and with the MSD cracks respectively. The central frame was then severed and the residual strength tests were conducted. The tests were performed by inserting the panels in a wide-body cylindrical fixture fabricated by Boeing and described in Sect. 8.2.1 (cf. Fig. 8.8).

Rosette strain gauges were installed back-to-back on the skins and tear straps, as shown in Fig. 10.12. Strain gauge readings recorded under the maximum cyclic pressure when the lead fatigue cracks reached their maximum lengths specified above were used to verify predicted stress distributions. Interestingly, unlike in the case of simple lap joints considered in Chap. 6, at all four locations the measured hoop bending stresses appeared negative on side A and positive on side B, the local bending factors ranging from 0.1 at location 1 to 0.33 at location 3.

One of the goals of the numerical analyses was to check the transition accuracy of the hierarchical modelling. Both the geometric and material nonlinearities were accounted for. Rivets in holes without cracks were modelled according to Fig. 10.11a and a crack with a length equal to the rivet diameter plus the MSD length was used to model the MSD cracks. Pressure loading was applied on all the internal surfaces. For the global model, symmetric boundary conditions were imposed on all the boundary edges to simulate a cylinder-like fuselage structure and uniform axial expansion was allowed at one longitudinal end. At the global level the panel of about the size of the test panel was modelled with the mesh density around the gauge locations progressively refined until the convergence was observed. For a sufficiently refined FE mesh the predicted stresses compared well with the measurements. With the global-local approach two hierarchical modelling levels were employed, a 1×1 bay panel being modelled at the local level with the kinematic boundary conditions extracted from the global model results. When the

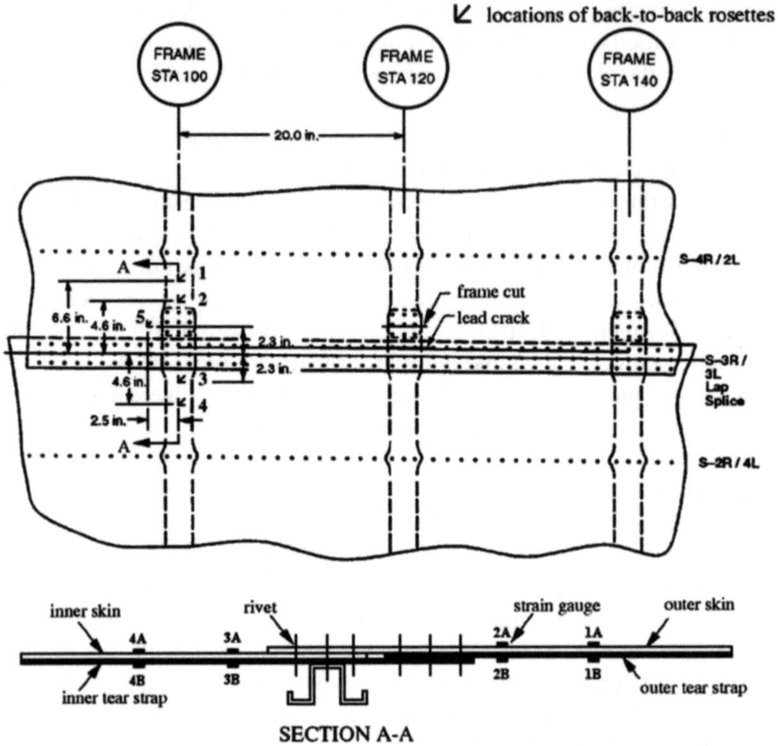

Fig. 10.12 Lead crack and gauge locations on a panel considered by Chen et al. (1999)

mesh density in the local model was the same as for the corresponding region in the global model with the refined mesh, the local model results agreed well with the global model predictions, and therefore with the strain gauge readings. However, when in the local approach a much higher mesh density, that is suitable for the stable tearing analysis, was applied, the results from the local model disagreed with the rest of numerical predictions and experimental measurements. In particular, the predicted membrane hoop stresses were much lower than those observed. The authors traced the discrepancy to the inadequacy of the idealized representation of the rivet-sheet connection according to Fig. 10.11a. The single point connection caused unrealistic distortion of the surrounding sheet elements, which led to premature yielding of the surrounding shell elements and reduced load transfer. This artificial effect was enhanced with refining the mesh. The local model with the dense mesh yielded a much better prediction if the rivet-sheet connection for the intact rivets was idealized according to Fig. 10.11b.

As already said in Sect. 10.2.2.1, the ψ_c-value that best correlates the residual strength in a stable tearing test simulation typically yields an underprediction of the load initiating the stable tearing and an overestimation of crack growth in the initial part of the stable crack growth. Such a disparity has been reported for both saw-cuts and fatigue precracks. For example, Chen et al. (1999) noted a considerable underprediction of the pressure level to initiate stable crack growth in the curved lap-jointed fuselage panels described above in this section. The analyses were performed using the FRANC3D/STAGS system. To allow for the residual plastic deformation in the crack wake due to cyclic loading, an elastic-plastic fatigue loading simulation prior to the stable tearing simulation was performed, as suggested by Newman (1976). To this end, in the first step the cracks were closed over a certain length and crack growth simulation was performed until the cracks reached the initial positions for stable tearing. The length of the MSD cracks enabled applying only two fatigue cycles. Allowing for the plasticity-induced crack closure increased the predicted initiation pressure by 150–210%, which made the calculated crack initiation loads to be within 6% of the test results.

An alternative option to improve the prediction of the stable tearing initiation pressure was adopted by Hsu et al. (2003) who assumed CTOA-values to initiate stable tearing (ψ_{ci}) higher than ψ_c. The ψ_{ci}-values were found by trial and error to correlate the measured crack initiation stresses for flat panels of 0.09 in. thick 2024-T3 sheets containing riveted lap joints with single and MSD saw-cut cracks along the upper rivet row. STAGS predictions of residual strength were performed in which fastener holes along the crack path with or without the MSD cracks were simulated as slits, as described earlier in this section (cf. Fig. 10.11c). The ψ_{ci} angle was set equal to 8° if MSD was present and 12° if not. Lead cracks also used ψ_{ci} of 8°. Once the crack was initiated from the saw-cut, the "normal" ψ_c was used in the simulation. Surprisingly enough, the same values of ψ_{ci} were found to correctly predict stress levels for the stable tearing initiation from precracks generated by cyclic pressure loading on curved fuselage panels with riveted joints (see Sect. 10.5.2). Because the profiles of a saw-cut and of a fatigue crack are very different, the above results can be a hint that the 3D CTOA behaviour during the stable tearing initiation is a primary reason of problems with predictions for the crack initiation stage.

A step towards the determination of the residual strength plot for a curved crack path and, at the same time, assessing a potential of tear straps for toughening a structure can be an investigation by Keesecker et al. (2003). In a residual strength test for an aluminium cylinder with two (external) circumferential tear straps, an initially longitudinal crack grew in a stable self-similar manner with increasing pressure until at the maximum pressure it bifurcated into branches running parallel to the tear straps. The crack bifurcation point location determined by the STAGS code under the assumption that the crack would grow along a path that provides the least crack growth resistance was consistent with the experimental observation. An extension of this concept for the case when the crack path is not prescribed *a priori* seems to be feasible. This could be achieved, however, at the cost of a largely increased computational effort because the crack turning angle must be found by trial and error looking for a direction requiring the lowest applied pressure to continue crack growth. In this way the crack preference for either growing through the tear strap or flapping ahead of the tear strap could be recognized. Like results by Kosai et al. (1999) addressed in Sect. 10.3, the STAGS code results by

Keesecker et al. (2003) corroborate the validity of the maximum tangential stress directional criterion for elastic-plastic conditions in a shell structure. It is shown that at the observed crack bifurcation point the directions perpendicular to the maximum principal stress at the crack tip on both the inner and outer surface are nearly aligned with the circumferential direction.

10.5 Comparisons Between Predicted and Measured Residual Strength of Fuselage Lap Joints for Self-Similar Crack Growth

10.5.1 Flat Panels

Anderson et al. (2004) validated through comparisons with test results residual strength analyses performed for large flat panels 1,219 mm wide and 2,285 mm long from Al 2024-T3 clad, 1.60 mm thick sheets containing typical fuselage riveted joints with MSD cracks in a critical rivet row. Two of the four joint geometries considered in this investigation represented longitudinal lap joints shown in Fig. 10.13. The doublers were made from 2024-T3 Alclad and the longerons were made from Al 7075-T6511. Three specimens were tested for each type of lap joint: one containing only a large crack and the other two containing multiple small cracks

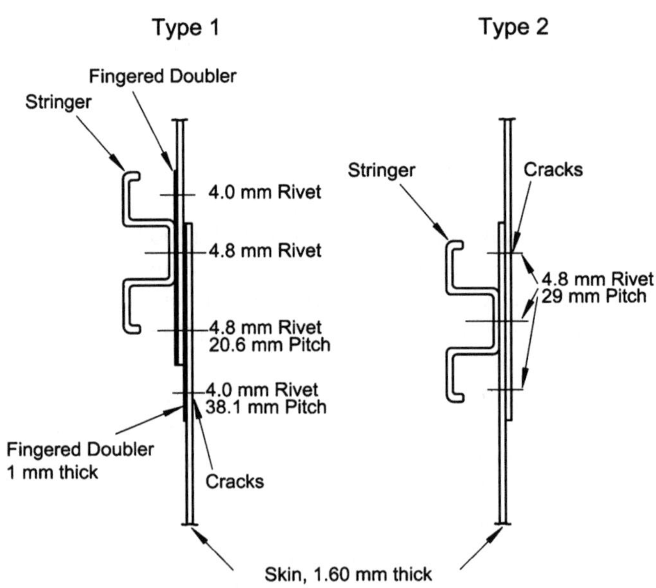

Fig. 10.13 Riveted lap joints geometries considered by Anderson et al. (2004)

in the fastener holes ahead of the large crack. The MSD crack lengths were 1.27 and 2.54 mm. The lead crack half length was 177 mm for the Type 1 panels and 163 mm for the Type 2 panels. The cracks were saw-cuts introduced prior to the riveting with the tip radius of 0.075–0.20 mm. The specimens were uniaxially pulled to fracture with guide plates to minimize lateral buckling effects.

The analyses were performed using three fracture criteria: PZL, T^*-integral resistance curve and CTOA. In the predictions based on the PZL criterion, FE solutions for global models with various lead crack lengths were used to create boundary conditions for local models that were then analyzed using the linear 2D-FEAM (cf. Sect. 10.4) to determine the β correction factors in Eq. 10.7. For panels with no MSD cracks $\beta_{MSD} = 0$ was assumed in Eq. 10.7, which implied that the lead crack plastic zone was touching the edge of the adjacent rivet hole. In agreement with the test data, the predicted second link-up for all panels occurred at the highest applied stress that was therefore referred to as the residual strength. Several values of the C-factor in Eq. 10.7 proposed in the literature have been allowed for in the predictions. That according to Ingram et al. (1999) appeared to provide most accurate results. For all four riveted joint geometries considered by Anderson et al. (2004) the predictions showed, in agreement with the experimental trend, that the size of the MSD rather than the size of the lead crack, was the dominant factor influencing the link-up and residual strength.

Again the two-stage hierarchical approach was applied for stable tearing and residual strength predictions based on the T^*-integral resistance curve criterion. The global models were analyzed using both a linear elastic and elastic-plastic FEM to generate the boundary conditions and bearing loads at the rivets needed in the local model analysis by the 2D-EPFEAM program. Stable tearing analyses for the panels without MSD cracks could not be performed since the latter approach did not have any provision to account for the case when, after the lead crack had met a fastener hole, the crack reinitiation occurred on the other side of the hole. The prediction accuracy was found to be highly affected by the T^*-integral resistance curve input data. Specifically, it was of key importance that the effectiveness of restraint on the out-of-plane displacements for laboratory specimens used to generate the resistance curve should be possibly the same as in the tests of the panels considered in the predictions. Another conclusion was that the elastic-plastic solution for the global model provided a considerably better correlation of the test data than the linear analysis. In the latter case, the prediction accuracy was worse than for the PZL criterion.

According to Anderson et al. (2004), the global-local procedure appears to be not feasible for the CTOA analysis using the STAGS code.[2] Hence, the global models used in the global-local approach for the PZL and T^* curve criteria were refined with the elements near the crack path of about 1 mm in size and translated into the STAGS format. The rivets were idealized according to Fig. 10.11c, which

---

[2]Chen et al. (1999) did apply the global-local approach utilizing the FRANC3D/STAGS code, see Sect. 10.4.

Table 10.1 Prediction errors (%) for the first link-up stress and residual strength in riveted flat panel analyses (Anderson et al. 2004)

		First link-up stress			Residual strength		
Panel type	MSD size (mm)	PZL	T^*	CTOA	PZL	T^*	CTOA
1	0	2	–	−1	−16	–	−12
	1.27	9	12	6	0	13	0
	2.54	6	1	3	0	7	−1
2	0	−1	–	−5	−15	–	−10
	1.27	−10	−6	2	−10	−2	−2
	2.54	−5	−1	2	2	−4	6

implied that a single slit was introduced to the model to simulate the fastener hole with or without MSD cracks. The values of CTOA for the stable tearing initiation (ψ_{ci}) were adopted, as described in Sect. 10.4. The ψ_c-value chosen for the crack growth predictions was obtained by a simulation of residual strength test results on laboratory specimens using the STAGS code. The CTOA criterion based predictions had an average error of about 3% for the first link-up stress and of 5% for the residual strength.

All three criteria investigated were able to predict σ_r within on average 10% of the experimental results for all four panel types, which can be considered a very plausible accuracy. The comparison of the errors obtained from all three methods (best prediction options) for the panels with the riveted lap joints is shown in Table 10.1. It cannot be overlooked that while the T^* curve and CTOA based methods are extremely time consuming and require an expert skill and sophisticated analyses to ensure the appropriate quality of predictions, the PZL criterion provides a quick and simple solution to assess the effect of MSD cracks with the use of a standard linear elastic FEM.

10.5.2 Curved Panels

Essential differences between a fuselage structure and a flat panel involve: a uniaxial uniform remote stress on a flat panel versus nonuniform biaxial stress in a fuselage structure; out of plane buckling for a flat panel versus fuselage skin bulging; free rotation of lap joints in a flat panel versus restrained rotation in a fuselage structure. In tests utilized to validate analytical procedures for residual strength predictions considered in this section large curved panels rather than full-scale barrels have been used. Special facilities were used to generate loading on the tested curved panels emulating actual load distribution in a fuselage barrel, e.g. Ilcewicz et al. (1992), Miller et al. (1994), Bakuckas (2002).

Comparisons between observed and computed results on residual strength for curved panels with longitudinal riveted lap joints have been provided in several works. Young et al. (1997) performed a residual strength pressure test and analysis

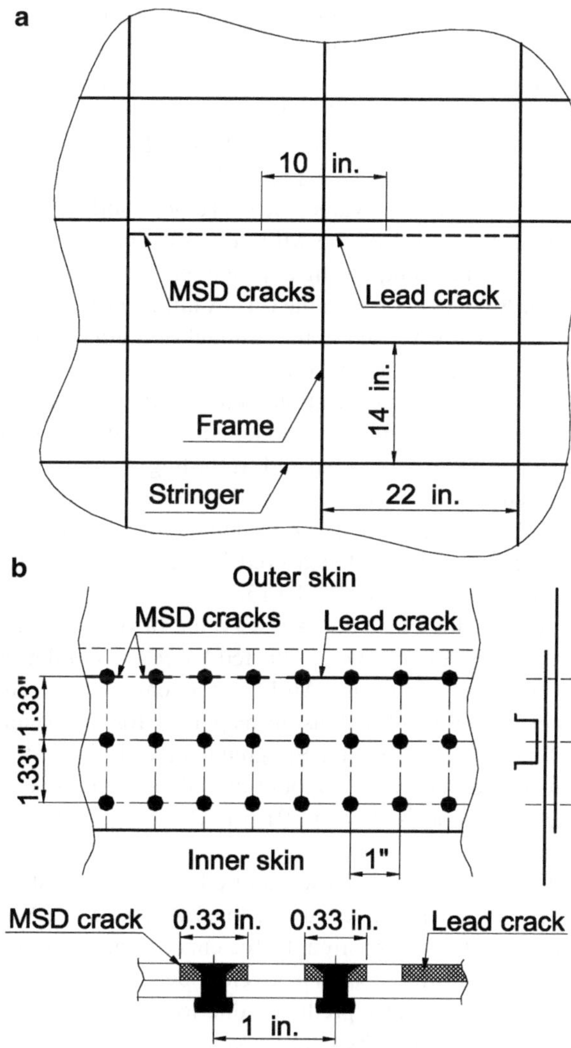

Fig. 10.14 Curved panel details: (**a**) configuration of lead crack and MSD; (**b**) rivet configuration in lap joint (Young et al. 1997)

of a fuselage panel with four stringers and three frames with a spacing shown in Fig. 10.14a. The panel overall dimensions included a 122 in. radius, a 72 in. length and a 63 in. arc width. The skin was 0.063 in. thick 2024-T3 Al alloy with the sheet rolling direction parallel to the stringers. The stringers and frames were also made from Al 2024-T3. Waffle tear straps from the same material, 0.04 in. thick, bonded to the skin were located under the stringers and frames. The stringers and frames were riveted to the skin and tear straps. The frames were connected to the stringers by riveted clips. A lap joint of a configuration shown in Fig. 10.14b consisted of three rows of 0.125 in. diameter countersunk rivets. The initial damage of the panel consisted of a 10 in. long lead crack centred on a severed frame and MSD cracks that extended 0.05 in. on each side of the fastener countersink, Fig. 10.14. The cracks were introduced in the outer skin critical rivet row by saw-cutting prior to the panel assembly. The test was performed using a special pressure-box test machine. The loading condition for the panel was a combination of internal pressure and axial tension as an equivalent to the bulkhead pressure load in a closed pressurized cylinder. No visible crack growth for pressure below 9.95 psi was observed. When pressure reached this value, the lead crack suddenly extended on each end and linked up with the series of the MSD cracks. The crack extended over the entire panel length in an instant, failing adjacent tear straps and frames.

The STAGS code with the CTOA failure criterion and the PSC core (cf. Fig. 10.7) implemented was used in the residual strength analysis. Due to symmetry about the middle frame centre line, only a half-length symmetry model for the panel could be used. The entire cross section including the middle frame was modelled using plate elements. The rivets were modelled as beams and their connections with the outer and inner skins was idealized according to Fig. 10.11b. In locations of the MSD cracks the elements in the outer skin were disconnected and the rivets were attached only to one side of the crack, as detailed in Sect. 10.4. A good correlation was found between predicted and measured axial and hoop strains at three locations around the lead crack tip region, Fig. 10.15.

A ψ_c-value of 5° and a PSC-value of 0.04 in. needed for the residual strength analyses by STAGS were selected based on simulations of tearing tests performed on laboratory C(T) and M(T) specimens of different widths. The predicted and experimental crack extension responses for the panel are compared in Fig. 10.16 as functions of pressure. In agreement with the test results, the analysis indicated that only a very small amount of stable tearing occurred with a transition to final fracture at a pressure of 11.01 psi. The analysis overestimated the residual strength by 11%.

Figure 10.17a, b show reported by Chen et al. (1999) comparisons between observed and predicted using the FRANC3D/STAGS software results on stable tearing and residual strength for curved panels. Information on the design of the panels and testing details is given in Sect. 10.4. Though the stable tearing initiation pressure is correctly predicted due to accounting for the residual plasticity in the crack wake induced by fatigue loading, as detailed in Sect. 10.4, the residual strength in Fig. 10.17a is considerably overestimated. ψ_c-values of 4.5°and 5.5° used in the analyses are within the scatter of the measured critical CTOA (5.5°±1°). The simulated plastic wake increased the predicted σ_r by 3–9%. The authors claimed that the occurrence of failure of other structural components was a possible cause for the low σ_r observed in the experiments. Specifically, they considered the breakage of the inner tear strap to be probable as the local fuselage model showed net section yielding due to the high stress concentration at the rivet holes. Figure 10.17b shows the predicted results for the fuselage model with the tear strap cut prior to the crack closure and residual strength analyses. The correlation with the experiment has improved, especially with ψ_c of 4.5°. In that case the computed σ_r was within 13% and 1% of the observed value for the panel without and with MSD respectively. Additional analyses were performed in which the inner tear strap stayed intact until

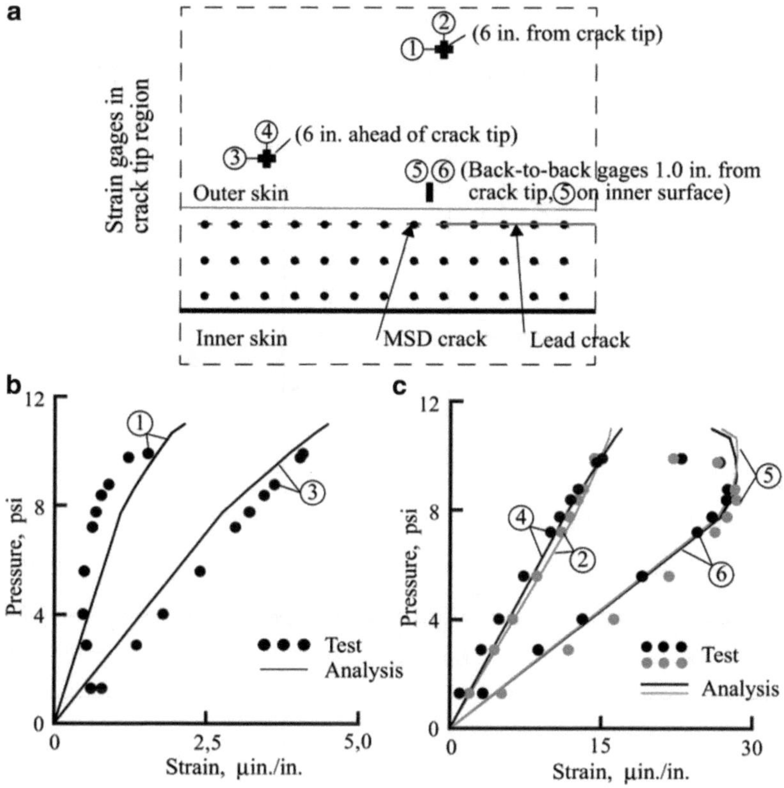

Fig. 10.15 Predicted and measured strains in curved panel skin: (**a**) strain gauge locations; (**b**) axial strain; (**c**) hoop strain (Young et al. 1997)

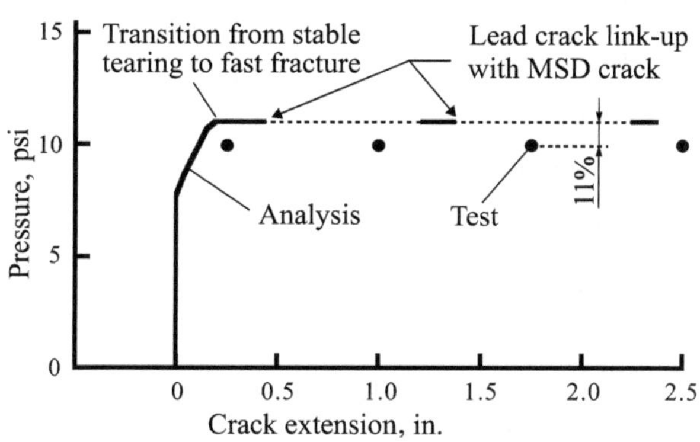

Fig. 10.16 Curved panel test-analysis correlation of crack extension as function of pressure (Young et al. 1997)

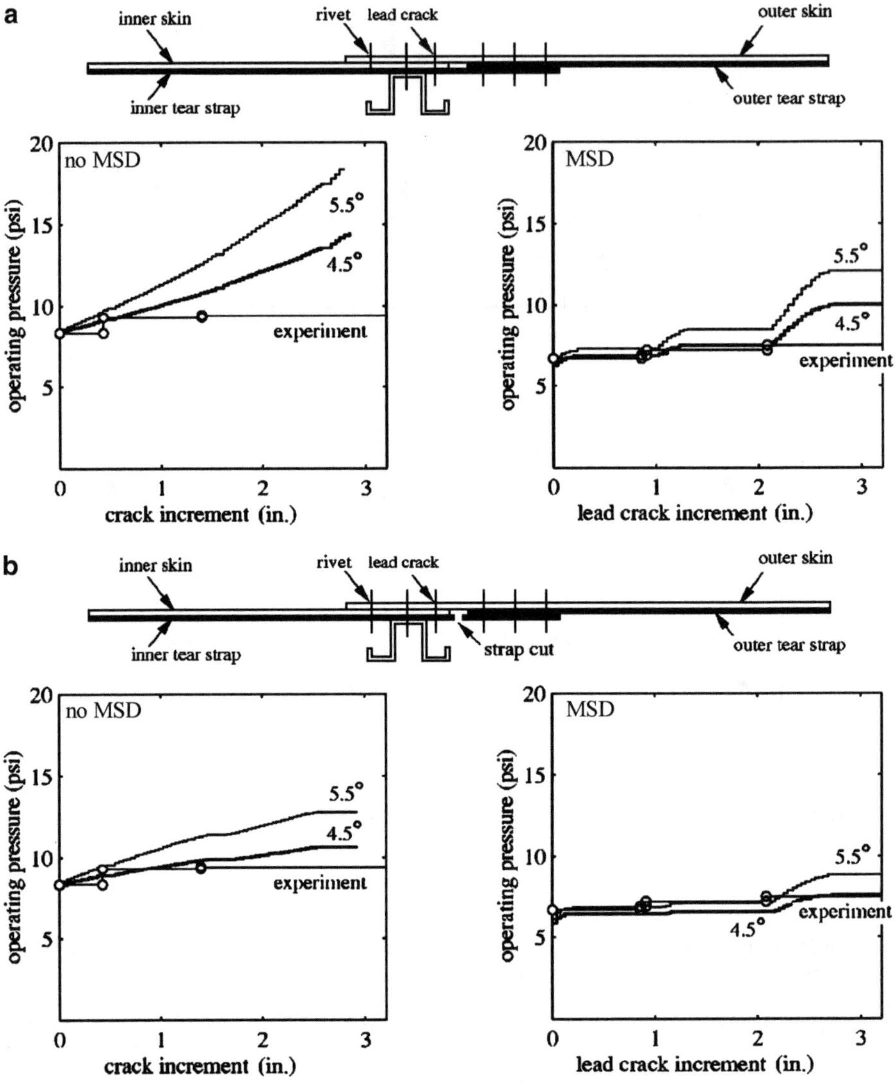

Fig. 10.17 Comparison between predicted stable crack growth and residual strength for curved fuselage panels without and with MSD cracks: (**a**) tear strap intact; (**b**) tear strap broken (Chen et al. 1999)

a certain amount of stable crack growth. Next, the simulation was continued with the broken tear strap until the residual strength was reached. The tear strap damage scenario was found to have a very mild influence on the prediction results.

Anderson et al. (2004) performed residual strength predictions for full-scale curved fuselage panels containing riveted longitudinal lap joints the same as in the Type 1 flat panel, Fig. 10.13. The curved panel test specimen is shown in Fig. 10.18. The panel overall size was about 120 in.×68 in. with a radius of 66 in., a stringer

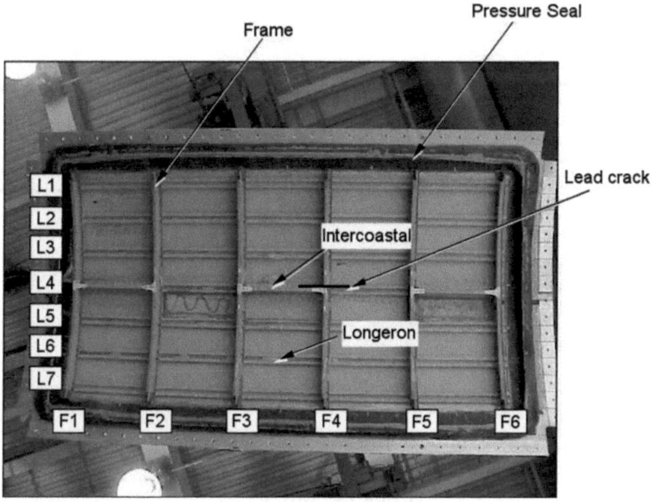

Fig. 10.18 Curved panel test specimen in experiments by Anderson et al. (2004) and Hsu et al. (2003)

spacing of 7.5 in. and a frame spacing of 19 in.. The skin and the doublers were made from 2024-T3 Alclad whilst Al 7075 was the frame, stringer and shear clip material. The CVP-1 panel contained only a large lead crack in the critical fastener row and the CVP-2 panel contained a lead crack and MSD cracks. The latter were introduced in fastener holes along the path of the lead crack up to a distance of one and a half bay width from frame F4 at which the lead crack was centred, Fig. 10.18. The 7 in. long lead cracks installed using a jeweller's saw and sharpened with diamond-coated wire were propagated to the predetermined length of about 25 in. via fatigue loading. The MSD cracks 0.05–0.15 in. in size were introduced in the same way after the fastener holes had been drilled. Frame F4 and the shear clip directly above the lead crack were artificially severed to simulate a two-bay skin crack. The testing was performed using the FASTER facility described in Sect. 8.2.4. The panel specimens were subjected to a combination of simulated cabin pressure and hoop and longitudinal loads simultaneously. The latter loads were assumed as for a pressurized cylinder which implied: total hoop load = (applied pressure) × (panel radius) × (length of pressurized area) and total longitudinal load = 0.5 × (applied pressure) × (panel radius) × (width of pressurized area). Internal pressure was applied using water as the medium, and the hoop and longitudinal loads were applied via servo-controlled actuators. The hoop load distribution between the skin and the frames was based on a geometrically nonlinear FE analysis. The loading system allowed the panel to pivot about the centre and remain free to expand tangentially, in the same way as a fuselage cylinder does in reality. During the residual strength tests the applied loads in terms of the simulated cabin pressure were incrementally applied until catastrophic failure occurred.

The residual strength analyses employing the PZL and CTOA failure criteria were performed using the STAGS program. The modelling of structural features

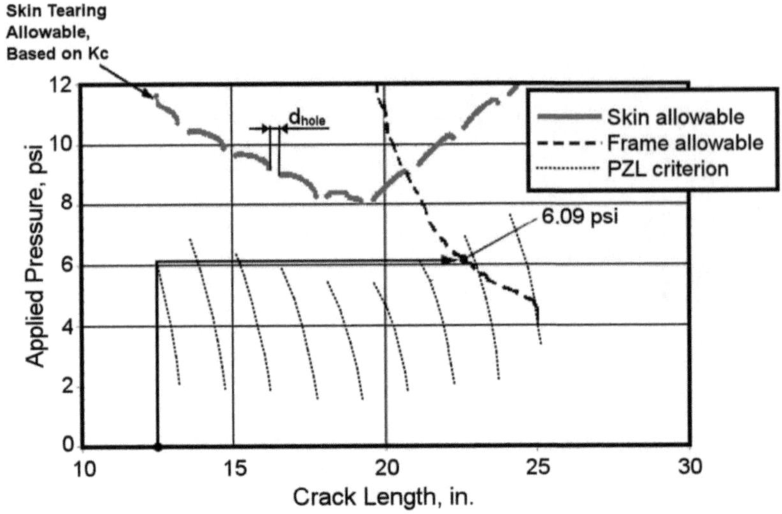

Fig. 10.19 Residual strength prediction for panel CVP-1 based on the PZL criterion (Anderson et al. 2004)

and, specifically, of the rivets was the same as in the case of the flat panels considered in Sect. 10.5.1. Although the lead crack was not located at the centre of the panel (cf. Fig. 10.18), the test results indicated that the cracks propagated symmetrically about the severed frame. Thus only one half of the panel was modelled. The SIFs of both the lead crack and the MSD cracks required for using the PZL criterion were computed based on the energy release rate given in the STAGS output (cf. Eq. 10.6). To obtain the correct elastic G (mode I), linear elastic material was assumed, but nonlinear geometric effects, such as skin bulging were accounted for. The stress in the resulting K_I-solution was the nominal hoop stress under unit pressure. Again the C-coefficient values in Eq. 10.7 according to several researchers have been considered in the predictions. In the case of panel CVP-1 (lead crack only), $\sigma_{LINK\text{-}UP}$ was computed as described earlier for the riveted flat panels.

Failure of a complex stiffened panel can be initiated in skin, stiffener, or fasteners. A failure in any of these components may result in failure in another component instantaneously, which can lead to total failure of the panel. Predictions based on the PZL criterion for panel CVP-1 obtained using the C-coefficient according to the original Swift concept are shown in Fig. 10.19, where the hoop stress has been converted to cabin pressure. The presentation is analogous with that in Fig. 10.3. Without considering the MSD cracks (rivet holes in the case of CVP-1), the skin fracture-allowable stress can be calculated using the K_C criterion, $K_C = 150\ \text{ksi}\sqrt{\text{in}}$ being adopted in Fig. 10.19. The link-up stress is shown in Fig. 10.19 as a group of semi parallel curves (dotted line), each starting from the right side of the corresponding fastener hole. It was found that the maximum stresses in the frame are located at the point (frame cap) where it connects to the stringer. The frame allowable stress curve (dashed line in Fig. 10.19), was derived from a separate FE analysis assuming that the frames would fail when the average stress in the frame cap reached the allowable stress of the material. The frame allowable stress was computed as $P_{\mathrm{allow}} = P_{\mathrm{apply}}(F_{\mathrm{TU}}/\sigma_{\mathrm{avg}})$, where P_{apply} is the applied cabin pressure, F_{TU} is the material allowable stress, and σ_{avg} is the average stress of the (usually two) elements at the frame cap in the direction parallel to the frame. As shown in Fig. 10.19, the lead crack (half length of 12.5 in.) propagated when applied pressure reached the level of 6.09 psi. The propagation was continued through the successive six fastener holes. The frame failure had occurred before the crack could be arrested by the next (seventh) fastener hole. Failure of the frame and skin simultaneously resulted in total failure of the panel. Again the C-coefficient according to Ingram's modification resulted in an improved prediction of σ_{r} compared to the original Swift concept, but the computed results were still overly conservative: the predicted 7.45 psi versus observed 11.14 psi for CVP-1, and the predicted 7.28 psi versus observed 9.16 psi for CVP-2. According to the authors, a likely reason for the week correlation with CVP-1 and CVP-2 was the effect of skin bulging. The SIF obtained for a unit load was scaled up linearly for other load levels. Since the bulging factor is a nonlinear function of applied pressure, the approach resulted in overestimated K-factors and conservative predictions for σ_{r} ensued. Indeed, the authors demonstrated that the PZL-criterion produced significant correlation with the experiment for two curved panels with circumferential riveted butt joints where skin bulging was not an issue.

Predictions based on the CTOA criterion were performed using the same methodology as for the flat panels (cf. Sect. 10.5.1). Again the PSC with a half height equal to the sheet thickness was incorporated in the STAGS code and the same ψ_{ci}-values were assumed.[3] The predicted results on skin tearing, Fig. 10.20a, b, indicate that the skin was capable of carrying higher loads once the crack had passed the adjacent frame (F5, about 20 in. from the central frame, see Fig. 10.18). This prediction was made on the assumption that the adjacent frames remain intact. As with the PZL criterion based analyses (cf. Fig. 10.19), the residual strength of the panel was defined as the intersecting point of the skin tearing curve and the frame allowable curve. Depending on the element size and the number of elements used for calculating the average stress at the frame cap (σ_{avg}), the error in predictions on σ_{r} was found to vary from -16% to $+5\%$.

10.6 Comparisons Between Observed and Predicted Effect of Tear Straps on Crack Path

Comparisons between predicted and observed curving of cracks propagating from a precrack along a lap joint in a pressurized fuselage are lacking in the open literature. Kosai et al. (1992) simulated crack flapping in an aircraft fuselage reinforced by stringers, floating frames and tear straps. The skins and straps were of 0.9 mm thick

[3]See the comment to the Hsu et al. (2003) study in Sect. 10.4.

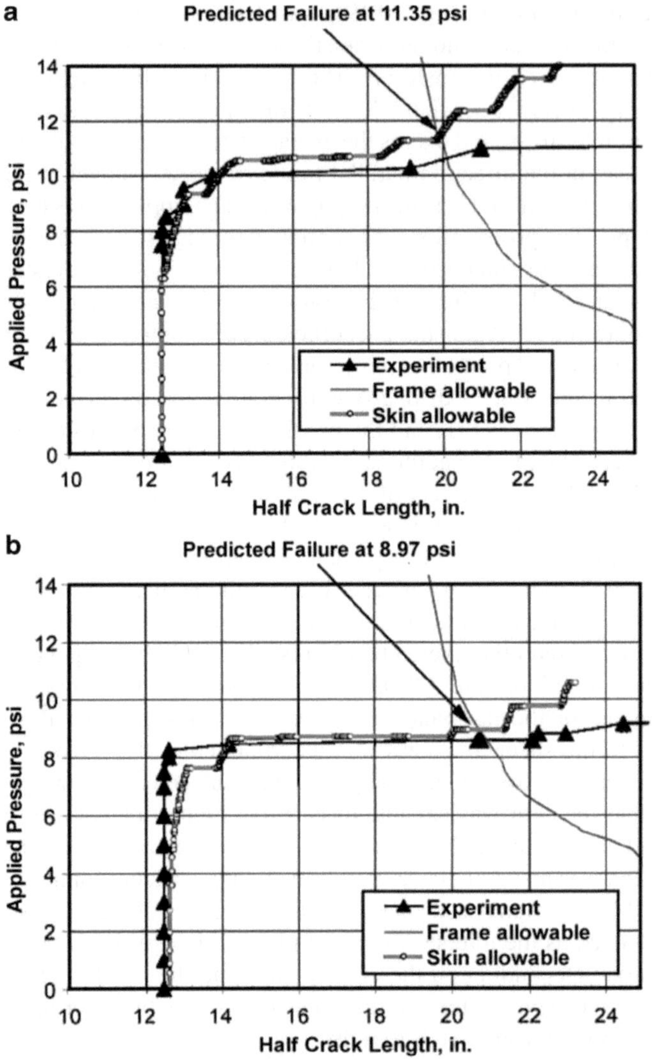

Fig. 10.20 Residual strength prediction for curved panels based on the CTOA criterion: (**a**) panel CVP-1; (**b**) panel CVP-2 (Anderson et al. 2004)

2024-T3 Al alloy sheet. The fuselage diameter of 3,759 mm, tear strap spacing of 254 mm and frame spacing of 508 mm was assumed. In order to account for the unsymmetrical deformation relative to the longitudinal direction for a crack in a longitudinal lap joint (cf. Sect. 10.3), one half rather than one quarter of the airplane fuselage was modelled with an implicit assumption that the axial crack centred at a frame will develop symmetrically relative to the frame in the other half of the fuselage. The stiffeners were modelled by beam elements and the straps were represented by increasing the skin thickness appropriately. An incremental elastic-plastic FE analysis was initiated at a half crack length of 178 mm and the crack growth criterion according to Eq. 10.12 was triggered as the crack tips were at a distance of 25.4 mm (equal to the tear strap width) from the tear strap edge. The K_{I} and K_{II} SIFs were computed using a linear elastic FE method accounting for geometric nonlinearity. Unfortunately, analysis details about the T-stress estimate method, determination of the r_{c} distance, MSD distribution and modelling, and accounting for the effect of MSD on the SIFs are lacking in the paper. The computations predicted a first kinking with an angle of 40° at a location of 25.4 mm before the tear strap. Subsequently the crack propagated in a straight line after which another kink of 40° occurred very close to the tear strap. The authors claim that their predicted results on crack flapping agree qualitatively with those observed by Maclin (1991), cf. Sect. 10.3.

Comparisons between measured and computed crack paths in fuselage structures have been reported only for cracks located in skin close to a stringer. As said earlier, for such a crack location a one-side flapping on the crack edge side opposite to the stringer can occur. Chen et al. (2002) evaluated the tangential stress criterion capability of predicting non-self similar crack growth observed in a full scale generic narrow-body fuselage panel with tear straps, stringers, stringer clips and frames under cyclic internal pressure. The initial saw-cut was centred on the midbay tear strap completely through both the skin and the tear strap. Geometrically nonlinear analysis was performed with pressure loading applied on the skin of the shell model and symmetric boundary conditions imposed on all the boundary edges to simulate a cylinder-like fuselage structure. Uniform axial expansion was allowed at one longitudinal end. The SIFs K_{I} and K_{II} were evaluated using the modified closure integral method, while a simple displacement correlation method was used to compute the T-stress. Crack growth trajectories predicted ignoring and allowing for the effect of fracture toughness orthotropy, i.e. according to Eqs. 10.12 and 10.14 respectively, are shown in Fig. 10.21 together with the measured results. It is seen that neither prediction option was capable to fully correlate with the experimental data. The simulations including fracture orthotropy match pretty well the earlier crack growth (note the nonsymmetrical measured crack paths on both sides of the centre tear strap) but fail to predict flapping as the crack grows further into the tear strap region. The reverse is true for the isotropic case predictions. Among possible reasons for the prediction inadequacy the authors mention different orientations of the material in the tear strap and in the skin (ignored in the simulations), approximations in modelling the structure (neglecting the 3D effects in the vicinity of the tear strap) and inaccuracies in the T-stress determination.

Kosai et al. (1999) applied the directional criterion according to Eq. 10.12 to predict crack growth trajectories observed in residual strength tests conducted by Miller et al. (1994) on full scale narrow body fuselage panels and referred to in Sect. 10.3. The numerical procedure used was the same as applied for the pressurized cylinder also considered in this work and was described in Sect. 10.3. A

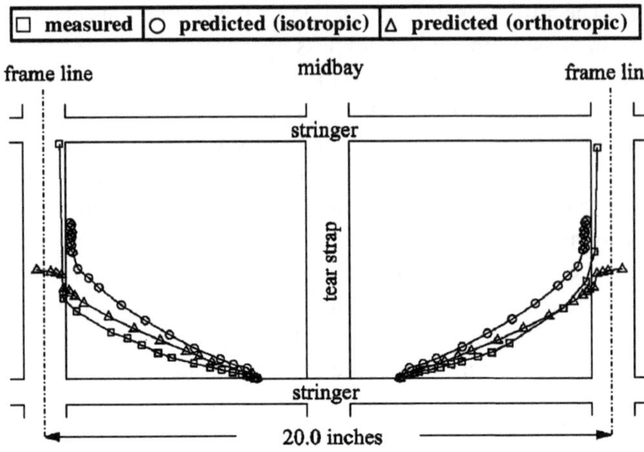

Fig. 10.21 Comparison between predicted and measured crack trajectories in a curved fuselage panel (Chen et al. 2002)

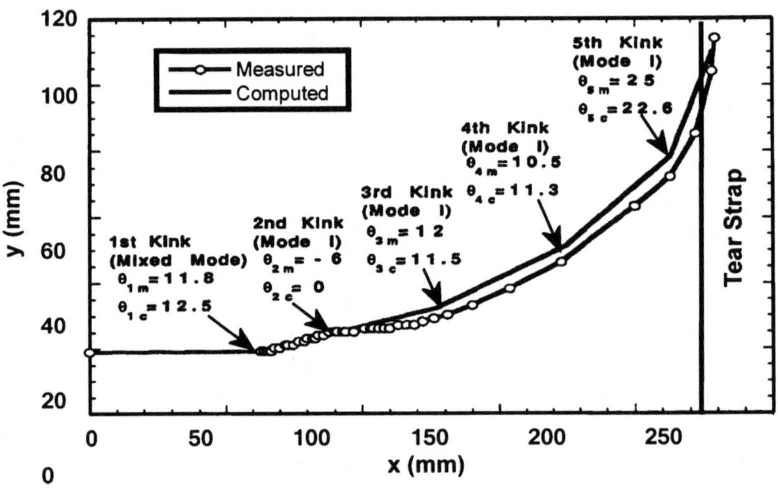

Fig. 10.22 Comparison between measured and computed crack growth trajectories for a fuselage panel representative for narrow-body Boeing aircraft (Kosai et al. 1999)

very good agreement between the measured and computed crack growth trajectories was obtained, as shown for one of the panels in Fig. 10.22, where the indices m and c refer to the measured and computed crack growth directions respectively. Interestingly, only the first kink occurred under mixed mode conditions at the crack tip, i.e. with a non-zero K_{II}, and the successive kinks occurred under pure mode I loading ($K_{II} = 0$) because the condition given by Eq. 10.13 was satisfied.

10.7 Structural Risk Analysis

Structural risk analysis is an aircraft fleet management tool that is used for decisions regarding the timing and extent of inspection, repair, or replacement maintenance actions on critical structural details. Because the major contributor to the risk of structural failure is fatigue, most methods of risk assessment involve modelling the effect of fatigue crack growth by some probabilistic method. As reviewed in Chap. 9, fracture mechanics provides deterministic models that predict crack growth under a given load sequence from an initial size to critical size. By introducing probabilistic descriptions of the factors that produce different initiating conditions and crack growth in the population, the results from deterministic approaches can be extended to quantify the degree of safety during an operating period. Typically, stochastic input data for the risk analysis are associated with a variability in material properties and crack growth rates, the initial fatigue quality of a component, inspection reliability and repair quality. A complementing of the deterministic damage tolerance analysis by incorporating stochastic elements is most commonly achieved by two methodologies, namely the conditional reliability method, which involves calculating the probability of failure (POF) using numerical integration, and the Monte Carlo simulation.

The best known analysis tool based on the conditional reliability method is a PROF (Probability of Fracture) software developed by the USAF for structural risk assessments (Berens et al. 1991, 1994; Hovey et al. 1998). A single execution of PROF produces the POF for a population of structural elements being potential cracking sites. All elements in the population experience equivalent load histories and have equivalent SIF correction coefficients. Fracture is assumed to occur when the applied maximum stress produces a SIF that exceeds the fracture toughness of the cracked detail. The stochastic input data to PROF include distributions of the initial crack size, maximum stress per flight, fracture toughness and, in the case of application after a maintenance action, the probability of crack detection function that characterizes the inspection system capability. These inputs are in a tabular form and it is crucial that the tails of the distributions should be well defined. Sensitivity analyses carried out by Berens et al. (1991) indicated that these tails were critical in the calculation of the POF.

Given an initial distribution of crack sizes at a reference time T_R, the programme estimates the distribution of crack sizes at $T_R+\Delta T$ flight hours for the expected stress sequence. Variability in crack growth rates is neglected, which implies a deterministic crack size versus flight hour relation derived from a crack growth prediction model. The calculation is repeated for all percentiles of the initial crack size distribution. If for a virgin aircraft T_R represents a start of the usage, the initial crack size input data is the EIFS distribution, which is dominated by the manufacturing quality. When the analysis is performed after a maintenance action, the so called equivalent repair crack size distribution is used as the initial crack size input. This equivalent repair crack size distribution is determined from the prior to maintenance crack size distribution changed in accordance with the probability of crack detection function and the repair quality. The POF calculated for a stress sequence and distribution of maximum stress per flight representing the mean usage severity corresponds to a detail in a randomly selected aircraft of the fleet. If POF were computed for stress sequences and maximum stress per flight distributions representative of the percentiles of the usage severity distribution, a distribution of POF would be generated. These POF values could be interpreted in terms of the population of individual aircraft usage.

Another problem, referred to as discrete source damage, that can be solved from a single PROF run is estimating the conditional POF of an intact structural element given partial failure in an adjacent structural member. A prototype for this analysis can be the appearance of a two-bay crack in fuselage skin. The conditional probability that the crack-stopping structures on either side of the crack will fail provides a measure of the ability of the aircraft to complete the mission. Due to the cracking, load can be redistributed from the cracked component to the intact component (or components) whose residual strength becomes, therefore, a function of the crack size. The overall POF at time T is determined as the product of the probability density function of crack sizes in the cracked element at T and the probability that the residual strength for a given crack size is exceeded at the maximum stress in flight. Such a change in the fracture criteria requires the residual strength versus crack size (deterministic) input data, in addition to the usual input data for a standard PROF run described earlier in this section.

There are many structural details for which conditions are not constant across an entire aircraft fleet and the effect of these conditions can be modelled through multiple runs of PROF. To this end, the total population of details is divided into sub-populations of equivalent geometry, material and stress. Conditional fracture probabilities are then calculated for each sub-population through a separate PROF run represented by the equivalent conditions. When the relative frequencies of occurrence (probabilities) of the conditions are also known, the conditional fracture probabilities can be combined to provide an overall (unconditional) fracture probability at time T as

$$\mathrm{POF}(T) = \sum \mathrm{POF}\,(T/C_i) \cdot P\,(C_i), \text{ with} \sum P\,(C_i) = 1 \tag{10.15}$$

where POF(T/C_i) is the probability of failure at T, given that condition C_i was used to predict crack growth, and $P(C_i)$ is the probability that condition C_i applies.

Facilitating this type computations was met by incorporating in PROF an Excel Workbook interface for a multi-run data management capability, which allows the programme to perform multiple runs in a single analysis (Hovey et al. 1998). Presented in the latter report is an exemplary application of the formulation according to Eq. 10.5 to risk analysis for a riveted lap joint with a single crack originating from the central rivet hole in the top rivet row according to two patterns, namely from both sides of the hole and from one side. Failure risks in the presence of MSD in a riveted joint cannot be currently reliably evaluated within the PROF methodology, mainly because of lack of good approaches to estimating the distribution of MSD cracks in the joint.

A probabilistic model by Sampath and Broek (1991) for MSD fatigue damage predictions in lap joints (cf. Sect. 9.5) was conceived to analyze the probability of detection of MSD cracks in lap joints. The objectives were to gain an insight into the cumulative probability of detection of cracks leading to failure (link-up) and to arrive at suggestions for a "safe" inspection interval. Similarly as the stochastic input data for the MSD analysis model considered in Sect. 9.5, the probability of detection curve is described by a curve fitting (Weibull) equation. If the predicted crack growth curve for a given location and the probability of detection curve are known, the probability of detection of a crack at that location at a certain inspection i (p_i) can be determined. The cumulative probability of detection after n inspections is $P_n = (1-p_1)\cdot(1-p_2)\cdot\ldots\cdot(1-p_n)$ with $p_n > \ldots > p_2 > p_1$ because in each consecutive inspection the crack is longer and, therefore, the probability of its detection higher. Sampath and Broek demonstrate that despite a considerable statistical variability of the MSD analysis results inherent in their model (cf. Sect. 9.5), the computed cumulative probability of detection appears relatively insensitive to statistical parameters and assumptions. For example, for an inspection interval of 3,000 flights, the cumulative probability of detection of the very largest crack is practically equal to 1 in most cases. For the same inspection interval, several dozen runs of the analysis with different statistical parameters yielded values of the cumulative probability of detection of the second largest crack between 0.92 and 1 though the predicted lives to failure widely differed (from 12 to 37 kcycles). Of significance was a precipitous drop in the computed probability of detection versus inspection interval curve when the inspection interval exceeded 3,000 flights. Sampath and Broek did not present any piece of validation of their conceptions.

The Monte Carlo approach is considered to be the most appropriate structural risk analysis tool suitable for MSD crack growth problem in lap joints. The Monte Carlo simulations involve performing a large number of deterministic analyses starting from initial conditions selected at random from the distribution appropriate to the input variable. This is repeated until a sufficiently large random sample of failed and unfailed items has been simulated. An item is failed if the limit state function is violated, i.e. $G(\hat{x}_i) \leq 0$, where $\hat{x}_i$ is a sample value for each random variable x_i. The POF can be directly determined from the failed and unfailed items proportion as

$$\mathrm{POF} = \frac{n[G(\hat{x}_i) \leq 0]}{N}$$

where n denotes the number of trials when the limit state is violated and N is the number of all trials.

It is obvious that the number N of trials required is related to the degree of accuracy and degree of confidence level required for the POF. For transport aircraft structure the POF per flight of 10^{-7} is generally acceptable. The number N of simulations for a 95% confidence level in the POF should be at least three times greater than the inverse of the cumulative failure probability per aircraft safe life. Because for transport aircraft the acceptable value of the latter probability is in the order of 0.001, the required number N of trials should exceed 3,000.

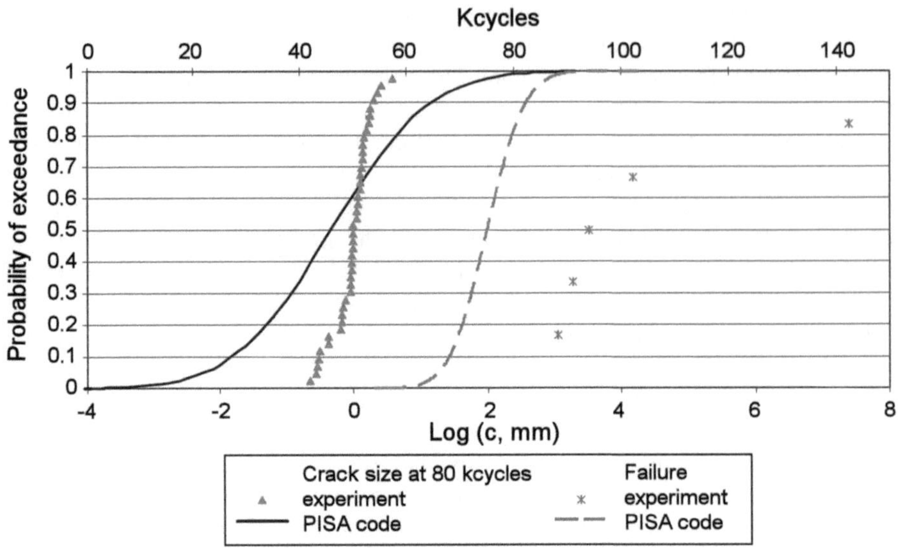

Fig. 10.23 Predicted and experimental results on probability distributions of the crack size at 80 kcycles and life to failure for a riveted lap joint (Cavallini and Lazzeri 2012)

Deterministic crack growth prediction models for MSD crack growth at rivet holes utilized in the Monte Carlo approach were addressed in Sect. 9.5. Though the risk analysis methodology applicable to MSD in mechanically fastened lap joints has been developed, its validation through experiments remains extremely limited. Figure 10.23 shows a comparison between probability distributions of the crack size at 80 kcycles and of the life to failure calculated by Cavallini and Lazzeri (2012) using their PISA code and experimental data for 300 mm wide flat panels from Al 2024-T3, 2 mm thick sheets containing three-row lap joints with countersunk rivets 4.8 mm in diameter. The CA loading in these fatigue tests was the same as in tests on simple coupons carried out to obtain the EIFS distribution used in the Monte Carlo simulations incorporated in the PISA software. The details of the Monte Carlo simulations have been described in Sect. 9.5. Figure 10.23 indicates that the predicted and experimental probability of exceedance curves for the crack size at 80 kcycles intersect for *c* of about 1 mm, which implies that the computations provide a lower than observed probability of occurrence for cracks above that size. On the other hand, the simulated results on the probability distribution of lives to failure are overly conservative.

Attempts to account for MSD in risk analysis have been made also in a "Structural Maintenance of Ageing Aircraft" (SMAAC) programme (Trey and Schmidt 2001), which is based on the Monte Carlo simulations addressed in Sect. 9.5. Comparisons between results computed using that tool and experimental observations have not been reported in the open literature.

10.8 Summary of This Chapter

Fracture behaviour of fuselage panels with MSD cracks in a rivet row depends on the structural configuration (rivet spacing in row, presence of crack stoppers, MSD pattern) and is relatively insensitive to the size of the MSD cracks. For a given configuration, the size of MSD rather than the size of a lead crack is the dominant factor influencing the link-up and residual strength.

Residual strength analyses for pressurized aircraft fuselage structures should account for essential characteristics of the very last stage of crack growth which include large scale yielding, a substantial amount of crack extension, and out-of-plane deformation (bulging at the crack edges). Hence, EPFM parameters are the most suitable fracture criteria.

The CTOA is currently deemed an EPFM parameter most appropriate to control stable tearing and fracture behaviour of flat or curved thin panels containing a single lead crack or a lead crack and MSD. However, the technique used to measure the critical CTOA remains in a stage of development, and the measured data are still subject to interpretation by the researchers. Available analysis results indicate that it is not feasible to simulate stable tearing for either simple laboratory specimens or complicated fuselage panels with an acceptable accuracy when a single critical CTOA-value is used. A probable reason for it seems to be variations of the critical CTOA along the crack front that occur in the initial phase of stable tearing in thin sheets due to crack tunnelling.

In the current stage of development, the specialized code employing the T^*-integral fracture criterion imposes various limitations, when applied to the analysis of complex fuselage structures, because the local model involved can only deal with a planar problem. In addition, the approach does not have any provision to account for the presence of rivets without cracks ahead of a lead crack.

Even if streamlined, single-process analytical tools are used, procedures for estimating residual strength of a complex structure employing the CTOA or T^*-integral fracture criterion remain extremely time consuming. Substantial computing resources to perform sophisticated FE analyses, expert skill and knowledge of fracture theories are required to ensure the quality of the predictions. The results are very sensitive to modelling techniques and the prediction accuracy is highly affected by various analyst's choices and assumptions.

Application of the PZL criterion requires only linear-elastic FE analyses. In the case of curved panels nonlinear geometric effects, such as skin bulging, must be accounted for. Compared to EPFM residual strength criteria, the PZL criterion provides a quick and simple solution to assess the effects of MSD cracks on the residual strength of a structure. For riveted flat panels, predictions on the link-up stress and residual strength obtained from the PZL criterion empirically modified (based on experiments on simple laboratory specimens) were not less accurate than those obtained from the T^*-integral based analysis. For the curved panels the PZL criterion considerably underpredicted the residual strength. This inadequacy was attributed to scaling up linearly with load the stress intensity factor obtained for a unit load only.

Only CTOA criterion based analyses were shown to provide acceptable prediction results for complicated curved panels. It cannot be, however, overlooked that this satisfactory accuracy was attained due to experimental results known in advance, which made it possible to select the best analysis option.

Comparative studies available in the literature are limited to a specific configuration when a lead crack is much larger than MSD cracks. MSD without a lead crack, namely the case when the largest crack size is of the same order of magnitude as the rest of cracks, as has been observed in aging aircraft, has been addressed neither in residual strength tests nor predictions.

The MSD cracks are typically simulated by saw-cuts at fastener holes. Laboratory tests on simple specimens indicate that the crack tip geometry and – in the case of a precrack generated by fatigue loading – also the fatigue load amplitude affect the stable tearing behaviour, especially in the initial phase. Hence, using in residual strength tests specimens with naturally generated MSD cracks would be a more realistic option.

Reported comparisons between observed and predicted residual strength for riveted fuselage panels are limited to cases when the initial configuration of cracks in a critical rivet row is the same for the analysis and the experiment. In real-work predictions it is necessary to estimate residual strength after a certain period of aircraft usage. In that case, the initial crack configuration in a rivet row is not known and needs to be determined using a crack growth prediction methodology. To this end, the EIFS concept could be utilized. EIFS-values randomly picked up from an appropriate statistical EIFS distribution must be randomly assigned to each rivet hole in a row and predictions of multiple fatigue crack growth until a preset number of loading cycles must be made. The resulting input data for the residual strength analysis would be generated as a variety of initial crack configurations in a row. Monte Carlo simulations will be, therefore, required both in the crack growth prediction and the residual strength prediction phase. Though conceivable, the above methodology is currently difficult to be effected due to its high degree of complexity, especially in the case of realistic, curved fuselage panels.

The occurrence of crack curving at a tear strap is conditioned by the structure design and the severity of MSD. The more severe MSD, the more preference for self-similar crack growth. The maximum tangential stress criterion with T-stress accounted for is considered to be most suitable for predicting the crack growth direction for pressurized stiffened cylinders. This criterion is strictly valid for LEFM conditions, but certain results suggest that it is applicable in the case of geometric and material nonlinearity. Concepts to predict the residual strength for non-self-similar crack growth, and therefore to evaluate the potential of tear straps to arrest a crack, are still under development.

The Monte Carlo approach is currently considered to be the most appropriate structural risk analysis tool suitable for the problem of MSD crack growth in lap joints with mechanical fasteners. However, its validation through experiments remains extremely limited.

References

Abbreviations

ICAF	International Committee on Aeronautical Fatigue
NLR	National Aerospace Laboratory, The Netherlands
NRC	National Research Council Canada
TU Delft	Delft University of Technology

Ahmed, A., Bakuckas, J.G., Jr., Tan, P.W., Awerbuch, J., Lau, A.C., Tan, T.-M.: Growth of multiple-site damage (MSD) in an undamaged fuselage lap joint curved panel. In: Proceedings of 7th Joint FAA/DoD/NASA Conference on Aging Aircraft, New Orleans, LA, 8–11 Sept 2003

Ahmed, A., Bakuckas, J.G., Jr., Averbuch, J., Lau, A.C., Tan, T.-M.: Evolution of multiple-site damage in the riveted lap joint of fuselage panel. In: Proceedings of 8th Joint NASA/FAA/DoD Conference on Aging Aircraft, Palm Springs, CA, Jan 31–Feb 3 2005

Aircraft Accident Report, Aloha Airlines, Flight 243, Boeing 737–200, N73711, near Manui, Hawaii, April 28, 1988, Report No. NTSB/AAR-89/03. National Transportation Safety Board, Washington, DC (June 1989)

Anderson, T.L.: Fracture Mechanics: Fundamentals and Applications, 3rd edn. CRC Press, Boca Raton (2005)

Anderson, B.L., Hsu, C-L., Carr, P.L., Lo, J.G., Yu, J-C., Duong, C.N.: Evaluation and verification of advanced methods to assess multiple-site damage of aircraft structure. DOT/FAA/AR-04/42,I. Available at the FAA page: actlibrary.tc.faa.gov (2004)

Änggård, J.: FE-analysis of riveted lap joint. TN 2000–15, The Aeronautical Research Institute of Sweden FFA (2000)

Atre, A.: A finite element and experimental investigation on the fatigue of riveted lap joints in aircraft applications. Ph.D. thesis, Georgia Institute of Technology, Atlanta (2006)

Bakuckas, J.G., Jr.: Comparison of boundary correction factor solutions for two symmetric cracks in a straight-shank hole. Technical Report DOT/FAA/AR-98/36 (1999)

Bakuckas, J.G., Jr.: Full-scale testing and analysis of fuselage structures containing multiple cracks. Final Report, DOT/FAA/AR-01/46 (July 2002)

Bakuckas, J.G., Jr., Bigelow, C.A., Tan, P.W.: Characterization of fatigue behaviour of aircraft fuselage structures. In: Guillaume, M. (ed.) Proceedings of 22nd ICAF Symposium, Fatigue of Aeronautical Structures as an Engineering Challenge, Lucerne, Switzerland, 7–9 May 2003, pp. 341–360. EMAS, Warley (2003)

A. Skorupa and M. Skorupa, *Riveted Lap Joints in Aircraft Fuselage*, Solid Mechanics and Its Applications 189, DOI 10.1007/978-94-007-4282-6,

Bakuckas, J.G., Jr., Mosinyi, B., Steadman, D., Awerbuch, J., Lau, A.C., Tan, T.-M.: Teardown inspection and extended fatigue test of fuselage lap joints from retired passenger service airplane. In: Lazaretti L., Salvetti S. (eds.) Proceedings of 24th ICAF Symposium, Durability and Damage Tolerance of Aircraft Structures; Metals vs. Composites, Naples, Italy, 16–18 May 2007, pp. 749–768. Publ. Pacini, Naples (2007)

Barret, R.T.: Fastener design manual. NASA Reference Publication 1228 (1990)

Barrois, W.: Stress and displacements due to load transfer by fasteners in structural assemblies. Eng. Fract. Mech. **10**, 115–176 (1978)

Berens, A.P., Hovey, P.W., Skinn, D.A.: Risk analysis for aging aircraft fleets. Analysis, vol. 1. WL-TR-91-3066, Flight Dynamics Directorate. Wright Laboratory. Wright-Patterson AFB, October 1991

Berens, A.P., Gallagher, J.P., Dhar, S.: Risk analysis for modeling damage at multiple sites. ASIAC-TR-94-14, UDR-TR-94-15, Flight Dynamics Directorate, Wright Laboratory. Wright-Patterson AFB, February 1994

Beuth, J.L., Hutchinson, J.W.: Fracture analysis of multi-site-cracking in fuselage lap joints. Comput. Mech. **13**, 315–331 (1994)

Blom, A.F.: Fatigue science and engineering – achievements and challenges. In: Rouchon, J. (ed.) Proceedings of 21th ICAF Symposium, Design for Durability in the Digital Age, Tolouse, France, 27–29 June 2001. 18th Plantema Memorial Lecture, pp. 3–63. Cépaduès Edition, Touluse (2001)

Bombardier, Y., Liao, M., Renaud, G.: Modelling of continuing damage for damage tolerance analysis. In: Komorowski, J. (ed.) Proceedings of the 26th ICAF Symposium, Structural Integrity: Influence of Efficiency and Green Imperatives, Montreal, Canada, 1–3 June 2011, pp. 231–247. Springer (2011)

Brenner, L., Hübsch, B.: Lebensdauer-Untersuchungen an Nietverbindungen. Report EF 30-TB 8, Dornier, Friedrichshafen (1969)

Broek, D., Yeong, D.Y., Thomson, D.: Testing and analysis of flat and curved panels with multiple cracks. In: Harris, Ch.E. (ed.) Proceedings of FAA/NASA International Symposium on Advanced Structural Integrity Methods for Airframe Durability and Damage Tolerance, Hampton, VA, 4–6 May 1994, NASA CP 3274, pp. 85–98 (1994)

Brown, A.M., Straznicky, P.V.: Simulating fretting contact in single lap splices. Int. J. Fatigue **31**, 375–384 (2009)

Brust, F.M., Nishioka, R., Atluri, S.N., Nakagaki, M.: Further studies on elastic-plastic stable fracture utilizing T*-integral. Eng. Fract. Mech. **22**, 1079–1103 (1985)

Cavallini, G., Galatolo, R., Lazzeri, R.: Short crack growth in the lap joint: experimental and numerical results. In: Rouchon, J. (ed.) Proceedings of 21th ICAF Symposium, Design for Durability in the Digital Age, Tolouse, France, 27–29 June 2001, pp. 85–98. Cépaduès Edition, Tolouse (2001)

Cavallini, C., Lazzeri, R.: A probabilistic approach to fatigue design of aerospace components by using the risk assessment evaluation. In: Agarwal, R. (ed.) Recent Advances in Aircraft Technology, ISBN: 978-953-51-0150-5, InTech, Available from: http://www.intechopen.com/books/recent-advances-in-aircraft-technology/a-probabilistic-approach-to-fatigue-design-of-aerospace-components-by-using-the-risk-assessment-eval (2012)

Chakherlou, T.N., Vogwell, J.: The effect of cold expansion on improving the fatigue life of fastener holes. Eng. Fail. Anal. **10**, 13–24 (2003)

Chakherlou, T.N., Mirzajanzadech, M., Saaedi, K.H.: Fatigue crack growth and life prediction of a single interference fitted holed plate. Fatigue Fract. Eng. Mater. Struct. **33**, 632–644 (2010)

Chen, Ch.-S., Wawrzynek, P.A., Ingraffea, A.R.: Crack growth simulation and residual strength predictions in airplane fuselages. NASA/CR-1999-209115. Langley Research Center, Hampton, March 1999

Chen, Ch-S, Wawrzynek, P.A., Ingraffea, A.R.: Prediction of residual strength and curvilinear crack growth in aircraft fuselage. J. Aircr. **40**, 1644–1652 (2002)

Das, G.K., Miller, M., Sovar, T.: Durability assessment of fuselage single shear lap joint with pads. In: Rouchon, J. (ed.) Proceedings of 21th ICAF Symposium, Design for Durability in the Digital Age, Tolouse, France, 27–29 June 2001, pp. 567–595. Cépaduès Edition, Tolouse (2001)

Dawicke, D.S.: Residual strength predictions using a crack tip opening angle criterion. In: Bigelow, C.E. (ed.) Proceedings of FAA-NASA Symposium on the Continued Airworthiness of Aircraft Structures, Atlanta, GA, 28–30 Aug 1996. DOT/FAA/AR-97/2, pp. 555–566 (1997)

Dawicke, D.S., Newman, J.C., Jr.: Evaluation of various fracture parameters for predictions of residual strength in sheets with multi-site damage. In: Gallagher, J. (ed.) Proceedings of 1st Joint DoD/FAA/NASA Conference on Aging Aircraft, Odgen, 8–10 July 1997, pp. 1307–1326 (1997)

Dawicke, D.S., Newman, J.C., Jr.: Residual strength predictions for multiple site damage cracking using a three-dimensional finite element analysis and a CTOA criterion. In: Panotin, T.L., Sheppard, S.D. (eds.) Fatigue and Fracture Mechanics. ASTM STP 1332, vol. 29, pp. 815–829. ASTM, Philadelphia (1998)

Dawicke, D.S., Sutton, M.A.: CTOA and crack tunnelling measurements in thin sheet 2024-T3 aluminium alloy. Exp. Mech. **34**, 357–368 (1994)

Dawicke, D.S., Newman, J.C., Jr., Bigelow, C.A.: Three-dimensional CTOA and constraint effects during stable tearing in a thin-sheet material. In: Reuter, W.G., Underwood, J.H., Newman, J.C., Jr. (eds.) Fracture Mechanics. ASTM STP 1256, vol. 26, pp. 223–242. ASTM, Philadelphia (1995a)

Dawicke, D.S., Newman, J.C., Jr., Sutton, M.A., Amstutz, B.E.: Influence of crack history on the stable tearing behaviour of a thin-sheet material with multiple cracks. In: Bigelow, C.A. (ed.) Proceedings of FAA-NASA 6th International Conference on the Continued Airworthiness of Aircraft Structures, Atlantic City, 27–28 June 1995. DOT/FAA/AR-95/86, pp. 193–212 (1995b)

de Koning, A.U.: A contribution to the analysis of slow stable crack growth. Euromech Colloquium 64, Stockholm (Aug 1975) (Also: NLR, Report MP 75035)

de Rijck, J.J.M.: Stress analysis of fatigue cracks in mechanically fastened joints. An analytical and experimental investigation. Ph.D. thesis, TU Delft, Delft (2005)

de Rijck, J.J.M., Homan, J.J., Schijve, J., Benedictus, R.: The driven rivet head dimensions as an indication of the fatigue performance of aircraft lap joints. Int. J. Fatigue **29**, 2208–2218 (2007)

de Wit, R., Fields, R.J., Low III, S.R., Harne, D.E., Foecke, T.: Fracture testing of large-scale thin-sheet aluminium. Report No. DOT/FAA/AR-95/11. U.S. Department of Transportation. Federal Aviation Administration (1996)

Deng, X., Hutchinson, J.W.: The clamping stress in a cold-driven rivet. Int. J. Mech. Sci. **40**, 683–694 (1998)

Eastin, R., Mowery, J.: Aircraft structural fail safety: a noble but problematic concept. In: Lazaretti, L., Salvetti, S. (eds.) Proceedings of 24th ICAF Symposium, Durability and Damage Tolerance of Aircraft Structures; Metals vs. Composites, Naples, Italy, 16–18 May 2007, pp. 55–72. Publ. Pacini, Naples (2007)

Eijkhout, M.T.: Fractographic analysis of longitudinal fuselage lap joint at stringer 42 of Fokker 100 full scale test article TA 15 after 126 250 simulated flights. Fokker Report RT2160. Fokker Aircraft Ltd, Amsterdam (1994)

Ekh, J., Schön, J.: Effect of secondary bending on strength prediction of composite, single shear lap joints. Compos. Sci. Technol. **65**, 953–965 (2005)

Ekh, J., Schön, J., Melin, L.G.: Secondary bending in multi-fastener, composite-to-aluminium single shear lap joints. Composites **Part B 36**, 195–208 (2005)

El Haddad, M.H., Topper, T.H., Smith, K.N.: Prediction of nonpropagating cracks. Eng. Fract. Mech. **11**, 537–584 (1979)

Erdogan, F., Sih, G.C.: On the crack extension of plates under plane loading and transverse shear. J. Basic Eng. **85**, 519–527 (1963)

ESDU data sheet No. 79031. Endurance of riveted lap joints (aluminium alloy sheets and rivets). Engineering Science Data Unit, London (1979)

Fawaz, S.A.: Fatigue crack growth in riveted joints. Ph.D. thesis, TU Delft, Delft (1997)

Fawaz, S.A.: Equivalent initial flaw size testing and analysis AFRL-VA-WP-TR-2000-3024. Air Force Research Laboratory, Wright-Patterson AFB (2000)

Fawaz, S.A.: Equivalent initial flaw size testing and analysis of transport aircraft skin splices. Fatigue Fract. Eng. Mater. Struct. **26**, 279–290 (2003)

Fawaz, S.A.: Using the world's largest stress intensity factor database for fatigue life predictions. In: Benedictus, R., Schijve, J., Alderliesten, R.C., Homan, J.J. (eds.) First International Conference on Damage Tolerance of Aircraft Structures. Paper No. 13, TU Delft, Delft (2007)

Fawaz, S.A., Andersson, B.: Accurate stress intensity factor solutions for corner cracks at a hole. Eng. Fract. Mech. **71**, 1235–1254 (2004)

Fawaz, S.A., de Rijck, J.J.M.: A thin-sheet, combined tension and bending specimen. Exp. Mech. **39**, 171–176 (1999)

Feeney, J.A., Mc Millan, J.C., Wei, R.P.: Environmental fatigue crack propagation of aluminium alloys at low stress intensity levels. Metall. Trans. **1**, 1741–1757 (1970)

Flugge, W.: Stress problems in pressurized cabins, NACA TN 2612, February 1952

FRANC3D, Concepts & users guide. Version 2.6. Available at http://www.cfg.cornell.edu/software/Current%20Docs/F3D_Concepts_Users_CFG_V2.6 (2003)

FRANC3D &STAGS, Thin shell tutorial. Version 2.6. Available at http://www.cfg.cornell.edu/software/Current%20Docs/F3D_Thin_Shell_Tutorial_CFG_V2.6 (2003)

Fredell, R.S.: Damage tolerant repair techniques for pressurized aircraft fuselages. Ph.D. thesis, TU Delft, Delft (1994)

Fung, C.-P., Smart, J.: Riveted lap joints. Part 1: A numerical parametric study. Proc. Inst. Mech. Eng. **Part G 211**, 13–27 (1997)

Furuta, S., Terada, H., Sashikuma, H.: Fatigue strength of fuselage joint structures under ambient and corrosive environment. In: Cook, R., Poole, P. (eds.) Proceedings of 19th ICAF Symposium, Fatigue in New and Ageing Aircraft, Edinburgh, Scotland, 18–20 June 1997, pp. 231–249. EMAS, Warley (1997)

Gadalińska, E., Kaniowski, J., Wojtas, A.: Stress measurements with X-ray diffractometry of aluminium alloys. Determination of the most optimized parameters of the measurement. In: Bos, M.J. (ed.) Proceedings of the 25th ICAF Symposium, Bridging the Gap between Theory and Practice, Rotterdam, The Netherlands, 27–29 May 2009, pp. 1285–1304. Springer (2009)

Giummarra, C., Zoker, R.: Improving the fatigue response of aerospace structural joints. In: Donne, C.D. (ed.) Proceedings of 23rd ICAF Symposium, Structure Integrity of Advanced Aircraft and Life Extension for Current Fleets. Lesson Learned in 50 Years After the Comet Accidents, Hamburg, Germany, 8–10 June 2005, pp. 445–456. EMAS, Warley (2005)

Goranson, U.G.: Fatigue issues in aircraft maintenance and repairs. Int. J. Fatigue **20**, 413–431 (1997)

Gruber, M.L., Wilkins, K.E., Worden, R.E.: Investigation of fuselage structure subject to widespread fatigue damage. In: Bigelow, C.A. (ed.) Proceedings of FAA/NASA Symposium on the Continued Airworthiness of Aircraft Structures, Atlanta, GA, 28–30 Aug 1996. DOT/FAA/AR-97/2, pp. 439–459 (1997)

Harish, G., Farris, T.N., Wang, H.L., Grandt, A.F.: Nucleation and growth of cracks in lap joints. In: 1999 USAF Aircraft Structural Integrity Program Conference, 30 Nov–2 Dec 1999, San Antonio, TX, pp. 1–14

Harris, C.E., Piascik, R.S., Newman, J.C., Jr.: A practical engineering approach to predicting fatigue crack growth in riveted lap joints. In: Rudd, J.L., Bader, R.M. (eds.) Proceedings of 20th ICAF Symposium, Structural Integrity for the Next Millennium, Bellevue, WA, USA, 14–16 July 1999, pp. 89–109. EMAS, Warley (1999) (Also NASA/TM-2000/210106, May 2000)

Harter, J.A.: AFGROW user's guide and technical manual. AFGROW for Windows 2K/XP, Version 4.0011.14. AFRL-VA-WP-TR-2006-XXXX. Air Force Research Laboratory, Wright-Patterson AFB, June 2006

Hartman, A.: A comparative investigation on the investigation on the influence of sheet thickness, type of rivet and number of rivet rows on the fatigue strength at fluctuating tension of riveted single lap joints of 24 ST Alclad sheet and 17 S rivets. Report NLR M.1943. NLR, Amsterdam (1954)

Hartman, A.: Some tests on the effect of fatigue loading on the friction in riveted light alloy specimens. Report NLR M. 2008. NLR, Amsterdam (1961)
Hartman, A.: Fatigue tests on single lap joints in clad 2024-T3 aluminium alloy manufactured by a combination of riveting and adhesive bonding. Report NLR M.2170. NLR, Amsterdam (1966)
Hartman, A.: The influence of manufacturing procedures on the fatigue life of 2024-T3 Alclad riveted single lap joints. Report NLR TR 68072 U. NLR, Amsterdam (1968)
Hartman, A.: The effect of secondary bending on the fatigue strength of (clad 2024-T3 and 7075-T6) aluminium alloy riveted joints. Report NLR TR 72007 U. NLR, Amsterdam (1972)
Hartman, A., Duyn, G.C.: A comparative investigation on the fatigue strength at fluctuating tension of several types of riveted lap joints, a series of bolted and some series of glued lap joints of 24 ST Alclad. Report NLR M.1857. NLR, Amsterdam (1952)
Hartman, A., Klaassen, W.: The fatigue strength at fluctuating tension of simple lap joints of clad 24 ST and 75 ST aluminium alloy with 2 rows of 17 S rivets. Report NLR M.2011. NLR, Amsterdam (1956)
Hartman, A., Schijve, J.: The effect of secondary bending on the fatigue strength of 2024-T3 Alclad riveted joints. Report NLR TR 69116U. NLR, Amsterdam (1969)
Hartman, A., Jacobs, F.A., de Rijck, P.: Tests on the effect of the size of the specimen on the fatigue strength of 2024-T3 Alclad double row riveted single lap joints. Report NLR M.2104. NLR, Amsterdam (1962)
Hartman, A., Jacobs, F.A., van der Vet, W.J.: Constant amplitude and programme fatigue tests on single lap joints in clad 2024-T3 and 7075-T6 aluminium alloy with two rows of rivets or huckbolts. Report NLR TN M.2147. NLR, Amsterdam (1965)
Hertel, H.: Ermüdungsfestigkeit der Konstruktionen. Springer, Berlin/Heidelberg/New York (1969)
Hoeppner, D.W.: The formation/nucleation of fatigue cracks in aircraft structural materials. In: Komorowski, J. (ed.) Proceedings of 26th ICAF Symposium, Structural Integrity: Influence of Efficiency and Green Imperatives, Montreal, Canada, 1–3 June 2011, pp. 219–227. Springer, Berlin/Heidelberg/New York (2011)
Hoeppner, D.W., Eliot III, Ch.D., Moesner, M.W.: The role of fretting fatigue on aircraft rivet hole cracking. Technical Report DOT/FAA/AR-96/10. FAA, Utah University Salt Lake City (1996)
Hoffer, K.: Permanent Fasteners for Light-Weight Structures. Aluminium Verlag, Düsseldorf (1984)
Homan, J.J.: Mechanically fastened joints. Lecture Notes ae4-740. TU Delft, Faculty of Aerospace Engineering, Delft (2008)
Homan, J.J., Jongebreur, A.A.: Calculation method for predicting the fatigue life of riveted lap joints. In: Blom, A.F. (ed.) Proceedings of 17th ICAF Symposium, Durability and Structural Integrity of Airframes, Stockholm, Sweden, 9–11 June 1993, pp. 175–190. EMAS, Warley (1993)
Hovey, P.W., Berens, A.P., Loomis, J.S.: Update of the probability of fracture (Prof) computer program for aging aircraft risk analysis. Modifications and user's guide, vol. 1. AFRL-VA-WP-TR-1999-3030. Air Vehicles Directorate, Air Force Research Laboratory, Air Force Material Command. Wright-Patterson AFB, Nov 1998
Hsu, Y.C., Forman, R.G.: Elastic-plastic analysis of an infinite sheet having a circular hole under pressure. J. Appl. Mech. **42**, 347–352 (1975)
Hsu, C.-L., Yu, J.: Initial quality of typical aircraft fuselage splice joints investigated in WFD programme. In: Proceedings of International Conference on Computational Engineering Science, Atlanta, GA, 5–9 Oct 1988 (1998)
Hsu, C.-L., Lo, J., Yu, J., Lee, X.-G., Tan, P.: Residual strength analysis using CTOA criteria for fuselage structures containing multiple site damage. Eng. Fract. Mech. **70**, 525–545 (2003)
Hurricks, P.L.: Mechanism of fretting. Wear **15**, 389–409 (1970)
Huth, H.: Experimental determination of fastener flexibility. LBF-Bericht Nr. 4980. Fraunhofer Institut für Betriebfestigkeit, Darmstadt (1983)

Huth, H.: Influence of fastener flexibility on the prediction of load transfer and fatigue life for multiple-row joints. In: Potter, J.M. (ed.) Fatigue in Mechanically Fastened Composite and Metallic Joints. ASTM STP 927, pp. 221–250. ASTM, Philadelphia (1987)

Ilcewicz, L.B., Smith, P.J., Horton, R.E.: Advanced composite fuselage technology. In: Davis, J.G., Bohon, H.L. (eds.) Proceedings of 3rd NASA Advanced Composites Technology Conference, Long Beach, CA, 8–11 June 1992. NASA CP-3178, pp. 97–156 (1992)

Ingram, J.E., Kwon, Y.S., Duffie, K.J., Irby, W.D.: Residual strength analysis of skin splices with multiple site damage. In: Harris, C.E. (ed.) Proceedings of 2nd Joint NASA/DoD/FAA Conference on Aging Aircraft, Williamsburg, 1–4 Sept 1998. NASA/CP/1999/208982, pp.427–436 (1999)

Ingram, J.E., Kwon, Y.S., Fawaz, S.A.: 3D finite element modeling of aircraft structural joints with multiple site damage. In: Blom, A.F. (ed.) Fatigue 2002. Proceedings of Eighth International Fatigue Congress, Stockholm, Sweden, 3–7 June 2002, pp. 133–140. EMAS, Warley (2002)

Ireman, T.: Three-dimensional stress analysis of bolted single-lap composite joints. Compos. Struct. **43**, 195–216 (1998)

James, M.A., Newman, J.C., Jr.: The effect of crack tunnelling on crack growth: experiments and CTOA analyses. Eng. Fract. Mech. **70**, 457–468 (2003)

Jarfall, L.: Shear loaded fastener installation. Int. J. Vehicle Des. **7**, 337–379 (1986)

Jaya, A., Tiong, V.H., Clark, G., Swift, S., Trathen, P.: A review on the effect of corrosion inhibitors on the fatigue performance of riveted joints. In: Jackson, P., Trasteli, Ch. (ed.) A Review of Australian and New Zeeland Investigation on Aeronautical Fatigue During the Period April 2009 to March 2011. Presented on 26th ICAF Symposium, 29 May–3 June 2011, Montreal, Canada, pp. 8–9 (2011)

Johnson, K.E., Ahmed, A., Bakuckas, J.G. Jr., Awebuch, J., Lau, A., Tan, T.-M.: Multiple-site damage nucleation and growth in a fuselage lap joint. In: Lazaretti, L., Salvetti, S. (eds.) Proceedings of 24th ICAF Symposium, Durability and Damage Tolerance of Aircraft Structures; Metals vs. Composites, Naples, Italy, 16–18 May 2007, pp. 988–1005. Publ. Pacini, Naples (2007)

Jones, R., Molent, L., Pitt, S.: Understanding crack growth in fuselage lap joints. Theor. Appl. Fract. Mech. **49**, 38–50 (2008)

Keesecker, A.L., Dávila, C.G., Johnson, E.R., Starnes, J.H.: Crack path bifurcation at a tear strap in a pressurized shell. Comput. Struct. **81**, 1633–1642 (2003)

Klaassen, W.: The fatigue diagram for fluctuating tension of single lap joints of clad 24 ST and 75 ST aluminum alloy with 2 rows of 17 S rivets. Report NLR M.1980. NLR, Amsterdam (1955)

Knops, H.A.J.: Numerical simulation of crack growth in pressurized fuselages. Ph.D. thesis, TU Delft, Delft (1994)

Kosai, M., Kobayashi, A.S., Ramulu, M.: Tear straps in airplane fuselage. In: Atluri S.N., Harris C.E., Hoggart A., Miller N., Sampath S.N. (eds.) International Workshop on Structural Integrity of Ageing Airplanes, Durability of Metal Aircraft Structures, Atlanta, GA, 31 Mar–2 Apr 1992, pp. 443–457. Atlanta Technical Publications, Atlanta (1992)

Kosai, M., Shimamoto, A., Kobayashi, A.S., Tan, P.W.: Axial crack propagation and arrest in a pressurized cylinder: an experimental-numerical analysis. Exp. Mech. **39**, 256–264 (1999)

Krasnowski, B.R., Reddy, D.J., Franada, B.G., Reid, L., Restis, J.: Fatigue strength and damage tolerance of thin sheet riveted lap/splice joints with cold-expanded holes. In: Rouchon, J. (ed.) Proceedings of 21st ICAF Symposium, Design for Durability in the Digital Age, Toulouse, France, 27–29 June 2001, pp. 195–207. Cépaduès Edition, Touluse (2001)

Kuo, A., Yasgur, D., Levy, M.: Assessment of damage tolerance requirements and analyses - Task I report, vol. II, ICAF Doc. 1583, AFVAL-TR-86-3003. AFVAL Wright-Patterson AFB (1986)

Le Telier, L., Repiton, F.: Full-scale testing and analysis of Falcon 7X curved panels with butt joints. In: Lazaretti, L., Salvetti, S. (eds.) Proceedings of 24th ICAF Symposium, Durability and Damage Tolerance of Aircraft Structures, Naples, Italy, 16–18 May 2007, pp. 340–358. Publ. Pacini, Naples (2007)

Leon, A.: Benefits of split mandrel cold working. Int. J. Fatigue **20**, 1–8 (1988)

Li, G., Shi, G.: Investigation of residual stress in riveted lap joints: experimental study. LTR-SMPL-2003-0099 Report. Institute for Aerospace Research, NRC, Ottawa (2003)

Li, G., Shi, G.: Effect of the riveting process on the residual stress in fuselage lap joints. Can. Aeronaut. Space J. **50**, 91–105 (2004a)

Li, G., Shi, G.: Investigation of residual stress/strain in lap joints with a single countersunk rivet. LTR-SMPL-2004-0130 Report. Institute for Aerospace Research, NRC, Ottawa (2004b)

Li, G., Shi, G., Bellinger, N.C.: Neutron diffraction measurement and FE simulation of residual strains and stress in fuselage lap joints. LTR-SMPL-2004-0003 Report. Institute for Aerospace Research, NRC, Ottawa (2004)

Li, G., Shi, G., Bellinger, N.C.: Study of the residual strain in lap joints. J. Aircraft **43**, 1145–1151 (2006)

Li, G., Shi, G., Bellinger, N.C.: Residual stress/strain in three-row, countersunk, riveted lap joints. J. Aircraft **44**, 1275–1285 (2007)

Li, G., Shi, G., Bellinger, N.C.: Stress in triple-row riveted lap joints under the influence of specific factors. J. Aircraft **48**, 527–539 (2011)

Liao, M., Shi, G., Xion, Y.: Analytical methodology for predicting fatigue crack distribution of fuselage splices. Int. J. Fract. **23**, 177–185 (2001)

Liao, M., Bombardier, Y., Renaud, G., Bellinger, N., Cheung, T.: Development of advanced risk assessment methodologies for aircraft structures containing MSD/MED. In: Bos, M.J. (ed.) Proceedings of the 25th ICAF Symposium, Bridging the Gap between Theory and Practice, Rotterdam, The Netherlands, 27–29 May 2009, pp. 811–837. Springer (2009)

Liu, Y., Mahadevan, S.: Probabilistic fatigue life prediction using an equivalent initial flaw size distribution. Int. J. Fatigue **31**, 476–487 (2009)

Maclin, J.R.: Performance of fuselage pressure structure. In: Harris, C.E. (ed.) Proceedings of 3rd International Conference on Aging Aircraft and Structural Airworthiness, Washington, DC, 19–21 Nov 1991, vol. 3160, pp. 67–74. NASA Conference Publications (1991)

Manning, S.D, Yang, J.N.: Advanced durability analysis, vol. 1 analytical methods. Technical Report AFFDL-TR-86-3017. U.S. Air Force Flight Dynamics Laboratory, Wright-Patterson AFB, Nov 1986

Mayville, R.A., Warren, T.J.: A laboratory study of fracture in the presence of lap splice multiple site damage. In: Atluri, S.N., Atluri, S.N., Sampath, S.G., Tong, P. (eds.) Structural Integrity of Aging Airplanes, pp. 263–273. Springer, Berlin (1991)

MIL-HDBK-5H, US Department of Defence. Military Handbook, Metallic Materials and Elements for Aerospace Vehicle Structures, October 2001

Miller, M., Gruber, M.L., Wilkins, K.E, Worden, R.E.: Full-scale testing and analysis of fuselage structure. In: Harris, Ch.E. (ed.) Proceedings of the FAA/NASA International Symposium on Advanced Structural Integrity Methods for Airframe Durability and Damage Tolerance, Hampton, VA, 4–6 May 1994. NASA CP 3274, pp. 481–496 (1994)

Molent, L., Jones, R.: Crack growth and repair of multi-site damage of fuselage lap joints. Eng. Fract. Mech. **44**, 627–637 (1993)

Moreira, P.M.G.P.: A contribution to the study of fatigue of riveted lap joints. MSc thesis, Facultate de Engenharia, Universidade de Porto, Porto (2004)

Morris, G.: Defining a standard formula and test-method for fastener flexibility in lap-joints. Ph.D. thesis, TU Delft, Delft (2004)

Müller, R.P.G.: An experimental and analytical investigation on the fatigue behaviour of fuselage riveted lap joints. The significance of the rivet squeeze force, and a comparison of 2024-T3 and Glare 3. Ph.D. thesis, TU Delft, Delft (1995)

Müller, R.P.G., Hart-Smith, L.J.: Making fuselage riveted lap splices with 200-year crack-free-lives. In: Cook, R., Poole, P. (eds.) Proceedings of 19th ICAF Symposium, Fatigue in New and Aging Aircraft, Edinburgh, Scotland, 18–20 June 1997, pp. 499–522. EMAS, Warley (1997)

Munroe, J., Wilkins, K., Gruber, M.: Integral Airframe Structures (IAS) – validated feasibility study of integrally stiffened metallic fuselage panels for reducing manufacturing costs. NASA/CR-2000-209337, May 2000

NASA, Fatigue crack growth computer program. "NASGRO" Version 3.0. Reference Manual. JSC-22267B (NASA, Lyndon B. Johnson Space Center, Texas, 2000)

Nesterenko, B.G., Nesterenko, G.I., Schmidt, H.J.: Fatigue and damage tolerance of fuselage skin longitudinal joints. In: Lazaretti, L., Salvetti, S. (eds.) Proceeding of 24th ICAF Symposium, Durability and Damage Tolerance of Aircraft Structures, Naples, Italy, 16–18 May 2007, pp. 1006–1018. Publ. Pacini, Naples (2007)

Nesterenko, B.G., Nesterenko, G.I., Basow, V.N.: Fracture behaviour of skin materials of civil airplane structures. In: Bos, M.J. (ed.) Proceedings of 25th ICAF Symposium, Bridging the Gap between Theory and Practice, Rotterdam, The Netherlands, 27–29 May 2009, pp. 661–683. Springer (2009)

Newman, J.C., Jr.: A finite element analysis of fatigue crack closure. In: Rice, J., Paris, P.C. (eds.) Mechanics of Crack Growth. ASTM STP 590, pp. 281–301. ASTM, Philadelphia (1976)

Newman, J.C., Jr.: A crack closure model for predicting fatigue crack growth under aircraft spectrum loading. In: Chang, J.B., Hudson, C.M. (eds.) Methods and Models for Predicting Fatigue Crack Growth under Random Loading. ASTM STP 748, pp. 53–84. ASTM, Philadelphia (1981)

Newman, J.C., Jr.: FASTRAN-II a fatigue crack growth structural analysis program. NASA Technical Memorandum 104159. NASA Langley Research Center, Hampton, Feb 1992

Newman, J.C., Jr.: Advances in fatigue and fracture mechanics analyses for aircraft structures. In: Rudd, J.L, Bader, R.M. (eds.) Proceedings of 20th ICAF Symposium, Structural Integrity for the Next Millenium, Bellevue, WA, 14–16 July 1999. 17th Plantema Memorial Lecture, pp. 3–40. EMAS, Warley (1999)

Newman, J.C., Jr., Raju, I.S.: Stress-intensity factor equations for cracks in three-dimensional finite bodies subjected to tension and bending loads. In: Atluri, S.N. (ed.) Computational Methods in Mechanics of Fracture, pp. 311–334. Elsevier, Amsterdam/New York (1986)

Newman, J.C., Jr., Ramakrishnan, R.: Fatigue analyses of riveted lap-splice joints in a narrow-body aircraft. In: Komorowski, J. (ed.) Proceedings of 26th ICAF Symposium, Structural Integrity: Influence of Efficiency and Green Imperatives, Montreal, Canada, 1–3 June 2011, pp. 289–302. Springer, Dordrecht/Heidelberg/London/New York (2011)

Newman, J.C., Jr., Harris, C.E., James, M.A., Shivakumar, K.N.: Fatigue life prediction of riveted lap splice joints using small crack theory. In: Cook, R., Poole, P. (eds.) Proceedings of 19th ICAF Symposium, Fatigue in New and Aging Aircraft, Edinburgh, Scotland, 18–20 June 1997, pp. 523–552. EMAS, Warley (1997)

Newman, J.C., Jr., James, M.A., Zerbst, U.: A review of the CTOA/CTOD fracture criterion. Eng. Fract. Mech. **70**, 371–385 (2003)

Niu, M.C.Y.: Airframe Structural Design. Practical Information and Data on Airframe Structures. Conmilit Press Ltd., Hong Kong (1988)

Niu, M.C.Y.: Airframe. Stress Analysis and Sizing, 2nd edn. Conmilit Press Ltd., Hong Kong (1999)

Oldersma, A.: Fatigue of riveted joints. A literature survey and statistical analysis of existing test data. Report NLR CR 92401 L. NLR, Amsterdam (1992)

Oldersma, A., Wanhill, R.J.H.: Variability of fatigue crack growth properties for 2024 T-3 aluminium alloy. Report NLR TP 96038 L. NLR, Amsterdam (1996)

Ottens, H.H.: Multiple crack initiation and crack growth in riveted lap joint specimens. In: Grandage, J.M., Jost, G.S. (eds.) Proceedings of 18th ICAF Symposium, Estimation, Enhancement and Control of Aircraft Fatigue Performance, Melbourne, Australia, 3–5 May 1995, pp. 487–506. EMAS, Warley (1995)

Ottens, H.H., Wanhill, R.J.H.: Review of aeronautical fatigue investigations in the Netherlands during the period March 2001–March 2003. Presented on 22nd ICAF Symposium, Fatigue of Aeronautical Structures as an Engineering Challenge, Lucerne, Switzerland, 7–9 May 2003, pp. 4–5 (2003)

Palmberg, B.: Load transfer and secondary bending in two mechanical joints. FOI Memo 82–0011. Swedish Defence Research Agency, Aeronautics Division, FFA, April 2002

Park, J.H., Atluri, S.N.: Fatigue growth of multiple cracks near a row of fasteners holes in a fuselage lap-joint. Comput. Mech. **13**, 47–63 (1993)

Pettit, R., Ingraffea, A.: Crack turning and arrest mechanism for integral structure. Final report, NASA Contract NAG-1-2013. Cornell University, Ithaca (1999)

Phillips, E.P., Britt, V.O.: Measurements of a fuselage skin strains and displacements near a longitudinal lap joint in a pressurized aircraft. NASA Technical Memorandum TM 104163. Langley Research Center, Hampton, Oct 1991

Piascik, R.S., Willard, S.A.: The characterisation of multi-site damage in the fuselage riveted lap splice joint. In: Cook, R., Poole, P. (eds.) Proceedings of 19th ICAF Symposium, Fatigue in New and Ageing Aircraft, Edinburgh, Scotland, 18–20 June 1997, pp. 93–114. EMAS, Warley (1997)

Piascik, R.S., Willard, S.A.: The growth of multi-site fatigue damage in fuselage lap joints. In: Harris, C.E. (ed.) Proceedings of 2nd Joint NASA DoD/FAA/NASA Conference on Aging Aircraft, Wiliamsburg, VA, 1–4 Sept 1998, NASA/CP/1999/208982, pp. 397–407 (1999)

Piascik, R.S., Willard, S.A., Miller, M.: The characterisation of widespread fatigue damage in fuselage structure. In: Harris, Ch.E. (ed.) Proceedings of FAA/NASA International Symposium on Advanced Structural Integrity Methods for Airframe Durability and Damage Tolerance, Hampton, VA, 4–6 May 1994. NASA CP 3274, pp. 563–580 (1994)

Potyondy, D.O.: A methodology for simulation of curvilinear crack growth in pressurized shells. Ph.D. dissertation, Cornell University, Ithaca (1993)

Potyondy, D.O., Wawrzynek, P.A., Ingraffea, A.R.: Discrete crack growth analysis methodology for through cracks in pressurized fuselage structures. Int. J. Numer. Methods Eng. **38**, 1611–1633 (1995)

Ralph, W.C., Johnson, W.S., Toivonen, P., Makeev, A., Newman, J.C., Jr.: Assessment of residual stresses and hole quality on the fatigue behavior of aircraft structural joints, vol. 2, Fastener hole drilling quality and fatigue lives. DOT/FAA/AR-07/56, V.2, Mar 2009

Ramulu, M., Kobayashi, A.S.: Dynamic crack turning – a photoelastic evaluation. Exp. Mech. **23**, 1–9 (1983)

Rankin, C.C., Brogan, F.A., Loden, W.A., Cabiness, H.D.: Structural analysis of general shells, STAGS, User Manual, Version 3.0. LMMS PO32594. Advanced Technology Center, Lockheed Martin, Palo Alto, Feb 1998

Rans, C.D.: The role of rivet installation on the fatigue performance of riveted lap joints. Ph.D. dissertation, Carleton University, Ottawa (2007)

Rans, C., Straznicky, P.V., Alderliesten, R.: Riveting process induced residual stress around solid rivets in mechanical joints. J. Aircraft **44**, 323–329 (2007)

Rans, C.D., Alderliesten, R.C., Straznicky, P.V.: Assessing the effects of riveting induced residual stresses on fatigue crack behaviour in lap joints by means of fractography. Int. J. Fatigue **31**, 300–308 (2009)

Renaud, G., Liao, M., Bombardier, Y.: Developement of an efficient methodology and tool to determine stress intensity correction factors for complex aircraft structures. In: Komorowski, J. (ed.) Proceedings of the 26th ICAF Symposium, Structural Integrity: Influence of Efficiency and Green Imperatives, Montreal, Canada. 1–3 June 2011, pp. 335–345. Springer (2011)

Rodman, G.A., Creager, M.: Split mandrel vs split sleeve coldworking: dual methods for extending the fatigue life of metal structures. In: Harris, Ch.E. (ed.) Proceedings of FAA/NASA International Symposium on Advanced Structural Integrity Methods for Airframe Durability and Damage Tolerance, Hampton, 4–6 May 1994, NASA-CP-3274, pp. 1078–1086 (1994)

Roebroeks, G.H.J.J.: Fiber metal laminates, recent developments and applications. Int. J. Fatigue **16**, 33–42 (1994)

Rudd, J.L., Gray, T.D.: Equivalent initial quality method. Technical Report AFFDL-TM-76-83. U.S. Air Force Flight Dynamics Laboratory, Wright-Patterson AFB (1976)

Rudd, J.L., Gray, T.D.: Quantification of fastener hole quality. J. Aircraft **15**, 143–147 (1978)

Samavedam, G., Thomson, D., Jeong, D.Y.: Evaluation of the fuselage lap joint fatigue and terminating action repair. In: Harris, Ch.E. (ed.) Proceedings of the FAA/NASA International

Symposium on Advanced Structural Integrity Methods for Airframe Durability and Damage, Hampton, VA, 4–6 May 1994, NASA CP 3274, pp. 653–664 (1994)

Sampath, S., Broek, D.: Estimation of requirements of inspection intervals for panels susceptible to multiple site damage. In: Atluri, S.N., Sampath, S.G., Tong, P. (eds.) Structural Integrity of Aging Airplanes, pp. 339–389. Springer, Berlin (1991)

Schijve J.: The fatigue strength of riveted joints and lugs. Technical Memorandum 1935. National Advisory Committee for Aeronautics, Washington, DC (1956)

Schijve, J.: The fatigue life of unnotched and notched 2024-T3 Alclad sheet material from different manufacturers. Report NLR TR 68093U. NLR, Amsterdam (1968)

Schijve, J.: Some elementary calculations on secondary bending in simple lap joints. Report NLR TR 72036. NLR, Amsterdam (1972)

Schijve, J.: Crack stoppers and ARALL laminates. Eng. Fract. Mech. **37**, 405–421 (1990)

Schijve, J.: Multiple-site-damage of riveted joints. In: Atluri S.N., Harris C.E., Hoggart A., Miller N., Sampath, S.N.: International Workshop on Structural Integrity of Ageing Airplanes, Durability of Metal Aircraft Structures, Atlanta, GA, 31 Mar–2 Apr 1992. Atlanta Technical Publication, Atlanta, pp. 2–27 (1992)

Schijve, J.: Fatigue life until small cracks in aircraft structures. Durability and damage tolerance. In: Harris, Ch.E. (ed.) Proceedings of the FAA/NASA International Symposium on Advanced Structural Integrity Methods for Airframe Durability and Damage Tolerance, Hampton, VA, 4–6 May 1994, NASA CP 3274, pp. 665–680 (1994)

Schijve, J.: Multiple-site damage in aircraft fuselage structures. Fatigue Fract. Eng. Mater. Struct. **18**, 329–344 (1995)

Schijve, J.: Some considerations on the correlation between the rivet squeezing force and the dimensions of the driven head. Memorandum M-847. TU Delft, Faculty of Aerospace Engineering, Delft (2005)

Schijve, J.: Riveted lap joints with staggered thickness in the overlap of the joint. Calculation of secondary bending. Doc. B2-06-02. TU Delft, Faculty of Aerospace Engineering, Delft (2006)

Schijve, J.: Fatigue of Structures and Materials, 2nd edn. Springer, Dordrecht/Heidelberg/London/New York (2009a) (with CD-Rom)

Schijve, J.: Fatigue damage in aircraft structures, not wanted, but tolerated? Int. J. Fatigue **31**, 998–1011 (2009b)

Schijve, J., de Rijk, P.: The fatigue crack propagation in 2024 T-3 Alclad sheet materials from seven different manufacturers. Report NLR M.2162. NLR, Amsterdam (1966)

Schijve, J., Jacobs, F.A.: Programmed fatigue tests on notched light alloy specimens of 2024 and 7075 material. Report NLR M.2070. NLR, Amsterdam (1960)

Schijve, J., Jacobs, F.A., Tromp, P.J.: The significance of cladding for fatigue of aluminium alloys in aircraft structures. Report NLR TR 76065 C. NLR, Amsterdam (1976)

Schijve, J., Jacobs, F.A., Meulman, A.E.: Effect of an anti-corrosion penetrant on the fatigue life in flight-simulation tests on various riveted joints. Report NLR TR 77103 U. NLR, Amsterdam (1977)

Schijve, J., Skorupa, M., Skorupa, A., Machniewicz, T., Gruszczyński, P.: Fatigue crack growth in the aluminium alloy D16 under constant and variable amplitude loading. Int. J. Fatigue **26**, 1–15 (2004)

Schijve, J., Campoli, G., Monaco, A.: Fatigue of structures and secondary bending in structural elements. Int. J. Fatigue **31**, 1111–1123 (2009)

Schmidt, H.-J., Brandbecker, B.: The effect of environmental conditions and load frequency on the crack initiation life and crack growth in aluminium structure. In: Bigelow, C.A. (ed.) Proceedings of FAA/NASA Symposium on Continued Airworthiness of Aircraft Structures, Atlanta, GA, 28–30 Aug 1996, DOT/FAA/AR-97/1, pp. 171–182 (1997)

Schra, L., Ottens, H.H., Vlieger, H.: Fatigue crack growth in simulated Fokker 100 lap joints under MSD and SSD conditions. Report NLR CR 95729 C. NLR, Amsterdam (1995)

Schütz, W.: Zeitfestigkeit einschnittiger Leichtmetall-Nietverbindungen. Bericht Nr. F-47. Laboratorium für Betriebsfestigkeit, Darmstadt (1963)

Schütz, D., Lowak, H.: Die Berücksichtigung des Einflusses der Sekundär-Biegung auf die Schwingfestigkeit von Fügungen. Bericht Nr FB-113. Laboratorium für Betriebsfestigkeit, Darmstadt (1974). Also available as: Schütz, D., Lowak, H.: The effect of secondary bending on the fatigue strength of joints. Report FB-113. Laboratorium für Betriebsfestigket, Darmstadt (1974), RAE Library Translation No. 185

Seshadri, B.R., Newman, J.C., Jr.: Residual strength analyses of riveted lap-splice joints. NASA/TM-2000-209856, NASA Langley Research Center, Hampton (2000)

Shah, R.C.: Stress intensity factors for through and part through cracks originating at fastener holes. In: Rice, J.R., Paris, P.C. (eds.) Mechanics of Crack Growth. ASTM STP 590, pp. 429–459. ASTM, Philadelphia (1976)

Silva, L.F.M., Gonçolves, J.P.M., Oliveira, F.M.F., de Castro, P.M.S.T.: Multiple-site damage in riveted lap joints: experimental simulation and finite element prediction. Int. J. Fatigue **22**, 319–338 (2000)

Simpson, A.: The use of modern fastening systems to enhance the fatigue life of thin sheet structures. In: Berkovits, A. (ed.) Proceedings of the 15th ICAF Symposium, Aeronautical Fatigue in the Electronic Era, Jerusalem, Israel, 21–23 June 1989, pp. 243–262. EMAS, Warley (1989)

Skorupa, M.: Load interaction effects during fatigue crack growth under variable amplitude loading – a literature review. Part I: Empirical trends. Fatigue Fract. Eng. Mater. Struct. **21**, 987–1006 (1998)

Skorupa, M., Korbel, A.: Modelling the secondary bending in riveted joint with eccentricities. Arch. Mech. Eng. **55**, 335–352 (2008)

Skorupa, M., Machniewicz, T., Schijve, J., Skorupa, A.: Application of the strip-yield model from the NASGRO software to predict fatigue crack growth in aluminum alloys under constant and variable amplitude loading. Eng. Fract. Mech. **74**, 291–313 (2007)

Skorupa, M., Skorupa, A., Machniewicz, T., Korbel, A.: An experimental investigation on the fatigue performance of riveted lap joint. In: Bos, M.J. (ed.) Proceedings of the 25th ICAF Symposium, Bridging the Gap between Theory and Practice, Rotterdam, 27–29 May 2009, pp. 449–473. Springer (2009)

Skorupa, M., Machniewicz, T., Schijve, J., Skorupa, A., Korbel, A.: Measurements of rivet flexibility and load transfer in a lap joint. In: Proceedings of the European Conference on Fracture ECF-18, Drezden, 30 Aug–3 Sept 2010. Paper C.05.1-1 (2010a) (Proceedings on CD-Rom)

Skorupa, M., Skorupa, A., Machniewicz, T., Korbel, A.: Effect of production variables on the fatigue behaviour of riveted lap joints. Int. J. Fatigue **32**, 996–1003 (2010b)

Skorupa, M., Skorupa, A., Machniewicz, T., Korbel, A.: Improving the fatigue performance of riveted joints in airframes. Final report on EUREKA project IMPERJA, No. E!3496. University of Science and Technology AGH, Kraków (2011) (in Polish, unpublished results)

Sławiński, G., Niezgoda, T., Szymczyk, E., Jachimowicz, J.: Numerical study of the influence of shape imperfections on residual stress fields in a rivet hole. J. KONES Powertrain Transp. **17**, 427–434 (2010)

Smith, C.R.: Fatigue resistance. Design considerations. Aircr. Eng. **1960**, 142–144 (1960)

Smith, C.R., Lindenau, G.D.: Riveted joints fatigue strength. Reprint from Fatigue of Aircraft Structures. ASTM STP 203, pp. 10–25. ASTM, Philadelphia (1957)

Smith, B.L., Saville, P.A., Mouak, A., Myose, R.Y.: Strength of 2024-T3 aluminium panels with multiple site damage. J. Aircraft **37**, 325–331 (2000)

Starikow, R.: Mechanically fastened joints. Critical testing of single overlap joints. Scientific Report FOI-R-0441-SE. Swedish Defense Research Agency, Aeronautics Division, FAA, Stockholm (2002)

Starnes, J.H., Newman, J.C. Jr., Harris, C.E., Piascik, R.S., Young, R.D, Rose, C.A.: Advances in structural integrity analysis methods for aging metallic airframe structures with local damage. RTO AVT Specialists' meeting on "Life Management Techniques for Aging Air Vehicles", Manchester, U.K., 8–11 Oct 2001, RTO-MP-079 (II), pp. 27-1–27-16 (2003)

Steadman, D., Carter, A., Ramakrishnan, R.: Characterisation on MSD in an in-service fuselage lap joint. In: 3rd Joint FAA/DoD/NASA Conference on Aging Aircraft, Albuquerque, N.M., Sept 1999 (Proceedings on CD-Rom)

Stefanescu, D., Edwards, L., Fitzpatrick, M.E.: Stress intensity factor correction for asymmetric through-thickness fatigue cracks at holes. Int. J. Fatigue **25**, 569–576 (2003)

Swift, T.: Development of the fail-safe design features of the DC-10. Damage Tolerance in Aircraft Structures. In: Rosenfeld, M.S. (ed.) A Symposium Presented at the Seventy-third Annual Meeting of the American Society of Testing and Materials. ASTM STP 486, pp. 164–214. ASTM, Philadelphia (1971)

Swift, T.: The effects of fastener flexibility and stiffener geometry on the stress intensity factor in stiffened cracked sheet. In: Prospects of Fracture Mechanics, pp. 419–436. Noordhoff International, Leyden (1974)

Swift, T.: Damage tolerance in pressurized fuselages. In: Simpson, D.L. (ed.) Proceedings of 14th ICAF Symposium, New Materials and Fatigue Resistant Aircraft Design, Ottawa, Canada, 8–12 June 1987. 11th Plantema Memorial Lecture, pp. 1–77. EMAS, Warley (1987)

Swift, T.: Repairs to damage tolerant aircraft. Technical report FAA-AIR-90-01. Presented to International Symposium on Structural Integrity of Aging Airplanes, Atlanta, GA, 20–22 Mar 1990 (1990)

Swift, T.: Widespread fatigue damage monitoring – issues and concerns. In: Proceedings of 5th International Conference on Structural Airworthiness of New and Aging Aircraft, Hamburg, 16–18 June 1993, pp. 133–150. Deutsche Gesellschaft fur Luft- und Raumfahrt, Bonn (1993)

Szolwinski, M.P.: The mechanics and tribology of fretting fatigue with application to riveted lap joints. Ph.D. dissertation, Purdue University, West Lafayette (1998)

Szolwinski, M.P., Farris, T.M.: Mechanics of fretting fatigue crack formation. Wear **198**, 93–107 (1996)

Szolwinski, M.P., Harish, G., McVeigh, P.A., Farris, T.N.: The role of fretting crack nucleation in the onset of widespread fatigue damage: analysis and experiments. In: Bigelow, C. (ed.) Proceedings of the FAA/NASA Symposium on Continued Airworthiness of Aircraft Structures, Atlanta, GA, 28–30 Aug 1996. DOT/FAA/AR-97/2, pp. 585–596 (1997)

Szymczyk, E., Sławiński, G., Jachimowicz, J., Derewońko, A.: Comparison of the riveting process of a rivet with and without a compensator. J. KONES Powertrain Transp. **16**, 415–422 (2009)

Tan, P.W., Bigelow, C.A., Bakuckas, J.G., Jr.: Widespread fatigue damage assessment approach. RTO AVT Specialists' meeting on "Life Management Techniques for Aging Air Vehicles", Manchester, U.K., 8–11 Oct 2001, RTO-MP-079 (II), pp. 28-1–28-9 (2003)

Terada, H.: A proposal on damage tolerant testing for structural integrity of aging aircraft – learning from JAL accident in 1985. In: Erdogan, F. (ed.) Fracture Mechanics. ASTM STP 1220, vol. 25, pp. 557–574. ASTM, Philadelphia (1995)

Terada, H.: Structural fatigue and joint degradation. Int. J. Fatigue **23**, 21–30 (2001)

Tong, Ch.W.: Literature review on aircraft structural risk and reliability analysis. DSTO-TR-1100 Report. DSTO Aeronautical and Maritime Research Laboratory, Melbourne (2001)

Tong, Ch.Y.: Literature review on aircraft structural risk and reliability analysis. Aeronautical and Maritime Research Laboratory, Defence Science and Technology Organisation, DSTO-TR-1100, Melbourne, Febr. 2001

Trego, A., Cope, D.: Evaluation of damage tolerance analysis tools for lap joints. Am. Inst. Aeronaut. Astronaut. (AIAA) J. **39**, 2250–2254 (2001)

Trey, H., Schmidt, H.J.: Programme for life extension and widespread fatigue damage evaluation to ensure continued structural integrity of airbus large transport category airplanes. RTO AVT Specialists' Meeting on "Life Management Techniques for Ageing Air Vehicles", Manchester, U.K. RTO-MP-079(II), 8–11 October 2001, pp. 29–1 – 29–15

van Dalen, F., Fraterman, N.J., Homan, J.J., van der Schee, P.A., Vogelesang, L.B.: Extension of fatigue life prediction method for riveted joints. Fokker Report S-186 (July 1992)

Vlieger, H.: Results of uniaxial and biaxial tests on riveted fuselage lap joint specimens. In: Harris, Ch.E (1994) Proceedings of FAA/NASA International Symposium of Advanced Structural Integrity Methods for Airframe Durability and Damage Tolerance, Hampton, VA, 4–6 May 1994. NASA CP 3274, pp. 911–930 (1994)

Vlieger, H., Ottens, H.H.: Results on uniaxial and biaxial tests on riveted fuselage lap joint specimens. Report NLR CR 97319 L. NLR, Amsterdam (1997)

Vlot, A.: Glare. History of the Development of a New Aircraft Material. Kluwer Academic, Dordrecht (2001)

Wang, G.S.: An elastic–plastic solution for a normally loaded centre hole in a finite circular body. Int. J. Press. Ves. Piping **33**, 269–284 (1988)

Wang, H.L.: Evaluation of multiple site damage in lap joint specimens, Ph.D. dissertation, Purdue University, West Lafayette, IN (1998)

Wang, L., Brust, F.W., Atluri, S.N.: The elastic-plastic finite element alternating method (EPFEAM) and the prediction of fracture under WFD conditions in aircraft structures. Part I: EPFEAM theory. Comput. Mech. **19**, 356–369 (1997a)

Wang, L., Brust, F.W., Atluri, S.N.: The elastic-plastic finite element alternating method (EPFEAM) and the prediction of fracture under WFD conditions in aircraft structures. Part II: Fracture and the T*-integral parameter. Comput. Mech. **19**, 370–379 (1997b)

Wang, L., Brust, F.W., Atluri, S.N.: The elastic-plastic finite element alternating method (EPFEAM) and the prediction of fracture under WFD conditions in aircraft structures. Part III: Computational predictions of the NIST multiple site damage experimental results. Comput. Mech. **20**, 199–212 (1997c)

Wang, X., Modarres, M., Hoffman, P.: Analysis of crack interactions at adjacent holes and onset of multiple-site damage in aging airframes. Int. J. Fract. **31**, 155–163 (2009)

Wanhill, R.J.H.: Some practical considerations for fatigue and corrosion damage assessment of ageing aircraft. Report NLR TP 96253 L. NLR, Amsterdam (1996)

Wanhill, R.J.H., Koolloos, M.F.J.: Fatigue and corrosion in aircraft pressure cabin lap splices. Int. J. Fatigue **23**(Suppl. 1), 337–347 (2001)

Wawrzynek, P.A.: Discrete modeling of crack propagation: theoretical aspects and implementation issues in two and three dimensions. Ph.D. dissertation, Cornell University, Ithaca, NY (1991)

White, P.: Review of methods and approaches for the structural risk assessment of aircraft. Air Vehicles Division, Defence Science and Technology Organisation, DSTO-TR-1916 (2006)

Wit, G.P.: Bending stresses in circumferential fuselage joints (3-D modeling). T.D. No. SCM-099, Fokker, Stress Office Technical Data Sheet (Aug 1993)

Wronicz, W., Kaniowski, J.: Numerical analysis of riveted lap joint used in aircraft structure. In: Niepokólczycki, A. (ed.) Fatigue of Aircraft Structures. Monographic Series. Issue 2010, pp. 106–116. Institute of Aviation Scientific Publications, Warsaw (2010)

Xiang, Y., Lu, Z., Liu, Y.: Crack growth-based fatigue life prediction using an equivalent initial flaw model. Part I: Uniaxial loading. Int. J. Fatigue **32**, 341–349 (2010)

Yang, J.N., Hsi, W.H., Manning, S.D., Rudd, J.L.: Stochastic crack propagation in fastener holes. J. Aircr. **22**, 810–817 (1985)

Young, R.D., Rouse, M., Ambur, D.R., Starnes, J.H.: Residual strength pressure tests and nonlinear analyses of stringer-and frame-stiffened aluminium fuselage panels with longitudinal cracks. In: Bigelow, C.A. (ed.) Proceedings of FAA/NASA Symposium on Continued Airworthiness of Aircraft Structures, Atlanta, GA, 28–30 Aug 1996. DOT/FAA/AR-97/2, pp. 171–182 (1997)

Zieve, P.B.: Low voltage electromagnetic riveter. In: Proceedings of FASTEC SME West '86, Anaheim, CA, 21–23 Oct 1986 (1986)

Index

A. Skorupa and M. Skorupa, *Riveted Lap Joints in Aircraft Fuselage*, Solid Mechanics and Its Applications 189, DOI DOI 10.1007/978-94-007-4282-6,

J

L

M

MIX
Papier aus verantwortungsvollen Quellen
Paper from responsible sources
FSC® C105338

If you have any concerns about our products,
you can contact us on
ProductSafety@springernature.com

In case Publisher is established outside the EU,
the EU authorized representative is:
Springer Nature Customer Service Center GmbH
Europaplatz 3, 69115 Heidelberg, Germany

Printed by Libri Plureos GmbH
in Hamburg, Germany